Asphaltene Deposition

Asphaltene Deposition

Fundamentals, Prediction, Prevention, and Remediation

Edited by
Francisco M. Vargas
Mohammad Tavakkoli

CRC Press
Taylor & Francis Group
Boca Raton London New York

CRC Press is an imprint of the
Taylor & Francis Group, an **informa** business

CRC Press
Taylor & Francis Group
6000 Broken Sound Parkway NW, Suite 300
Boca Raton, FL 33487-2742

First issued in paperback 2021

© 2018 by Taylor & Francis Group, LLC
CRC Press is an imprint of Taylor & Francis Group, an Informa business

No claim to original U.S. Government works

ISBN-13: 978-0-367-78121-7 (pbk)
ISBN-13: 978-1-138-03523-2 (hbk)

Visit the Taylor & Francis Web site at
http://www.taylorandfrancis.com

and the CRC Press Web site at
http://www.crcpress.com

Contents

Preface

Asphaltene deposition is a challenging problem for the oil industry in many regions around the world, due to the tendency of this petroleum fraction to precipitate and deposit in the wellbore, near wellbore region, and surface facilities, as a result of changes in pressure, temperature, and composition. The current tendency to produce from deeper waters or to enhance oil recovery by injection of light components, such as hydrocarbon gases or carbon dioxide under miscible conditions, gives a clear indication that this problem is likely to worsen even more. Wellbore plugging caused by asphaltene deposition is a multimillion-dollar problem, that not only requires expensive and time consuming intervention methods, but it is also responsible for significant downtime and associated monetary losses.

Despite the significant efforts and resources devoted by a number of companies, universities and research institutions, to better understand and solve this important flow assurance problem, the available knowledge, tools, and technologies are somewhat limited or not readily applicable. There are various reasons why the development of an ultimate solution to asphaltene deposition has been hindered:

- **Ambiguous definition of asphaltenes:** They constitute a solubility class of the crude oil that precipitates upon addition of aliphatic solvents, such as *n*-pentane or *n*-heptane, and that is soluble in aromatic solvents, such as toluene or benzene. As a result of this vague definition, a very wide distribution of species present in the crude oil is classified as asphaltenes. Moreover, the lack of standardization in the separation, quantification and characterization of these asphaltenes, makes the interpretation of results even more challenging, especially when comparing the results of different laboratories or research groups.
- **Some instruments and experimental methods are inadequate:** Some of the commercial instruments that are widely used to detect and quantify asphaltene precipitation and deposition are not sensitive enough, which results in imprecise data and/or incorrect interpretation, e.g. incorrect onset of asphaltene precipitation. Moreover, certain testing protocols are based on improper assumptions or methods that are unrealistic or irrelevant to the systems and conditions of interest.
- **Many contradicting theories and observations in the literature:** For someone new to this area, it can be overwhelming to fully understand the theories and properly interpret the reported observations in the literature, because it is not uncommon to find contradictions among the different sources. The two reasons presented above, along with the complex multistep process, and competing phenomena affecting asphaltene precipitation, aggregation, and deposition, may contribute to the significant variations in the conclusions of different investigations.

- **Mechanisms of inhibition are not fully understood:** Mitigation of asphaltene deposition via chemical injection is a very popular method that many companies have attempted in the past decade, with some mixed results in the field. In some cases, the chemicals are not effective and they can even make the deposition problem worse. Improper chemical screening and lack of understanding of the mechanisms of inhibition are two of the reasons behind these unsuccessful field trials.
- **Effective simulation tools are not widely available:** The utilization of simulation tools to predict the phase behavior and the occurrence of some of the flow assurance problems, such as wax deposition or gas hydrate formation, is common in the oil industry. However, when it comes to asphaltene deposition predictions, the available commercial solutions are limited. New models have been developed that have shown very promising results, but their availability is still somewhat limited.

Asphaltene Deposition: Fundamentals, Prediction, Prevention, and Remediation is the product of over ten years of active investigation, summarizing relevant literature resources, conducting novel experimental and computational work, and solving real field cases, which required us to address the challenges described above. With this book we intend to provide a quick immersion method for anyone interested in learning more about this important topic for the first time, and a comprehensive resource for advanced experimental and modeling methods and case studies for experienced professionals.

Throughout the various chapters of the book, the reader will be able to learn about:

- the motivation for improving our understanding and tools to tackle the potential flow assurance problem caused by asphaltene deposition;
- the chemical and physical properties of asphaltenes, and their separation and quantification methods;
- experimental methods to determine asphaltene precipitation;
- advanced modeling techniques to predict asphaltene precipitation at reservoir conditions;
- experimental methods to probe and quantify asphaltene deposition;
- advanced modeling methods to predict asphaltene deposition under realistic field conditions;
- methods to mitigate asphaltene deposition and some challenges associated to chemical inhibition;
- real case studies that have been successfully conducted by implementation of the methods presented in this book; and
- current research tendencies and some areas for future investigation.

With this book, we aim to consolidate and present in a systematic manner the state-of-the-art laboratory techniques, and the most advanced simulation methods, to better understand and predict the occurrence and the magnitude of asphaltene deposition, as well as some of the best practices to develop an effective mitigation strategy. In turn, we hope that this knowledge and these tools will contribute to the worldwide implementation of cost-effective strategies to manage this problem.

Francisco M. Vargas, Ph.D.

Mohammad Tavakkoli, Ph.D.

Editors

Dr. Francisco M. Vargas is an assistant professor in the department of chemical and biomolecular engineering at Rice University and the chief technology adviser at Ennova LLC. His expertise in petroleum thermodynamics and flow assurance is the result of a number of years combining fundamental research with industrial experience, which has allowed him to develop some unique technologies to effectively address problems faced by the oil and gas industry. Francisco was the manager of the Flow Assurance Research and Development Program of Abu Dhabi National Oil Company (ADNOC) and he worked for the Flow Assurance Team of Chevron Energy Technology Company. He also held academic positions at the Petroleum Institute, in Abu Dhabi, and Tecnologico de Monterrey, in Mexico. He is actively involved in the organization of and presentation at technical meetings of the American Institute of Chemical Engineers (AIChE) and the Society of Petroleum Engineers (SPE).

Dr. Mohammad Tavakkoli is the chief executive officer at Ennova LLC. He has over ten years of experience in the successful design and execution of projects related to petroleum characterization, phase behavior, and flow assurance, and is the co-author of numerous journal articles and conference presentations in these areas. Besides his technical contributions, Mohammad has been actively involved with professional societies. He has been the programming chair of the Upstream Engineering and Flow Assurance Forum (UEFA) of the AIChE and co-organizer of the technical sessions in flow assurance of the Offshore Technology Conference (OTC).

Contributors

M. I. L. Abutaqiya
Department of Chemical and
  Biomolecular Engineering
Rice University
Houston, Texas

A. Chen
Department of Chemical and
  Biomolecular Engineering
Georgia Institute of Technology
Atlanta, Georgia

R. Doherty
Department of Chemical and
  Biomolecular Engineering
Rice University
Houston, Texas

S. Enayat
Department of Chemical and
  Biomolecular Engineering
Rice University
Houston, Texas

J. Hu
Department of Chemical and
  Biomolecular Engineering
Rice University
Houston, Texas

A. T. Khaleel
Abu Dhabi National Oil Company
  (ADNOC)
Abu Dhabi, United Arab Emirates

and

Department of Chemical and
  Biomolecular Engineering
Rice University
Houston, Texas

J. Kuang
Department of Chemical and
  Biomolecular Engineering
Rice University
Houston, Texas

F. Lejarza
Department of Chemical and
  Biomolecular Engineering
Rice University
Houston, Texas

P. Lin
Department of Chemical and
  Biomolecular Engineering
Rice University
Houston, Texas

P. Pourreau
Department of Chemical and
  Biomolecular Engineering
University of Houston
Houston, Texas

N. Rajan Babu
Department of Chemical and
  Biomolecular Engineering
Rice University
Houston, Texas

S. Rezaee
Department of Chemical and
  Biomolecular Engineering
Rice University
Houston, Texas

C. Sisco
Department of Chemical and
  Biomolecular Engineering
Rice University
Houston, Texas

E. Song
Department of Chemical
 Engineering
Massachusetts Institute of
 Technology
Cambridge, Massachusetts

M. Tavakkoli
Ennova LLC
Stafford, Texas

F. M. Vargas
Department of Chemical and
 Biomolecular Engineering
Rice University
Houston, Texas

F. Wang
Department of Chemical and
 Biomolecular Engineering
Rice University
Houston, Texas

J. Yarbrough
Department of Chemical and
 Biomolecular Engineering
Rice University
Houston, Texas

J. Zhang
Department of Chemical and
 Biomolecular Engineering
Rice University
Houston, Texas

Nomenclature

Variables	Definitions
P	pressure
T	temperature
V	volume
n	moles
R	universal gas constant
V^{ig}	ideal gas volume
Z	compressibility factor
A	Helmholtz energy
G	Gibbs energy
U	internal energy
H	enthalpy
S	entropy
δ	solubility parameter
$\hat{V}$	molar volume
ρ	mass density
$\tilde{\rho}$	number density
ρ_h	hydrocarbon density
ρ_{ra}	reduced density
ρ_{sc}	density at standard conditions
$\Delta\rho_p$	density changes due to changes in pressure
$\Delta\rho_T$	density changes due to thermal expansion
ρ_{do}	dead oil density
ρ_w	water density
γ	aromaticity
M_i	molecular weight of component i
z_i	overall molar composition of component i
c_i	mass or mole concentration of component i
c_i^{eq}	solubility (mass or molar concentration units) of component i
n_{ik}	molar amount of component i in phase k
x_{ik}	molar composition of component i in phase k
φ_{ik}	volume fraction of component i in phase k
$\hat{f}_{ik}$	fugacity of component i in phase k
$\hat{\phi}_{ik}$	fugacity coefficient of component i in phase k
γ_{ik}	activity coefficient of component i in phase k
μ_{ik}	chemical potential of component i in phase k
$\overline{Z}_{ik}$	partial molar compressibility factor in phase k
K_i	partition coefficient of component i
β_k	molar fraction of phase k
B^{res}	residual thermodynamic property
B^{ex}	excess thermodynamic property

$\overline{B}_i$	partial molar thermodynamic property
$\dot{b}$	reduced molar thermodynamic property
m	PC-SAFT parameter, chain length
σ	PC-SAFT parameter, segment diameter
ε/k_b	PC-SAFT parameter, interaction energy
κ^{AB}	PC-SAFT or CPA parameter, association volume
ε^{AB}/k_b	PC-SAFT or CPA parameter, association energy
γ	aromaticity
T_c	cubic EOS parameter, critical temperature
P_c	cubic EOS parameter, critical pressure
ω	cubic EOS parameter, acentric factor
N_C	total number of components in a mixture
N_P	total number of phases in a mixture
k_p	precipitation kinetic parameter
k_{ag}	aggregation kinetic parameter
k_d	deposition kinetic parameter
a	surface deposition coefficient
b	entrainment coefficient
C	dimensionless concentration of the precipitated asphaltene particle
C_{ag}	dimensionless concentration of aggregated asphaltene particles
C_f	dimensionless concentration of asphaltene in the oil-precipitant mixture
C_{eq}	dimensionless thermodynamic equilibrium concentration of asphaltene
$\bar{t}$	residence time
r	radius of capillary tube or wellbore
L	axial length of the capillary tube or wellbore
$\overline{R}$	geometric spacing between two asphaltene aggregates
c	velocity vector of the particle
$f(c)$	distribution function of the particle
$\overline{r}$	position vector of the particle
F	external force on the particle
U_z	average axial velocity of the fluid
D_e	asphaltene particle diffusivity
D_{ax}	asphaltene axial dispersion coefficient
Da_p	precipitation damköhler number
Da_{ag}	aggregation damköhler number
DL	dilution ratio
$\Delta\Gamma$	change in half band-half width
Pe	Peclet number
$R_{\text{mass.tr}}$	rate of mass transfer into the boundary layer
R_{int}	rate of asphaltene depletion because of the deposition process
ScF	scaling factor
$K_{i,j}$	collision kernel between two asphaltene aggregates
N_{agg}	number of asphaltene molecules per nanoaggregate
R_p	rate of asphaltene precipitation

R_{ag}	rate of asphaltene aggregation
R_d	rate of asphaltene deposition
K_0	solvation coefficient
D_p	particle diffusivity
D_t	turbulent particle diffusivity
D_B	Brownian particle diffusivity
Sc_t	turbulent schmidt number
d_i	diameter of asphaltene aggregate
m_A	mass fraction of destabilized asphaltene per unit mass of oil
k_b	Boltzmann constant
f^{eq}	local equilibrium distribution function
v_x	velocity field in x direction
v_y	velocity field in y direction
m_D	mass fraction of the deposited asphaltene in oil
δ	boundary layer thickness
β	particle–particle collision efficiency
$\alpha_{i,j}$	collision frequency
μ	viscosity of the medium
μ^*	low-pressure gas mixture viscosity
μ_0	dilute gas contribution to viscosity
μ_f	friction contribution to viscosity
μ_c	characteristic critical viscosity
μ_{mix}	mixture viscosity
$[\mu]$	intrinsic viscosity
ϕ	volume fraction of the suspended particles in mixture
ϕ_m	maximum packing volume fraction
φ_o	volume fraction of oil in the oil–heptane mixture
Ω	LBM collision operator
σ	LBM collision frequency
τ	LBM relaxation factor
ξ	viscosity-reducing parameter
a_i	viscosity correlation polynomial coefficients
v_f	kinematic viscosity of the fluid
v_t	eddy diffusivity
r_{cr}	critical size of asphaltene particles in the oil–precipitant mixture
F	resonance frequency
D	dissipation
ρ_q	specific density
v_q	shear wave velocity
t_q	thickness of the quartz crystal
f_0	fundamental resonance frequency
$E_{dissipated}$	energy dissipated during a single oscillation
E_{stored}	energy stored in the oscillating system
ρ_f	density of the liquid
ϕ	porosity

K	permeability
S_{wi}	initial water saturation
S_0	oil saturation
A	Hamaker constant
C_{12}	van der waals constant for a pair of molecules
ε	dielectric constant
n	refractive index
h	Planck constant
v_e	main adsorption frequency in the UV region
A_{SLR}	composite Hamaker constant predicting sticking tendency of the asphaltene-rich phase on the surface
A_{SRL}	composite Hamaker constant predicting spreading tendency of the asphaltene-rich phase on the surface
A^{Blank}	area between the curve of the control and the horizontal baseline
A^{Disp}	area between the curve of the treated sample set and the horizontal baseline
E_D	energy of dispersion
E_P	energy of permanent-dipole
E_H	energy of electron-exchange
E	energy of vaporization
B_o	oil formation volume factor
δ_{mom}	momentum boundary layer
δ_{lam}	laminar boundary layer
Υ	characteristic constant for porous media
q_o	deposition mass flux in the absence of shear removal
k_{sr}	shear removal
CED	cohesive energy density
E_{coh}	cohesive energy
F_{RI}	function of the refractive index
m/z	mass to charge ratio of an ion
P_{RI}	refractive index at the onset of asphaltene precipitation
R_m	molar refractivity
M_n	number average molecular weight
M_W	weight average molecular weight
γ_o	oil gravity (°API)
PDI	polydispersity index
P_c	critical pressure
P_r	reduced pressure
τ	shear stress
R_s	solution GOR
SG	specific gravity
σ_{hw}	hydrocarbon–water surface tension
T_{amb}	ambient temperature
T_c	critical temperature
T_p	pour point temperature

T_r	reduced temperature
TC	total carbon
TH	total hydrogen
UC	unbounded carbon
μ_{ob}	oil viscosity at the bubble point
μ_{od}	dead oil viscosity
ΔV	voltage difference

Abbreviations   Definitions

μPBR	packed bed microreactor
AAPD	average absolute percent deviation
ACA	advancing contact angle
ADEPT	asphaltene deposition tool
ADT	asphaltene dispersion test
AN	acid number
AOP	asphaltene onset pressure
ART	acoustic resonance technique
ASIST	asphaltene instability trend
BGK	bhatnagar, gross, and krook model
BP	bubble pressure
BPR	back pressure regulator
CCD	charge-coupled device
CFD	computational fluid dynamics
CNP	carbon nanoparticles
CPA	cubic plus association EOS
CQD	carbon quantum dots
CSP	corresponding state principle model
CT	coiled tube
CTHRC	champion technologies hydrate rocking cell
CVD	chemical vapor deposition
DBE	double bond equivalent
DBF	deep-bed filtration
DCM	dichloromethane
DLA	diffusion limited aggregation
DP	4-dodecylphenol
DPE	dispersive performance efficiency
DSC	differential scanning calorimetry
DSSC	dye-sensitized solar cell
DV	differential vaporization
EDTA	ethylenediaminetetraacetic acid
EOR	enhanced oil recovery
EOS	equation of state
ESI	electrospray ionization
ETFE	ethylene tetrafluoroethylene

FIA	fluorescent indicator adsorption
FT-ICR	Fourier transform ion cyclotron resonance
FTIR	Fourier transform infrared spectroscopy
GC	gas chromatography
GOR	gas-to-oil ratio
GPC	gel permeation chromatography
GQD	graphene quantum dots
HP	4-hexylphenol
HPHT	high-pressure, high-temperature
HPLC	high-pressure liquid chromatography
HPM	high pressure microscopy
HSP	hansen solubility parameter
ICP-MS	inductively coupled plasma mass spectroscopy
ICP-OES	inductively coupled optical emission spectroscopy
ID	inner diameter
IFT	interfacial tension
LAOP	lower AOP
LBC	Lohrenz-Bray-Clark method
LBM	Lattice-Boltzmann method
LED	light emitting diode
LLE	liquid–liquid equilibrium
LPG	liquid petroleum gasses
LST	light scattering technique
MALDITOF-MS	matrix-assisted laser desorption ionization-time-of-flight-mass spectroscopy
mol%	mole percent
MS	mass spectrometry
NIR	near-infrared
NMR	nuclear magnetic resonator
NOA	Norland optical adhesive
NTU	nephelometric turbidity units
OBM	oil-based mud
OD	outer diameter
OOIP	original oil in place
OP	4-octylphenol
OSDC	organic solid deposition and control device
PAH	polycyclic aromatic hydrocarbon
PBM	population balance model
PC-SAFT	perturbed-chain statistical associating fluid theory EOS
PFA	perfluoroalkoxy
PIONA	n-paraffin, isoparaffin, olefin, naphthene, and aromatic
PL	photoluminescent
PNA	poly-nuclear aromatics

PR	peng-robinson EOS
PR-FT	peng-robinson EOS version of the friction theory
PSD	particle size distribution
PTFE	polytetrafluoroethylene
PV	pore volume
PVDF	polyvinylidene fluoride
PVT	pressure–volume–temperature
py-gc-ms	pyrolysis-gas chromatography-mass spectroscopy
QCM	Quartz crystal microbalance
QCR	Quartz crystal resonator
RCA	receding contact angle
RCF	relative centrifuge force
Re	Reynold's number
RPM	revolutions per minutes
SARA	saturates–aromatics–resins–asphaltenes
SBS	styrene butadiene styrene
Sc	schmidt number
SCSSV	surface-controlled subsurface safety valve
SDS	solid detection system
SDV	styrene-divinylbenzene
SEC	size exclusion chromatography
SEM	scanning electron microscope
Sh	Sherwood number
s-MWCNT	sulfonated multiwalled carbon nanotube
SRK	soave-redlich-kwong EOS
STO	stock tank oil
TBP	true boiling point
TEM	transmission electron microscopy
TGA	thermogravimetric analysis
THF	tetrahydrofuran
TLC-FID	thin-layer chromatography with flame ionization detection
UAOP	upper AOP
UV	ultraviolet
VLE	vapor–liquid equilibrium
VLLE	vapor–liquid–liquid equilibrium
vol%	volume percent
VPO	vapor pressure osmometry
WAT	wax appearance temperature
wt%	weight percent
XANES	X-ray absorption near-edge structures
XPS	X-ray photoelectron spectroscopy
XRD	X-ray powder diffraction

1 Introduction

M. Tavakkoli and F. M. Vargas

CONTENTS

Fossil fuels such as oil and gas satisfy a large portion of the energy demand all over the world. The U.S. Energy Information Administration (2016) predicts that the world energy consumption would grow by 56% between 2010 and 2040. Because we are running out of the *easy oil*, the petroleum industry faces the need to produce oil and gas in unconventional and complex conditions, deep waters, and difficult-to-access formations. One of the major challenges in this pursuit is to implement a holistic flow assurance program, that is, to guarantee the continuous and economic production and the flow of oil and gas to the refinery.

Asphaltenes constitute the heaviest fraction of the oil, which can deposit during oil production, damage the formation, and clog wellbores and production facilities. Asphaltene deposition can significantly decrease the rate of oil production. It can cause unnecessary well shut-in time and cleaning costs of up to several millions of dollars. For instance, in oil fields in the Gulf of Mexico, the average expenses associated with the asphaltene deposition problem is around US $70 million per well when well shut-in is required (González 2015). If the deposition occurs in the surface-controlled subsurface safety valve, the cost increases to US $100 million per well. Downtime losses can reach up to US $500,000 per day, based on a production of 10,000 barrel per day and oil price of US $50 per barrel. If the well is lost as a result of a severe asphaltene deposition problem, a cost around US $150 million is necessary to replace the well with a side track (González 2015).

Asphaltene has been an active research area during last few decades; however, oil companies still suffer from asphaltene deposition problems, and this has motivated both academia and the industry to actively research the asphaltene deposition problem and provide a proper solution.

1.1 ASPHALTENE DEFINITION

Asphaltenes are a polydisperse mixture of the heaviest and most polarizable fraction of crude oil (Tavakkoli et al. 2014). Boussingault (1837) used the term *asphaltene* to describe the black, alcohol-insoluble, and essence of turpentine-soluble material obtained from crude oil distillation residues. The study of asphaltenes first started in the 1930s when researchers realized that asphaltenes are widely distributed throughout the nature. Asphaltenes can be found in crude oil, Bitumen, tar-mat, and asphalts. In modern operations, asphaltenes are defined in terms of their solubility, being completely miscible in aromatic solvents, such as benzene, toluene, or xylenes, but insoluble in light paraffinic solvents, such as n-pentane or n-heptane. Depending on the normal alkane used to precipitate asphaltenes, the obtained asphaltenes have a significant difference in molecular weight, structure, and other properties. This difference is a result of the polydisperse nature of asphaltenes. Figures 1.1a and b show asphaltenes separated from the same crude oil sample using two different precipitants, that is, normal pentane and normal heptane (Buckley and Wang 2006). However, these figures show purified asphaltenes separated from an oil sample in the laboratory. During oil production, separated asphaltenes are not pure and look like a second liquid phase as shown by Figure 1.1c.

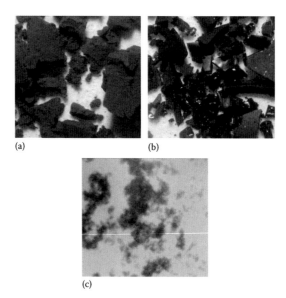

(a)   (b)

(c)

FIGURE 1.1 Purified asphaltenes separated from same crude oil sample in the laboratory using n-pentane (a) and n-heptane (b). (Reprinted from Buckley, J.S. and Wang, J.X., Personal communication, 2006.) (c) Asphaltene sample separated from a different source without any purification.

Because of the current definition of asphaltene, which is based on its solubility and not a unique chemical composition, researchers have faced many problems trying to identify this fraction of the crude oil (Vargas 2010). Despite decades of research, asphaltene chemical composition and molecular structure and their effect on the mechanisms of asphaltene stabilization are not completely understood (Vargas 2010). Chemical analyses of asphaltenes indicate that they contain fused ring structures, naphthenic rings, and some aliphatic chains that contain hydrogen and carbon; small quantities of heteroatoms, such as nitrogen, sulfur, and oxygen; and trace metals, such as vanadium and nickel (Dickie et al. 1969; Mansoori 2009). Most of these heteroelements are thought to be contained in the rings because they resist oxidation. There are two main structures for an asphaltene molecule proposed in the literature: *Island* and *Archipelago* structures. Recent experimental studies using laser desorption have shown the dominance of island structures (Borton et al. 2010), consistent with other mass spectral measurements (Tang et al. 2015). Chacon et al. (2017) concluded that the more easily ionizable asphaltenes are enriched in island-type molecules and the less ionizable asphaltene fractions are likely a mixture of island and archipelago molecules. Figure 1.2 presents some of the hypothetical island and archipelago structures for an asphaltene molecule reported in the literature (Murgich et al. 1999; Rogel 2000; Speight and Moschopedis 1982).

Asphaltene molecular weight has been subject to long-standing discussions as well (Badre et al. 2006). The reason is that asphaltenes aggregate even at low concentrations in good asphaltene solvents such as toluene and form asphaltene clusters called *nano-aggregates* (Indo et al. 2009). It is believed that six to eight asphaltene molecules stack together and form the nano-aggregates (Indo et al. 2009). Asphaltene molecular weight measurements using the vapor pressure osmometry (VPO) and the size exclusion chromatography (SEC) have shown large values for the asphaltene molecular weight, which is most likely due to the formation of asphaltene clusters of molecules.

Measurements of molecular diffusion for asphaltenes using the time-resolved fluorescence depolarization technique have indicated that asphaltene molecules are monomeric with average molecular weight of 750 g/mol and a range of 500–1000 g/mol (Groenzin and Mullins 1999, 2000). These values for asphaltene molecular weight have been confirmed by other techniques used to measure asphaltene molecular diffusion, such as Taylor dispersion (Wargadalam et al. 2002), nuclear magnetic resonance (Freed et al. 2007), and fluorescence correlation spectroscopy (Schneider et al. 2007).

Figure 1.3 illustrates typical asphaltene molecular weight distribution for asphaltenes in the liquid phase and also for precipitated asphaltenes (Creek 2005). As can be seen, asphaltenes form larger aggregates in precipitated phase, and therefore, the peak for the distribution function shifts to the higher values of asphaltene molecular weight for the precipitated asphaltenes compared to the asphaltenes in the liquid phase.

1.2   ASPHALTENE DEPOSITION PROBLEM

Asphaltene is known as the *cholesterol of petroleum* because of its ability to pre-cipitate, deposit, and as a result, interrupt the continuous production of oil from underground reservoirs (Kokal et al. 1995). Asphaltenes may precipitate out of the crude oil with changes in temperature, pressure, or composition. These precipi-tates may then adhere to surfaces and form deposit layers. The asphaltene deposi-tion problem has been observed in all stages of oil production and processing, in near wellbore formations, production tubings, surface facilities, and refinery units.

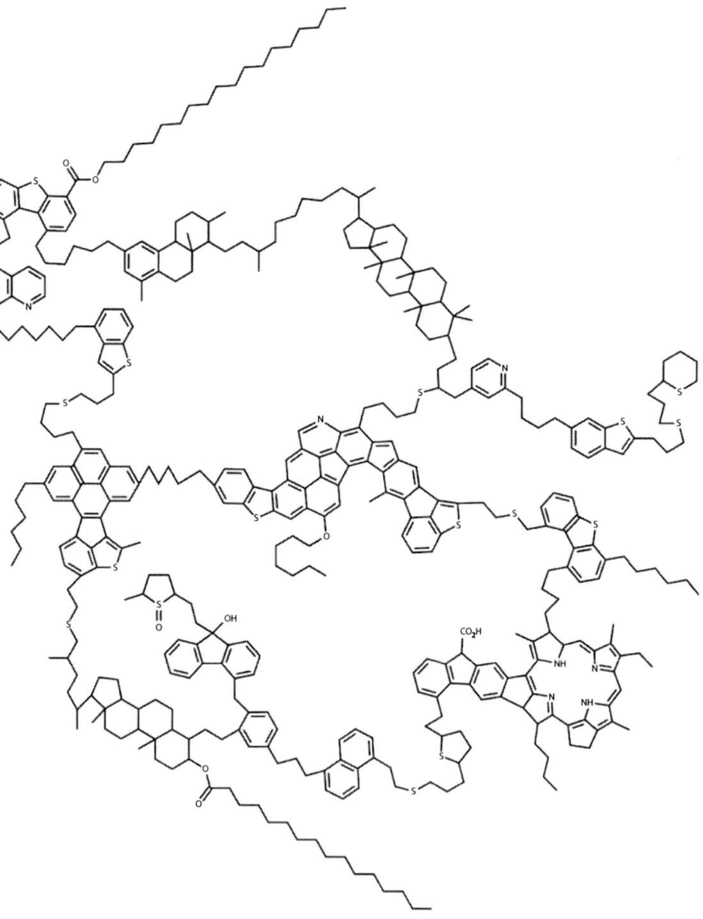

(a)

FIGURE 1.2 Different hypothetical structures for an asphaltene molecule: (a) shows Archipelago structure.   (*Continued*)

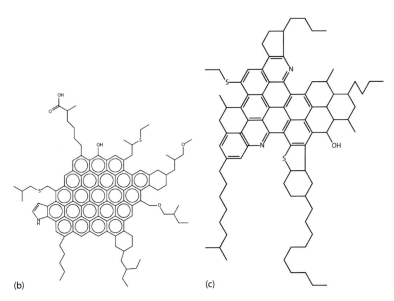

(b)   (c)

FIGURE 1.2 (Continued) Different hypothetical structures for an asphaltene molecule: (b) and (c) show the Island structure. (Reprinted from Murgich, J. et al., *Energy Fuels*, 13, 278–286, 1999; Rogel, E., *Energy & Fuels*, 14(3), 566–574, 2000; Speight, J.G. and Moschopedis, S.E., *Chemistry of Asphaltenes*, American Chemical Society, Washington, DC, 1982.)

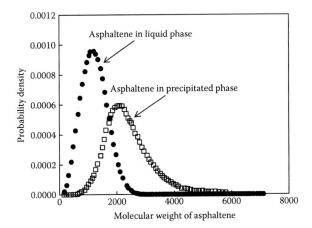

FIGURE 1.3 Typical asphaltene molecular weight distribution for asphaltenes in liquid phase and for precipitated asphaltenes. (Reprinted from Creek, J.L., *Energy Fuels*, 19(4), 1212–1224, 2005.)

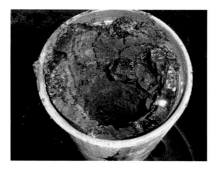

FIGURE 1.4 Asphaltene deposition in production tubing in an oil field in the Gulf of Mexico. (Reprinted from Yen, A. and Squicciarinim, M., *225th ACS National Meeting and Exposition*, New Orleans, LA, 2003.)

Figure 1.4 shows asphaltene deposition in production tubing in an oil field in the Gulf of Mexico (Yen and Squicciarini 2003).

Figure 1.5 illustrates a pressure–temperature (P–T) diagram for a certain oil sample. The oil flows from the reservoir, where it is normally a one-phase liquid at high-pressure, high-temperature conditions, and enters the production tubing. The pressure and temperature start to decrease as the oil moves upward toward the surface. As the pressure decreases, the oil swells because of the expansion of light hydrocarbon fractions in the oil. Asphaltenes, which are the heaviest fraction of the oil, are insoluble in the light hydrocarbons, and therefore, the oil becomes a poor solvent for the asphaltenes when the oil light fractions expand. As a result, asphaltenes leave the oil phase, precipitate, and form a separate phase. Asphaltenes appear as a separate phase for the first time at the asphaltene onset pressure (AOP), point A in Figure 1.5.

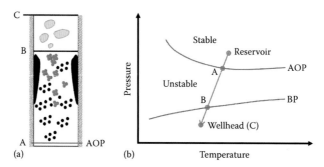

FIGURE 1.5 (a) Schematic of asphaltene precipitation, aggregation, and deposition in wellbore and (b) a typical pressure–temperature (P–T) diagram for an oil sample being produced from an underground reservoir. AOP, asphaltene onset pressure; BP, bubble point. (Reprinted from Ge, Q. et al., *9th International Conference on Heat Transfer, Fluid Mechanics, and Thermodynamics*, Malta, 2012.)

As the oil flows from point A to point B in Figure 1.5, more asphaltenes precipitate because of the more expansion of the oil light components. At point B, which is the oil bubble point (BP), light hydrocarbons start to evaporate, leaving the liquid phase. Therefore, the remaining oil becomes a better solvent for asphaltenes, and as a result, some of the asphaltenes redissolve into the oil phase. In some cases, when the oil reaches the wellhead, all the asphaltenes have been already redissolved in the oil phase and there is no sign of asphaltene precipitates. Based on this discussion, the maximum amount of asphaltene precipitation happens at the bubble point. Also, it is worth mentioning that the oil may reach its asphaltene onset pressure while it is still flowing inside the reservoir and before entering the production tubing. This mostly happens for the crude oils at their late stage of primary depletion or for the reservoirs under miscible gas injection such as carbon dioxide and nitrogen injections. These gases may get dissolved into the oil phase, make the oil a poor solvent for asphaltenes, and finally cause asphaltene precipitation and potential deposition inside the porous media.

Vargas et al. (2014) proposed a conceptual mechanism for asphaltene precipitation, aggregation, and aging that is illustrated in Figure 1.6. As previously mentioned in Section 1.1, asphaltenes aggregate even at a low concentration in a good asphaltene solvent, such as toluene, and form asphaltene clusters called nano-aggregates (Indo et al. 2009; Mullins 2010). These nano-aggregates are composed of less than 10 asphaltene molecules and have a size of about 3 nm. Based on the scanning electron microscopy (SEM) observations, Vargas et al. (2014) proposed that the precipitation of asphaltenes can lead to the formation of intermediate particles (i.e., primary particles) with a size of a few hundred nanometers. These primary particles can then aggregate and form bigger clusters. Finally, some sort of rearrangement takes place in which the precipitated phase becomes more compact and acquires a solid-like appearance. This last step is called the *aging process*. Vargas et al. (2014) showed that the formation of clusters shortly after precipitation is a fully reversible process. However, once the material reaches the final step of the mechanism, redissolution is no longer readily achieved. This aged solid phase could be the main reason for asphaltene deposition problem.

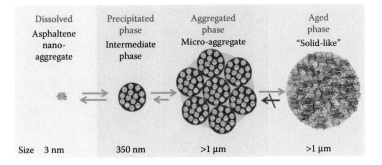

FIGURE 1.6 Proposed multistep mechanism for asphaltene precipitation, aggregation, and aging. The dual green arrows show the reversibility of the process. (Reprinted from Vargas, F. M. et al., *Offshore Technology Conference*, doi:10.4043/25294-MS, 2014.)

1.2.1 FIELD CASES

The asphaltene deposition problem has no geographical boundaries. It has been reported in all parts of the world, including the Middle East, North Africa, North Sea, and North and South Americas (Haskett and Tartera 1965; Tuttle 1983; Thawer et al. 1990). Also, asphaltene deposition may occur even in an oil field that contains a petroleum fluid with very low asphaltene content. For instance, although the asphaltene content of the crude oil from the Hassi Messaoud oil field in Algeria is only 0.06 wt%, severe asphaltene deposition was reported in the production tubings (Haskett and Tartera 1965).

There are several cases reported in the literature that show asphaltene deposition problems in different oil fields. The asphaltene deposition study in oil wells in the Hassi Messaoud field is one of the well-known cases (Haskett and Tartera 1965). A large number of deposition profiles in different wells and at different production rates were measured and analyzed. It was reported that in some cases the deposit thickness reached up to two-thirds of the tubing radius (Haskett and Tartera 1965). Figure 1.7 presents the asphaltene deposition profile in one of Hassi Messaoud oil wells.

In the Marrat oil field in west Kuwait, asphaltene deposition has been a serious problem in oil production tubings (Al-Kafeef et al. 2005). This problem significantly aggravated after the Iraqi invasion in 1990. The wells were open to the atmosphere for a long time, and as a result, the reservoir pressure decreased significantly. The caliper log was used to find the asphaltene deposit thickness in one of the oil wells in the Marrat oil field. It was found that almost 55% of the wellbore cross-sectional area was blocked by asphaltene deposits (Al-Kafeef et al. 2005).

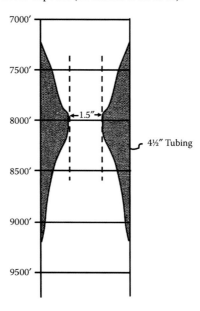

FIGURE 1.7   Asphaltene deposition profile in one of Hassi Messaoud oil wells. (Reprinted from Haskett, C.E. and Tartera, M., *JPT*, 17, 387–391, 1965.)

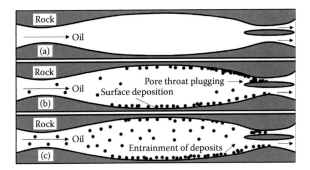

FIGURE 1.8   Asphaltene deposition in porous media: (a) Asphaltenes are dissolved in the oil phase; (b) asphaltenes precipitate as a result of changes in oil pressure, temperature, or composition. They deposit in the porous media and plug the pore throats; and (c) some of the deposits are entrained by the oil flow. (Reprinted from Wang, S. and Civan, F., *J. Energy Resour. Technol.*, 127, 310, 2005.)

Asphaltenes may also deposit in porous media, plugging the pore space, and decreasing the rate of oil production. Experimental investigations have shown that asphaltenes in crude oil may affect the rock properties. Most of these effects are the result of asphaltene adsorption on the rock surface (Clementz 1982). The main mechanisms that dominate asphaltene deposition in porous media are surface deposition, pore throat plugging, and entrainment of deposits, as shown in Figure 1.8 (Wang and Civan 2005).

In the Bangestan oil reservoir in the south of Iran, asphaltene deposition in the porous media has been a severe problem (Jamshidnezhad 2005). Although the asphaltene content of the oil is very low, 0.66 wt%, the skin formed by asphaltenes in the near wellbore region significantly decreased the rate of oil production (Jamshidnezhad 2005). Figure 1.9 shows monthly production from the Bangestan reservoir. It can be observed that the production rate considerably decreases at several points as a result of the asphaltene deposition in porous media (Jamshidnezhad 2005).

1.3   OBJECTIVES OF THE BOOK

The objective of this book is to provide the reader with a quick and deep enough immersion into the complex problem associated with asphaltene precipitation and deposition during oil production. The book shall provide the reader with a fundamental framework to better understand the chemistry, stabilization theories, and mechanistic approaches to understand and predict asphaltene behavior at high temperatures and pressures. The focus of this book is to introduce the asphaltene deposition problem and possible solutions in a practical way. The book teaches the reader the four main following items:

1. The state-of-the-art techniques for experimental determination of asphaltene precipitation and deposition.
2. Advanced modeling methods recently developed to predict asphaltene precipitation and deposition at high-pressure, high-temperature conditions.

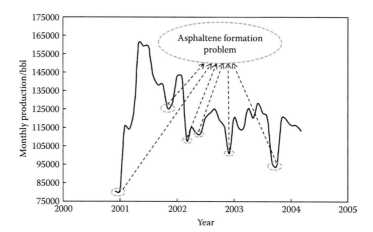

FIGURE 1.9 Oil production rate from Bangestan reservoir located in the south of Iran; asphaltene deposition caused significant decreases in the production rate. (Jamshidnezhad, M., *Canadian International Petroleum Conference*, Petroleum Society of Canada, 2005.)

3. Different strategies for mitigation and remediation of the asphaltene deposition problem along with the corresponding challenges.
4. Applications of the three previous items by introducing several case studies from different regions around the world, such as the Gulf of Mexico, the Middle East, and so on.

1.4 STRUCTURE OF THE BOOK

The structure of the book and the content of each chapter are briefly explained in this section.

Chapter 1: Introduction

In this chapter, we learn general information about asphaltenes and their definition and properties. A quick overview of asphaltene deposition and its impact on oil industry are also provided. Finally, the objective and what the reader should learn by reading this book is discussed at the end of the chapter.

Chapter 2: Crude Oil and Asphaltene Characterization

In Chapter 2, general properties of a crude oil sample are defined and the experimental techniques to measure these properties are briefly discussed. Different methodologies for characterization of a crude oil sample along with advantages and disadvantages of each technique are then presented. Various experimental methods for asphaltene characterization, including recent novel techniques, are explained in this chapter as well.

Asphaltene chemical structure and physical properties obtained from these experimental methods are also discussed.

Chapter 3: Experimental Investigation of Asphaltene Precipitation

There are several techniques available in the literature for experimental determination of asphaltene precipitation. Some of them are limited to ambient conditions and some could be applied to high-pressure, high-temperature conditions. These techniques along with their advantages and disadvantages are discussed in Chapter 3. Also, recent experimental studies including the results of the kinetics of asphaltene precipitation and aggregation and the reversibility of asphaltene precipitation investigations are also presented in this chapter.

Chapter 4: Asphaltene Precipitation Modeling

Modeling of asphaltene phase behavior has been an active research area for more than three decades. The developed modeling methods fall into two main categories: Solubility models and Colloidal models. Using equations of state under the solubility models category to model asphaltene phase behavior has shown promising results and has been recently applied in a number of commercial packages. In Chapter 4, first, the aforementioned modeling methods are explained. Then, implementation of the Perturbed Chain form of the Statistical Associating Fluid Theory (PC-SAFT) equation of state for modeling asphaltene phase behavior is described. Finally, a workout example for asphaltene phase behavior modeling using the equation of state is presented step by step.

Chapter 5: Experimental Determination of Asphaltene Deposition

Chapter 5 reviews the work conducted on experimental determination of asphaltene deposition in the oil reservoirs as well as in the wellbores and pipelines. Researchers have been actively involved in this area of asphaltene research for around a decade. Although several novel experimental setups have been recently developed to study asphaltene deposition, there is still a need for a straightforward and inexpensive experimental technique. The recent attempts to develop such an experimental technique are presented in this chapter.

Chapter 6: Modeling Methods for Prediction of Asphaltene Deposition

Modeling methods for prediction of asphaltene deposition in wellbores and pipelines as well as in porous media are discussed in Chapter 6. Modeling of asphaltene deposition in wellbores and pipelines has been investigated by researchers for around a decade, whereas modeling of asphaltene deposition in porous media has a longer history. However, researchers are still working on the development of a mechanistic model to properly explain and predict the phenomena of asphaltene deposition. These recent developments are also discussed in this chapter. Finally, two workout examples, one for asphaltene deposition in the wellbore and the other one for asphaltene deposition in porous media, are presented step by step.

Chapter 7: Strategies for Mitigation and Remediation of Asphaltene Deposition

The current mitigation and remediation strategies used to treat asphaltene deposition along with challenges of each methodology are explained in

Chapter 7. Assessment of chemical additives is a crucial step in the chemical treatment of asphaltene deposition and the currently available techniques show mixed results in the field. Therefore, new technologies for the proper assessment of chemical additives are presented in Chapter 7. Finally, the best practices for mitigation and remediation of asphaltene deposition problem, including the development of a new generation of asphaltene inhibitors, are discussed at the end of this chapter.

Chapter 8: Case Studies and Field Applications

Several case studies are presented and discussed in Chapter 8. Through these case studies, the reader may appreciate how the knowledge gained by reading this book can help to better understand, analyze, and solve problems related to asphaltene deposition. These case studies fall into four main categories:

1. Modeling of asphaltene phase behavior and comparison against experimental data.
2. Prediction of the occurrence and the magnitude of asphaltene deposition using computer models and comparison against field data.
3. Chemical inhibition of asphaltene deposition.
4. Interrelation of asphaltene deposition and other flow assurance problems.

Chapter 9: Conclusions, Current Research, and Future Directions

The lessons learned from this book are discussed in Chapter 9. Also, the current research directions, as well as the unresolved challenges for the topics explained in Chapters 2 through 8, are described in this chapter. Moreover, the recently developed research areas in the field of asphaltenes, as well as the strategies and the challenges associated with each research area, are presented in the final chapter of this book.

REFERENCES

Al-Kafeef, S. F., F. Al-Medhadi, and A. D. Al-Shammari. 2005. A simplified method to predict and prevent asphaltene deposition in oilwell tubings: Field case. *SPE Production & Facilities* 20 (2): 126–132. doi:10.2118/84609-PA.

Badre, S., C. C. Goncalves, K. Norinaga, G. Gustavson, and O. C. Mullins. 2006. Molecular size and weight of asphaltene and asphaltene solubility fractions from coals, crude oils and bitumen. *Fuel* 85 (1): 1–11. doi:10.1016/j.fuel.2005.05.021.

Borton, D., D. S. Pinkston, M. R. Hurt, X. Tan, K. Azyat, A. Scherer, R. Tykwinski, M. Gray, K. Qian, and H. I. Kenttämaa. 2010. Molecular structures of asphaltenes based on the dissociation reactions of their ions in mass spectrometry. *Energy & Fuels* 24 (10): 5548–5559. doi:10.1021/ef1007819.

Boussingault, M. 1837. Memoire Sur La Composition Des Bitumens. *Annales de Chimie et de Physique* 64: 141–151.

Buckley, J. S., and J. X. Wang. 2006. Personal communication.

Chacon, M. L., S. Rowland, and R. Rodgers. 2017. Asphaltenes are a continuum of island and archipelago structural motifs. In *Petroleum Phase Behavior and Fouling Conference*, Le Havre, France, June 11–15.

Clementz, D. M. 1982. Alteration of rock properties by adsorption of petroleum heavy ends: Implications for enhanced oil recovery. *Society of Petroleum Engineers*. doi:10.2118/10683-MS.

Creek, J. L. 2005. Freedom of action in the state of asphaltenes: Escape from conventional wisdom. *Energy & Fuels* 19 (4): 1212–1224. doi:10.1021/ef049778m.

Dickie, J. P., M. N. Haller, and T. F. Yen. 1969. Electron microscopic investigations on the nature of petroleum asphaltics. *Journal of Colloid and Interface Science* 29 (3): 475–484.

Freed, D. E., N. V. Lisitza, P. N. Sen, and Y. Q. Song. 2007. Molecular composition and dynamics of oils from diffusion measurements. In *Asphaltenes, Heavy Oils, and Petroleomics*, Springer-Verlag New York.

Ge, Q., Y. F. Yap, F. M. Vargas, M. Zhang, and J. C. Chai. 2012. Numerical modeling of asphaltene deposition. In *9th International Conference on Heat Transfer, Fluid Mechanics, and Thermodynamics*, Malta.

González, F. A. 2015. Asphaltene deposition economic impact. Reservoir performance global community of practice lead BP.

Groenzin, H., and O. C. Mullins. 1999. Asphaltene molecular size and structure. *The Journal of Physical Chemistry A* 103 (50): 11237–11245. doi:10.1021/jp992609w.

Groenzin, H., and O. C. Mullins. 2000. Molecular size and structure of asphaltenes from various sources. *Energy & Fuels* 14 (3): 677–684. doi:10.1021/ef990225z.

Haskett, C. E., and M. Tartera. 1965. A practical solution to the problem of asphaltene deposits-Hassi Messaoud field, Algeria. *Journal of Petroleum Technology* 17 (4): 387–391. doi:10.2118/994-PA.

Indo, K., J. Ratulowski, B. Dindoruk, J. Gao, J. Zuo, and O. C. Mullins. 2009. Asphaltene nanoaggregates measured in a live crude oil by centrifugation. *Energy & Fuels* 23 (9): 4460–4469. doi:10.1021/ef900369r.

Jamshidnezhad, M. 2005. Prediction of asphaltene precipitation in an Iranian south oil field. In *Canadian International Petroleum Conference*. Petroleum Society of Canada. Retrieved from https://www.onepetro.org/conference-paper/PETSOC-2005-015.

Kokal, S. L., S. G. Sayegh, and others. 1995. Asphaltenes: The cholesterol of petroleum. In *Middle East Oil Show*. Society of Petroleum Engineers. Retrieved from https://www.onepetro.org/conference-paper/SPE-29787-MS.

Mansoori, G. A. 2009. A unified perspective on the phase behaviour of petroleum fluids. *International Journal of Oil, Gas and Coal Technology* 2 (2): 141–167.

Mullins, O. C. 2010. The modified yen model. *Energy & Fuels* 24 (4): 2179–2207. doi:10.1021/ef900975e.

Murgich, J., J. A. Abanero, and O. P. Strausz. 1999. Molecular recognition in aggregates formed by asphaltene and resin molecules from the Athabasca oil sand. *Energy & Fuels* 13 (2): 278–286. doi:10.1021/ef980228w.

Rogel, E. 2000. Simulation of interactions in asphaltene aggregates. *Energy & Fuels* 14 (3): 566–574. doi:10.1021/ef990166p.

Schneider, M. H., A. B. Andrews, S. Mitra-Kirtley, and O. C. Mullins. 2007. Asphaltene molecular size by fluorescence correlation spectroscopy. *Energy & Fuels* 21 (5): 2875–2882. doi:10.1021/ef700216r.

Speight, J. G., and S. E. Moschopedis. 1982. On the molecular nature of petroleum asphaltenes. In *Chemistry of Asphaltenes*, J. W. Bunger and N. C. Li (Eds.), 195: 1–15. Washington, DC: American Chemical Society.

Tang, W., M. R. Hurt, H. Sheng, J. S. Riedeman, D. J. Borton, P. Slater, and H. I. Kenttämaa. 2015. Structural comparison of asphaltenes of different origins using multi-stage tandem mass spectrometry. *Energy & Fuels* 29 (3): 1309–1314. doi:10.1021/ef501242k.

Tavakkoli, M., S. R. Panuganti, V. Taghikhani, M. R. Pishvaie, and W. G. Chapman. 2014. Understanding the polydisperse behavior of asphaltenes during precipitation. *Fuel* 117: 206–217.

Thawer, R., D. C. A. Nicoll, and G. Dick. 1990. Asphaltene deposition in production facilities. *SPE Production Engineering* 5 (4): 475–480.

Tuttle, R. N. 1983. High-pour-point and asphaltic crude oils and condensates. *Journal of Petroleum Technology* 35 (6): 1–192.

U.S. Energy Information Administration. 2016. International Energy Outlook 2016 with Projections to 2040.

Vargas, F. M., M. Garcia-Bermudes, M. Boggara, S. Punnapala, M. I. L. Abutaqiya, N. T. Mathew, S. Prasad, A. Khaleel, M. H. Al Rashed, and H. Y. Al Asafen. 2014. On the development of an enhanced method to predict asphaltene precipitation. In *Offshore Technology Conference*. doi:10.4043/25294-MS.

Vargas, F. M. 2010. Modeling of asphaltene precipitation and arterial deposition. Rice University. Retrieved from https://scholarship.rice.edu/handle/1911/62084.

Wang, S., and F. Civan. 2005. Modeling formation damage by asphaltene deposition during primary oil recovery. *Journal of Energy Resources Technology* 127 (4): 310. doi:10.1115/1.1924465.

Wargadalam, V. J., K. Norinaga, and M. Iino. 2002. Size and shape of a coal asphaltene studied by viscosity and diffusion coefficient measurements. *Fuel* 81 (11): 1403–1407.

Yen, A., and M. Squicciarini. 2003. Characterization of asphaltenes from chemica treated oil, *225th ACS National Meeting and Exposition*, March 23–27, 2003, New Orleans, LA.

2 Crude Oil and Asphaltene Characterization

R. Doherty*, S. Rezaee*, S. Enayat,
M. Tavakkoli, and F. M. Vargas

CONTENTS

---

* R. Doherty and S. Rezaee contributed equally to this work.

Crude oil is one the most important substances used by modern society, and probably, the least understood. However, what it is known is that its behavior can be related to its chemical composition and physical properties. The first part of this chapter focuses on the description of the chemical and physical properties of crude oil and offers a review of the experimental methods involved in their determination. The properties discussed include density and API gravity, viscosity, molecular weight, water content, refractive index, and solubility parameter. The correlation between these properties is also highlighted. This part also gives an overview of the main methods to classify crude oils according to their boiling point, chemical structure, and the different polarities of its components, including the advantages and disadvantages of the methods used for each classification.

Because asphaltenes, the heaviest part of the crude oil, are responsible for the million-dollar losses caused by their deposition on wells and pipes, the second part of the chapter is dedicated to exploring more closely their properties, structure, and composition. Other properties, such as aromaticity, molecular weight, polydispersity, and wettability that help to understand, predict, and prevent their deposition are described as well.

2.1 CRUDE OIL

The word *petroleum* comes from the Latin words *petra*: *rock* and *oleum*: *oil*. Unrefined petroleum is better known as crude oil. This is a liquid mixture composed mainly of carbon and hydrogen but also containing some other elements such as oxygen, nitrogen, sulfur, and some trace heavy metals such as nickel, vanadium, copper, cadmium, and lead (Speight 2014). In fact, it is believed that these heavy metals acted as catalysts in the formation of petroleum (Riazi 2005). Because the diversity of its origin, its color, odor, and flow properties vary widely (Ceric 2012; Speight 2014). Nowadays, the most accepted theory about the origin of crude oil is that of an organic origin. According to this theory, dead organic material, including algae, zooplankton, and small aquatic animals, accumulated on the bottom of oceans, riverbeds, and swamps, mixing with clay, silt, and sand. Over time, more sediment piled up on top, increasing the heat and the pressure. In the absence of oxygen, anaerobic bacteria transformed the organic layer into a dark and waxy substance known as kerogen. Left alone, the kerogen molecules cracked and broke up into shorter and lighter molecules composed almost solely of carbon and hydrogen atoms. It has been reported that the geologic time needed to produce petroleum is 1 million years, under a maximum pressure of 17 MPa and a maximum temperature of 373 K–393 K (Riazi 2005; Buryakovsky et al. 2005).

Crude oil did not originate in the place it is found today, but much deeper. Through time, it migrated near the surface of Earth and stopped in the sediments layer in which water was present and above which was a dense leak proof cover formed by marl, clay, and oil shell. This accumulation of crude oil is called a *reservoir*. Reservoirs can be found beneath land or the ocean floor, and they are *not* equally distributed around the globe.

The latest research in 2016 show that the largest proved reserves, or the quantities of petroleum determined by analysis of geological and engineering data, are found in

Venezuela (300 billion of bbl), Saudi Arabia (266 billion of bbl), Canada (170 billion of bbl), Iran (158 billion of bbl), and Iraq (143 billion of bbl). The United States is positioned in 11th place with 36.5 billion of bbl ("International Energy Statistics" 2017; "The World Factbook—Central Intelligence Agency" 2017).

Crude oil has been used since the times of the Sumerians (6000 years B.C.) in construction, in ornamental works, and as a fossil fuel. Other civilizations used it as sealer for ships, for plaster and coating production, moisture protection, road construction, mummification, lightening, and even as disinfection agent in medicine. After the Greek and Roman Empires went into decadence, its use and importance were greatly reduced; however, in the middle of the nineteenth century, its importance as an energy source gained force, and, now, it is still the most used one around the globe (Ceric 2012). According to the U.S. Energy Information Administration (2017), the total consumption of energy in the United States was of 97.7 quadrillions BTU in 2015. Out of this amount, 81% was produced by the three major fossil fuels (crude oil: 36%, natural gas: 29%, and coal: 16%).

2.1.1 PROPERTIES

Crude oil is a mixture of widely varying constituents and proportions. Consequently, its physical properties also can vary widely. In appearance, for example, its color can go from yellow to black, passing for greenish, reddish, and dark brown (Donaldson et al. 1985). The color depends on the way the incident light interacts with the molecules and molecular bonds in the fluid, and it is transmitted or reflected, giving it different hues. In a similar way, other properties can vary according to the composition of crude oil. A summary of the most important properties is presented next.

2.1.1.1 Density and API Gravity

Density (ρ) is defined as the mass per unit of volume (e.g., kg/m^3). It is a state function that depends on both temperature and pressure. The density of crude oil, as for other liquids, decreases as the temperature increases and the effect of pressure is usually negligible at moderated pressures—less than few bars (Riazi 2005). At higher pressures, the density of crude oil will increase because an increase in pressure will lead to a decrease in volume.

At the standard conditions for the petroleum industry (60°F or 288.65 K, 1 atm) the density of crude oil can vary from 800 kg/m^3 to more than 1000 kg/m^3 (Speight 2014). Density can also be reported as specific gravity (SG), which is the ratio of the density of a liquid at temperature, T, to the density of water at the same temperature (at standard conditions, the density of water is 999 kg/m^3).

In 1921, a more practical measurement to express the density of crude oil was created by the American Petroleum Institute (Huc 2010; Groysman 2014). The API gravity (°API) is a measurement of how heavy a crude oil is in relation to water and it is defined as:

$$\text{API gravity} = \frac{141.5}{SG \ (\text{at} \ 60°F)} - 131.5 \qquad (2.1)$$

In this scale, water has an API gravity of 10°; crude oils with an API gravity larger than 10 are lighter and will float on water, whereas if it is less than 10, the crude oil is heavier and sinks.

The broad difference in density is used to divide the crude oil into four grades: light (ρ < 870 kg/m^3; API > 31.1°), medium (870 < ρ < 920 kg/m^3; 22.3° < API < 31.1°), heavy (920 < ρ < 1000 kg/m^3; 10° < API < 22.3°), and extra heavy (ρ > 1000 kg/m^3; API < 10°) (Huc 2010; Ancheyta and Speight 2007; Jafarinejad 2016).

Density can be determined by the ASTM D287 2012, ASTM D1217 2015, and ASTM D1298 2012. The first two methods make use of a hydrometer, which is a floating cylinder graduated by API gravity units. Also, both are based on the principle that the gravity of a liquid varies directly with the depth of immersion of a body floating it. The API gravity can be read from the hydrometer scale when this is left freely floating inside a cylinder full of crude oil at a determined temperature.

In the third method, density is measured with a Bingham pycnometer. The pycnometer is equilibrated to the desired temperature and filled with the sample. Then, it is weighted, and the weight of the sample is obtained by subtracting the weight of the empty pycnometer from the total weight. The density can be calculated by dividing this weight by the volume. If the specific gravity is needed, the same procedure can be done with pure water at the same temperature.

For measurements at higher pressures and temperatures, different and more sophisticated equipment is used. As an example, Wang et al. (2016) have reported the use of an Anton Paar DMA HPM high-pressure and high-temperature (HPHT) densitometer for density measurements of up to 137.4 MPa and 684.3 K. Table 2.1 shows the properties of some of the oils mentioned in this book.

2.1.1.2   Viscosity

Viscosity of a fluid is a measure of its resistance to gradual deformation by shear stress or tensile stress; in other words, it is the measure of a fluid's internal resistance

---

TABLE 2.1

Properties of Some of the Oils Mentioned in This Book

Crude Oil	Density (g/cm^3; Tamb)	Viscosity (cP; Tamb)	Water Content (wt%)	Molecular Weight (g/mol)
A	0.884	56.0	0.07	273.0
C	0.885	22.0	0.03	245.0
P1	0.878	21.3	0.04	243.1
P60	0.911	338.9	0.06	339.9
P7	0.876	15.3	0.03	251.3
SB	0.867	60.5	0.03	176.1
SE	0.893	33.8	0.18	220.4
CN	0.899	55.0	0.08	278.9
S4	0.826	5.4	0.03	NA

Source: Courtesy of Vargas Research Laboratory, Rice University, Houston, TX.

to flow. It gives a notion of the *thickness* of the fluid. Knowing this parameter is useful for from designing the appropriate extraction process and transport of crude oil, to the planning and execution of measures for the remediation of spills in case of an accident.

Newton's viscosity law states that the shear stress between adjacent layers of a fluid is proportional to the gradients of velocity between the two layers (Riazi 2005). Figure 2.1 shows a representation of a fluid moving along a solid boundary at a laminar flow.

$$\tau_\alpha \frac{dv_x}{d_y} \text{ or } \tau = -\mu \frac{dv_x}{d_y} = -v_f \frac{d(\rho v_x)}{dy} \tag{2.2}$$

where:

$$v_f = \frac{\mu}{\rho}$$

τ is the shear stress (in N/m²)
v_x is the velocity (m/s)
dv_x/d_y is the velocity gradient or share rate (s⁻¹)
ρ is the density of the fluid (kg/m³)
μ is the dynamic viscosity (Pa s)
v_f is the kinematic viscosity (m²/s) (Bird et al. 2002)

Dynamic viscosity (also known as *absolute viscosity*) is a measurement of a fluid's resistance to flow when an external force is applied, while the kinematic viscosity is a measurement of the resistive flow of a fluid under only the action of gravity. In other words, dynamic viscosity provides information on the force needed to make the fluid flow at a certain rate, whereas kinematic viscosity tells how fast the fluid is moving when a determined force is applied.

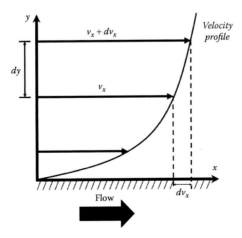

FIGURE 2.1 Laminar flow of a fluid along a solid boundary. (Adapted from Bird, R. B. et al., *Transport Phenomena*, 2nd ed., John Wiley & Sons, New York, 2002.)

Kinematic viscosity is determined experimentally by the ASTM D445 (2017). According to this method, the viscosity is determined at different temperatures, usually 298.15 K (77°F) and 373.15 K (212°F), by measuring the time that takes for a fixed volume of oil to flow under gravity through a calibrated viscometer. Other examples of viscometers can be found in the ASTM D445 (2017). The kinematic viscosity is then calculated by multiplying the time by the calibration constant of the viscometer. Dynamic viscosity can also be calculated from this result by multiplying the kinematic viscosity by the density of crude oil at the same temperature.

It is important to remember at this point that crude oil is a non-Newtonian fluid; that is, its viscosity varies according to the applied stress. There is more sophisticated equipment to determine the dynamic viscosity as a function of the stress such as rotational viscometers (i.e., cone and plate, Stabinger viscosity). Several of these methods are specified in the ASTM D4287 (2014); ASTM D5481 (2013); ASTM D7042 (2016); ASTM D7395 (2007). In addition to these standards, ASTM D7152 (2011) provides a methodology to calculate the viscosity of blends of petroleum products.

Oil viscosity is a strong function of many thermodynamic and physical properties such as oil composition, gas-to-oil ratio (GOR), temperature, and pressure and saturation pressure. As expected, viscosity increases with decreases in crude oil API gravity and decreases with temperature. Also, the gas dissolved decreases the viscosity. Above saturation pressure, viscosity increases almost linearly with pressure. Viscosity is usually determined by experimental methods at the temperature of the reservoir; however, many correlations have been developed to predict the viscosity of crude oil at the reservoir conditions of temperature and pressure (Alomair et al. 2016; Riazi and Al-Otaibi 2001; Al-Rawahi et al. 2012).

2.1.1.3   Molecular Weight

Molecular weight is a fundamental physical property that can be used with other physical properties to characterize pure hydrocarbons and mixtures of them. The information about molecular weight is necessary for the application of many correlation-based methods that are used for determining the composition of the heavier fractions of petroleum.

There are several experimental techniques available in the literature for molecular weight determination of crude oil samples. The most frequently used techniques are: (1) Freezing Point Depression; and (2) Thermoelectric Measurement of Vapor Pressure.

2.1.1.3.1   *Freezing Point Depression*

The molecular weight is measured by the freezing point depression or cryoscopic method, from the lowering of the freezing point, which a solution shows with respect to the pure solvent. This is made possible by the three fundamental concepts, found experimentally, and have historically led to the freezing point depression method (Biltz 1899):

1. A solution freezes at a lower temperature than the solvent.
2. The depression of the freezing point is proportional to the concentration of the solution.
3. Equimolecular solutions in the same solvent show an equal depression of the freezing point.

The solvent that is normally used for measuring the molecular weight of a crude oil sample using the freezing point depression method is benzene.

2.1.1.3.2 Thermoelectric Measurement of Vapor Pressure

This method has been published as the standard method ASTM D2503 (1992) for measurement of hydrocarbons molecular weight. It can be applied to petroleum fractions with molecular weights up to 3000 g/mol; however, the precision of the method has not been established for the molecular weight values above 800 g/mol. The method should not be applied to oils with boiling points lower than 493.15 K (ASTM D2503 1992).

In this method, a weighed portion of the sample is dissolved in a known quantity of appropriate solvent. A drop of this mixture and a drop of the solvent are suspended, side by side, on separate thermistors in a closed chamber, which is saturated with solvent vapor. Because the vapor pressure of the solution is lower than that of the solvent, solvent condenses on the sample drop and causes a temperature difference between the two drops. This change in temperature is measured and used to determine the molecular weight of the sample using a previously prepared calibration curve (ASTM D2503 1992).

Other techniques for determination of molecular weight are the boiling point method and the method based on the principle of lowering of solubility. The boiling point technique is similar to the freezing point depression and is based on these fundamental statements: A solution boils at higher temperature than the solvent; the rise in the boiling point is proportional to the concentration; and finally, the equimolecular solutions in the same solvent show the same rise in boiling point. The experimental technique on the principle of lowering of solubility works based on the relative lowering of the capacity to dissolve a second liquid, which a solvent experiences on adding a foreign substance (Biltz 1899).

2.1.1.4 Water Content

The determination of the amount of water in crude oil and petroleum products has always been important. Rather than paying the cost of crude oil for a mixture of oil and water, contracts have been based on net dry oil. This is calculated by removing the amount of water and sediment, which is determined by analyzing a sample of the oil from the total gross standard volume. Knowledge of the water content of petroleum products is important in refining, purchase and sale, and transfer of products and is useful in predicting the quality and performance characteristics of the products (Nadkarni 2000). There are three different methods for determining the water content of crude oil and petroleum products: centrifugation, distillation, and Karl Fischer titration.

2.1.1.4.1 Centrifugation

The oldest and most widely used method to determine the water content of a crude oil sample is the centrifuge test. In this method, known volumes of crude oil and solvent are placed in a graduated centrifuge tube and heated up to 333 K. After centrifugation, the volume of the higher density water and sediment at the bottom of the tube is read. For some waxy crude oils, temperatures of 343 K or higher may be required to completely melt the wax crystals. In this case, waxes will not be measured as sediment (ASTM D4007 2002; ASTM D2709 1996). This test is detailed in several standard methods, including ASTM D4007 (2002) and ASTM D2709 (1996). Even if all details of this method are followed carefully, the total water content will be underestimated for most types of crude oil samples.

2.1.1.4.2 Distillation

During the past few years, the determination of the water content by distillation test (also called the *Dean and Stark test*), has been used more frequently because it is more accurate compared to the centrifuge test. Also, it is the accepted referee method when parties cannot agree on centrifuge results. In the test, which is described in detail in ASTM D4006 (2005), the sample is heated under reflux conditions with a water immiscible solvent, which co-distills with the water in the sample. Condensed solvent and water are continuously separated in a trap; the water settles in the graduated section of the trap, and the solvent returns to the distillation flask (ASTM D4006 2005).

2.1.1.4.3 Karl Fischer Titration

The third method for water content determination is based on the titration of the sample with Karl Fischer reagent, described in ASTM D4928 (2000) and ASTM D6304 (2007). This method has been used for many years to determine the water content of liquid petroleum products, but it has not been used for crude oil samples until the past few years. In this technique, a small amount of sample is injected into the titration vessel of a Karl Fischer apparatus in which iodine for the Karl Fisher reaction is coulometrically generated at the anode. When all the water has been titrated, excess iodine is detected by an electrometric end-point detector, and the titration is terminated. Based on the stoichiometry of the reaction, 1 mol of iodine reacts with 1 mol of water; and therefore, according to Faraday's Law, the quantity of water is proportional to the total integrated current (ASTM D4928 2000; ASTM D6304 2007).

The Karl Fischer method has several advantages over the other water determination methods, the most important of which is increased accuracy. Also, samples can be analyzed in less than 5 minutes, compared to at least 30 minutes for the centrifuge test and several hours for the distillation test. The Karl Fischer text can be used in the back of a pickup truck for field use, on an offshore platform for production use, and in a laboratory for quality control and other applications.

2.1.1.5 Refractive Index

One of the fundamental physical properties that can be used with other physical properties to characterize pure hydrocarbons and their mixtures is the Refractive Index (Nadkarni 2000). Refractive index is the ratio of the velocity of light in air at a specific wavelength to its velocity in the substance under study (ASTM D1747 2004). The relative index of refraction is defined as the sine of the angle of incidence divided by the sine of the angle of refraction, as light passes from air into the substance. If absolute refractive index (referred to vacuum) is needed, this value should be multiplied by the factor 1.00027; 1.00027 is the absolute refractive index of air. The refractive index of liquids varies inversely with both wavelength and temperature (ASTM D1218 2002).

The refractive index is an important optical parameter of crude oils that is used in various calculations related to their compositions. It is also used in a number of industrial applications such as determination of the asphaltene precipitation onset and the solubility parameter measurement (Speight 2014; Otremba 2000; Buckley 1999; Buckley et al. 1998).

The most widely used technique to measure the refractive index of crude oils is based on the ASTM D1218 (2002). There are commercially available refractometers, which are used to measure the refractive index of transparent, light-colored, or strongly colored hydrocarbons in the range of 1.33–1.50 (or above) based on the ASTM D1218. In this method, the refractive index is measured by the critical angle method using monochromatic light from a sodium lamp. The instrument is previously calibrated using certified liquid standards (ASTM D1218 2002). In the standard D1218, the test temperature varies from 293 K to 303 K. There is another method, ASTM D1747 (2004), which covers the measurement of refractive index of transparent and light-colored viscous hydrocarbon liquids and melted solids that have refractive indices in the range between 1.33 and 1.60, and at temperatures from 353 K to 373 K. Temperatures lower than 353 K can be used if the melting point of the sample is at least 10 K below the test temperature (ASTM D1747 2004).

There are also other techniques in the literature, which are not commercially available, for measuring the refractive index of a crude oil system such as the use of a capillary tube interferometer (El Ghandoor et al. 2003). In this technique, the oil-filled capillary tubes are illuminated by a thick helium-neon (HeNe) laser sheet and the obtained transverse interference patterns are projected on a screen, which its plane is perpendicular to the plane of the laser sheet. The patterns show bell-shaped fringes. The characteristic bell shape is mathematically demonstrated by calculating the optical path difference between a reference ray and a ray that passes through the oil sample. The refractive indices of the crude oil samples are then determined by measuring the detection angles of the fringes (El Ghandoor et al. 2003).

2.1.1.6 Solubility Parameter

The solubility parameter (δ) is a numerical value that indicates the relative solvency behavior of a specific solvent. To understand the solubility parameter, it is necessary to explore the relationship among solubility, van der Waals forces, and the cohesive energy density.

Van der Waals forces are the attractive forces (or intermolecular forces) that hold molecules together. The three most common types are London forces, dipole-dipole forces, and hydrogen bonding. London forces are the weakest of the intermolecular forces; they are the product of the temporary charges spontaneously developed in all compounds. Dipole-dipole are forces of medium intensity found in molecules with permanent or even partial charges (dipoles). Hydrogen bonding is the strongest of the three intermolecular forces, and it is a case of the dipole-dipole forces between atoms of hydrogen and fluorine, oxygen or nitrogen.

For a solute to be dissolved in a solvent, all intermolecular attractions should be broken, including the solvent-solvent and solute-solute forces. Once these forces do not exist, a new intermolecular attraction should be formed between solvent and solute molecules (solvation). This is accomplished best when the attractions between molecules of both components are similar.

A similar situation happens when a liquid is evaporated. When a liquid is heated to the boiling point, the applied heat will increase first the temperature of the liquid, but at the boiling point, the temperature will remain constant and all heat will be used to increase the kinetic energy of the molecules to a point in which the attraction forces will be broken. This energy is the heat of vaporization, and because it is a direct indication of the energy necessary to separate the molecules of a liquid, it is also an indication of the van der Waals forces associated with that liquid.

This energy is also known as cohesive energy (E), and when it is expressed as the quotient between the heat of vaporization and the molar volume, it receives the name of cohesive energy density (CED) (Burke 1984; Hansen 2007; Krevelen and Nijenhuis 2009).

Hildebrand defined the solubility parameter as the square root of the cohesive energy density (Barton 1991). In the following equation, E is the heat of vaporization, the product RT is the energy of the same material as an ideal gas at the same temperature, and $\hat{V}$ the molar volume (Kitak et al. 2015). This approximation is more detailed explained in Section 4.2.2.1.

$$\delta = \sqrt{\text{CED}} = \sqrt{\frac{E - RT}{\hat{V}}} \tag{2.3}$$

The Hildebrand solubility parameter was developed for simple liquid mixtures and does not account for the molecular association caused by polar interactions, including hydrogen bonding.

Many other researchers have added some other considerations to the development of this equation or have proposed other methodologies to include a vaster number of substances. For example, Hansen (1967) introduced a model that included the three forces when calculating the cohesion energy (Burke 1984). Other proposed methods use the group contribution of the molecular structure, (Fedors 1974; Krevelen and Nijenhuis 2009; Stefanis and Panayiotou 2008), some others use the refractive index (Karger et al. 1978; Wang and Buckley 2001), and others have used parameters such as density and viscosity (Correra et al. 2005). Also, some methodologies have been developed to calculate the solubility parameter at high temperature and pressure (Wang et al. 2016).

2.1.1.7 Correlation between Properties

Measuring each of the crude oil properties is time consuming, but the use of correlations between these properties helps to reduce the experimental time. On the other hand, the result of these correlations can be used both as a baseline and to verify the experimental data. Some of these correlations are described next.

2.1.1.7.1 Solubility Parameter

The solubility parameter is often needed for the modeling of complex systems such as asphaltenic crude oils, however, its experimental determination can be quite challenging. To simplify its determination, Wang and Buckley (2001) proposed a correlation between the solubility parameter and the refractive index of a non-polar hydrocarbon.

$$\delta = 52.042 F_{RI} + 2.904 \tag{2.4}$$

where:

δ is the solubility parameter in MPa$^{0.5}$

F_{RI} is a function of the refractive index (n) and is calculated by:

$$F_{RI} = \frac{n^2 - 1}{n^2 + 2} \qquad (2.5)$$

Vargas and Chapman (2010) proposed a one-third rule for a wide range of hydrocarbons according to the Lorentz–Lorenz model (A relationship between molar refractivity, R_m, and molecular weight, M_w).

$$R_m = \left(\frac{n^2 - 1}{n^2 + 2}\right)\frac{M_w}{\rho} = F_{RI}\frac{M_w}{\rho} \qquad (2.6)$$

Based on "One-Third" rule, the F_{RI} divided by the mass density, ρ, is a constant approximately equal to one-third. There is a deviation from one-third value for light and heavy hydrocarbons. The ratio of F_{RI}/ρ is approximately constant in a temperature range of 283 K–343 K. Also, it is possible to apply one-third rule in the Wang-Buckley correlation to find the relation between the solubility parameter and the mass density. For the light components, it is better to use the Lorentz–Lorenz expansion in Equation 2.7 to correlate the solubility parameter (in MPa$^{0.5}$) to the mass density (in g/cm^3).

$$\delta = 2.904 + 26.302\rho - 20.5618\rho^2 + 12.0425\rho^3 \qquad (2.8)$$

$$\frac{F_{RI}}{\rho} = \frac{1}{\rho^\circ} = 0.5054 - 0.3951\rho + 0.2314\rho^2 \qquad (2.9)$$

2.1.1.7.2 Viscosity

As mentioned previously, oil viscosity plays an important role in oil production, transportation, and oil recovery processes. Different correlations have been proposed to predict oil viscosity.

Beggs and Robinson (1975) proposed a correlation for the viscosity as a function of API gravity and temperature:

$$\log(\mu_{od} + 1) = a_1 + a_2\gamma_0 + a_3\log T \qquad (2.10)$$

where:

γ_0 is the oil gravity (°API)

T is the temperature (°F)

a_1, a_2, and a_3 are constants

μ_{od} is the viscosity of the dead oil (cP)

Later, (Egbogah and Ng 1990) found a significant difference between the experimental and calculated viscosity by the Beggs and Robinson correlation. For this reason, they added a new parameter—the pour point temperature—to the correlation which decreased the absolute error and modified the viscosity correlation.

$$\log(\mu_{od}+1) = a_1 + a_2 T_p + a_3 SG + (a_4 + a_5 T_p + a_6 SG)\log(T - T_p) \qquad (2.11)$$

where:

μ_{od} is the viscosity of dead oil (cP)
SG is the specific gravity (°API)
T is the temperature of interest (K)
T_p is the pour point temperature (K)

Despite Equation 2.10 provides better results than Equation 2.11, the results are still not accurate enough, probably because oil behaves like a non-Newtonian liquid at high temperatures.

The most accurate viscosity model was proposed in by Alomair et al. (2016) as a function of the temperature and the density:

$$\ln\mu_{od} = \hat{a} + \frac{\hat{b}}{T^2} + \hat{c}\left(\rho_{od}^2\right)\ln\rho_{od} \qquad (2.12)$$

where:

μ_{od} (in cP) and ρ_{od} (in kg/m³) are the viscosity and density of the dead oil
$\hat{a}, \hat{b}, \hat{c}$ are constants that are defined based on two different temperature ranges: normal temperature 293 K–373 K and high temperature 373 K–433 K

The average absolute error for this model is 8%.

In this model, the density is calculated according to the Standing method as shown in Equation 2.12. $\Delta\rho_p$ and $\Delta\rho_T$ are the changes in density resulting from changes in pressure and thermal expansion (Alomair et al. 2016).

$$\rho = \rho_{sc} + \Delta\rho_p + \Delta\rho_T \qquad (2.13)$$

$$\Delta\rho_p = \left[0.167 + (16.181)10^{0.0425\rho_{sc}}\right]\left(\frac{P}{1000}\right) \qquad (2.14)$$

$$\Delta\rho_T = \left[0.013 + 152.4\left(\rho_{sc} + \Delta\rho_p\right)^{-2.45}\right](T - 520) \qquad (2.15)$$
$$- \left[8.1\left(10^{-6}\right) - (90.06)10^{-0.764\left(\rho_{sc} + \Delta\rho_p\right)}\right]$$

where:

P is the pressure (in bar)
T is the temperature (in °R)
ρ_{sc} is the measured density at standard conditions (in kg/m³)

A more common way to find the density is the cubic equation of state (EOS) as explained in Section 4.2.3.1. It is also possible to calculate the viscosity of oil by applying the Vazquez and Beggs (1980) correlation.

$$\mu_o = \mu_{ob} \left(\frac{P}{BP} \right)^{\left[2.6 P^{1.187} 10^{(-3.9 \times 10^{-5} P - 5)} \right]} \tag{2.16}$$

The bubble point oil viscosity (μ_{ob}) can be calculated with the Chew and Connally (1959) correlations.

$$\mu_{ob} = A\mu_{od}^B \tag{2.17}$$

$$A = 0.2 + \frac{0.8}{10^{(0.00081 R_s)}} \tag{2.18}$$

$$B = 0.43 + \frac{0.57}{10^{(0.00072 R_s)}} \tag{2.19}$$

Dead oil viscosity (μ_{od}) can be calculated by the Glaso (1980) correlation:

$$\mu_{od} = \left(\frac{3.141 \times 10^{10}}{T^{3.444}} \right) \log \gamma_o^{[10.313 \log T - 36.447]} \tag{2.20}$$

where:
   μ_o is the oil viscosity in cP
   BP is the bubble point pressure in Psia
   T is the temperature in °F
   γ_o is the oil gravity in °API
   R_s is the solution GOR in scf/STB

2.1.1.7.3 Molecular Weight
Some other correlations between properties are reported in the literature, such as the one between the molecular weight and the viscosity (ASTM D2502 2014). In this method, the molecular weight of crude oil is calculated from kinematic viscosity measurements at 311 K (100°F) and 372 K (210°F). As limitations, this method can be only used for samples with mean relative molecular masses in the range from 250 to 700 g/mol and cannot be applied to crude oils that represent extremes of composition or with extremely narrow molecular weight.

2.1.2 CRUDE OIL FRACTIONATION
Crude oil contains different hydrocarbon molecules ranging from very light components to very heavy ones, such as highly asphaltenic crudes (Powers 2014). The study of the crude oil components, species by species, is not feasible because of the

large amount of hydrocarbon types. Instead, a hydrocarbon group type analysis is commonly employed (Powers 2014).

Crude oil fractionation is important because there is a wide variation in the properties of the lightest crude oil to the heaviest one. For example, heavy oil has higher viscosity, asphaltene content, and metals in comparison to conventional oil.

There are three main methods of crude oil fractionation: based on the boiling point, based on chemical structure, and based on polarity.

2.1.2.1   Based on Boiling Point

The most common and oldest way to separate the crude oil into its fractions is based on the different boiling temperature of its components. This process is better known as *fractional distillation.*

In a distillation tower (Figure 2.2), the crude oil is heated up until the evaporation of the most volatile components. At the top of the column, the lighter products such as butane, other liquid petroleum gasses (LPG), gasoline blending components, and naphtha are recovered. The components with a midrange boiling point stay in the middle of the distillation tower such as jet fuel, kerosene, and distillates. At the bottom of the column, at a temperature of 811 K (1000°F), the heaviest products such as residual fuel oil are recovered (Speight 2014).

The true boiling point (TBP) curve can be obtained based on the (ASTM D2892 2016). In this method, each component is separated at the same time. There is a large distillation column with 15 to 18 trays and a large reflux ratio of 5:1. The reboiler temperature and the distilled volume are recorded to plot the TBP curve. This method is time consuming and expensive in practice.

Another distillation method is the ASTM D86 (2016). This method is faster than the ASTM D2892 because there is no reflux. The sample is heated, and the vapors are collected and condensed while they are produced. The vapor temperature and the cumulative volume of liquid collected are recorded.

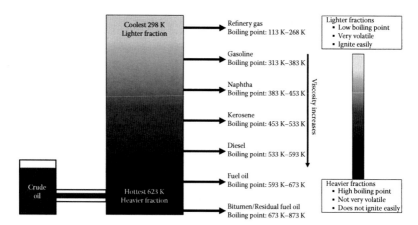

FIGURE 2.2   Crude oil distillation products. (Adapted from Aitani, A. M., *Encyclopedia of Energy*, 715–729, 2004.)

Because hydrocarbons start to crack when the temperature is higher than 613 K, another distillation method, the ASTM D1160 (2015), is used for heavy fractions. In this method, vacuum is applied (1–50 mmHg) instead of increasing the temperature to higher than 613 K.

The separation based on the boiling points fails when there are chemical reactions or extractive separations based on polarity (Hay et al. 2013).

2.1.2.2 Based on the Chemical Structure

The hydrocarbons present in crude oil are categorized into:

- Paraffins, that is, saturated hydrocarbons with straight or branched chains, but without any ring structure.
- Cycloparaffins (naphthenes), that is, saturated hydrocarbons containing one or more rings, each of which may have one or more paraffin side-chains (more correctly known as alicyclic hydrocarbons).

Paraffins and naphthenes together are called the saturates.

- Aromatics, that is, hydrocarbons containing one or more aromatic nuclei such as benzene, naphthalene, and phenanthrene ring systems that may be linked up with (substituted) naphthalene rings or paraffin side-chains.
- High-molecular weight and highly branched aromatic hydrocarbons are known as asphaltene molecules, which are normally between 4 and 20 wt% of crude oil.

The characterization based on the type of hydrocarbons is better known as n-paraffin, isoparaffin, olefin, naphthene, and aromatic (PIONA) analysis (Lundanes and Greibrokk 1994). Because olefins are not usually present in crude oil; this analysis for crude oil is known as PNA analysis. This technique was developed to study the hydrocarbon mixture thermodynamic properties, and it is based on two approaches.

The first approach uses a micropacked multi-column gas chromatography (GC) system. The aromatic fractions are separated by a polar-based column; the saturates with boiling point lower than 473 K are divided into paraffins and cycloalkanes (naphthenes) by using a molecular-sieve (13Å) column, and the branched alkanes are separated from n-alkanes by using a molecular-sieve (5Å) column. This method is time consuming (the analysis time for each test is 1–5 hours), but it is possible to separate the paraffins/naphthenes/aromatics (PNA) and paraffins/isoparaffins/naphthenes/aromatics (PIPNA) (Lundanes and Greibrokk 1994).

The second approach is based on the use of a single high-resolution GC capillary column. In this method, the different fractions are identified and quantified by the use of detectors such as flame ionization detection (FID), electron ionization (EI), and mass spectrometry (MS) (Lundanes and Greibrokk 1994). The fluorescent indicator adsorption detector (FIA) is also used for the determination of aromatic (A), olefinic (O), and saturate (S). PIONA analysis is performed by capillary GC, which uses a non-polar bonded capillary column. The hydrocarbons are separated into as many peaks as possible. Each peak can be identified based on the number of carbons

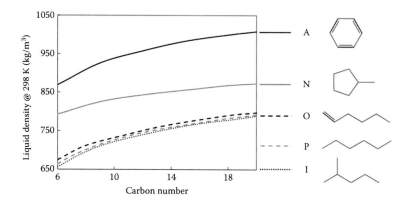

FIGURE 2.3 Crude oil fractionation based on PIONA analysis. (From Hay, G. et al., *Energy Fuels*, 27, 3578–3584, 2013.)

and hydrocarbon type. The disadvantages of this method are its long time of analysis and vulnerability to human error (Shimadzu excellence in science 2017).

Hay et al. (2013) developed a new PIONA characterization technique, based on the molecular structure of component groups. In this method, it is assumed that a group with the same carbon number has an average value of physical properties with different thermodynamic properties. This assumption introduces some error, as in the case of 1-butene, 2-butene, and iso-butene. Even if these three compounds have the same four carbons, their physical properties vary significantly (Hay et al. 2013). As an example (Aitani 2004), Figure 2.3 shows how the density changes in the PIONA molecular structure versus different carbon numbers at 298 K. Also, it shows the different groups of PIONA molecular structure with six carbon atoms.

2.1.2.3 Based on Polarity

Saturates, **a**romatics, **r**esins and **a**sphaltenes (SARA) analysis of crude oil is a physical fractionation method based on the solubility and polarity of the hydrocarbon components in various solvents and adsorbent materials. It is the most commonly used method for crude oil characterization (Powers 2014). It separates the crude oil components based on their solubility and polarity into four parts. Each of these subfractions group compounds that share similar characteristics (YingShi et al. 2012). Saturates are *n*-paraffins, iso-paraffins, and cyclo-paraffins. Aromatic hydrocarbons are derivatives of benzene. Resins and asphaltenes are made of relatively high-molecular-weight, more polar, polycyclic and aromatic ring compounds with some aliphatic side chains. Jewell et al. (1972) started working on SARA analysis in 1972.

There are three main approaches to separate crude oil into SARA fractions. The first is the thin-layer chromatography with flame ionization detection (TLC-FID) or Iatroscan. The TLC-FID method uses silica-coated rods to carry out the chromatographic separation of the components, which are later identified by a flame ionization detector (Pearson and Gharfeh 1986). There are several advantages related to

TLC-FID: Asphaltenes do not have to be removed from the sample and, in fact, they are also quantified; it uses a very small amount of sample (about 10 μL of sample per test); and the results are highly reproducible. However, a significant volume of volatile material, that contains saturates and aromatics, can be lost during the analysis meaning that this test is not viable for analyzing medium or light crude oils (Aske 2002; Jiang et al. 2008; Fan et al. 2002). Two more disadvantages are that loading the sample into a small spot can cause overloading and channeling at the origin of the silica-coated rods, and because resins have polar substituents like asphaltene, some of them might exist along with the asphaltene precipitated part. This causes some challenges to analyze the signals and to differentiate between the resins and asphaltenes signals (Pearson and Gharfeh 1986). Because of the possibility of having resins inside the asphaltenes fraction, the reported asphaltenes content may be different from the other methods.

The second method is a clay-gel adsorption-based chromatographic method described in the ASTM D2007 (2007). This system uses two columns filled with clay and silica gel. After asphaltene extraction, the remaining part called *maltene* (saturates, aromatics, and resins) is added to the columns. Resins are adsorbed onto attapulgus clay, whereas the aromatics are adsorbed onto the silica gel. The remaining saturates elute directly and are collected in a flask. The polar solvents toluene and toluene/acetone are used to dissolve and separate the aromatics and resins. This method requires large quantities of solvent and crude oil sample is time consuming and difficult to automate (Aske 2002). Fan and Buckley (2002) showed that there were some losses in the ASTM method that belonged to saturates fraction.

The third method is based on a high-pressure liquid chromatography (HPLC) analysis, which uses two columns of NH_2-bonded silica to separate the de-asphalted crude oil (also known as *maltene fraction*) based on the polarity and solubility of the different components.

Asphaltenes are removed from the oil according to the IP-143 method (ASTM D6560 2005) and the remaining components are separated by HPLC. Saturates pass through the column without adsorption and are detected by a refractive index detector. Next, the aromatics fraction is detected by an UV detector. The resins stay adsorbed in the column and have to be eluted with a more polar solvent. A detector recognizes the different maltene fractions, and the various peaks are then related to the corresponding saturate, aromatics, and resins fractions by a calibration curve. Although HPLC techniques are faster and more reproducible, the operational costs are higher compared to a simple chromatography method (Fan and Buckley 2002). Also, a specific method is needed for each type of sample because there is no universal HPLC method for all kinds of hydrocarbons (Woods et al. 2008).

Nowadays, there are different methods proposed for SARA analysis based on the three main methods described. Sieben et al. (2017) proposed a new method for SARA analysis named microfluidic SARA analysis (Maze). The asphaltene separation is based on the ASTM D7996 (2015). Then, the maltene separation is done by a miniaturized chromatographic column. Figure 2.4 shows the schematically process of SARA analysis by Maze.

The maltene passes through the mini-column and the mobile phase, which the solvent washes the column continually. Saturates content is calculated by measuring

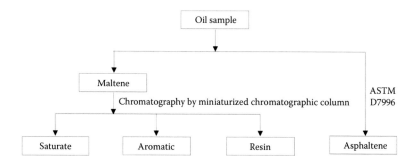

FIGURE 2.4   Maze method for SARA analysis. Part 1. Asphaltene content measurement by the microfluidic chip. Part 2. Maltene measurement by miniaturized chromatographic column. (Sieben, V. J. et al., *Energy & Fuels*, 31(4), 3684–3697, 2017.)

the refractive index of the sample. For resins and aromatics, the optical absorbance is measured by a spectrometer to find the content of each fraction. The advantages of this new method are that it is automated and the results are repeatable and reproducible (Sieben et al. 2017).

Another chromatographic-based method is known as automated chromatographic columns system for maltenes analysis. In this method, the asphaltene separation is done based on the IP-143 method (ASTM D6560 2005). Maltene analysis is performed in an automated chromatography column. This system consists of two chromatographic columns: one of them containing silica gel and the other one containing Attapulgus clay as shown in Figure 2.5.

The process is carried out in two steps: during the first step, saturates, aromatics, and resins are separated by selective adsorption into the clay and the silica gel columns, and during the second step, these components are desorbed and recovered. Both steps are described here.

During the adsorption step, all the valves on the solid brown line are open (Figure 2.5a), and the rest of them remained closed during the whole adsorption process. During this step, a solution with known concentration of a mix of saturates, aromatics, and resins in *n*-heptane is prepared and placed in the solution container (the maltene amount is calculated by subtracting oil weight from the asphaltene content). The HPLC pump is then turned on, and the solution is fed into the clay column in which only resins is adsorbed. Then, the solution passes through the silica column where the aromatic compounds are adsorbed. The solution continues its path and is collected in the effluent container. After the whole solution passes through the columns, the effluent, which contains the saturates fraction, is collected. Heptane evaporates from the solution, at approximately 371.15 K, to quantify the amount of saturates (Rezaee et al. 2017a).

After completing the first step, the columns are washed with the appropriate solvent(s) to recover resins and aromatics fractions separately. For the clay column

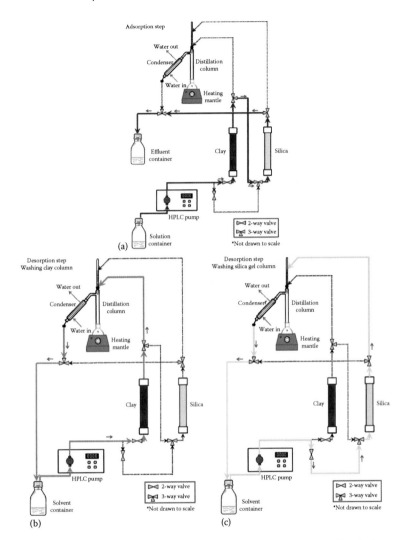

FIGURE 2.5 Automated chromatographic column: (a) Adsorption process, (b) resin desorption step, and (c) aromatic desorption step. (From Rezaee, S. et al., Crude oil characterization, fractionation and SARA analysis, In Preparation, 2017a.)

wash, the valves on the solid line in Figure 2.5b are open, and the rest of them remain closed. For the silica gel column wash, the valves on the solid line in Figure 2.5c are open, and the rest of the valves are closed.

The solvent container is filled with dichloromethane (DCM) (50 vol%) and acetone (50 vol%) for the clay wash and with DCM (62 vol%) and hexane (38 vol%) for the silica wash. The solvent is then pumped through the columns. When passing through

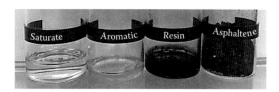

FIGURE 2.6   SARA separation by automated chromatographic columns. (From Rezaee, S. et al., Crude oil characterization, fractionation and SARA analysis, In Preparation, 2017a.)

the columns, the solvent dissolves the resins and aromatics from the clay and the silica gel media, respectively. The solution continues its way to the distillation system.

The solution of solvent and resins leaving the clay column is fed into the distillation column from the top side and is collected in a round-bottomed flask. The heating mantle maintains a temperature of around 321 K (118°F). At this temperature, the solvent evaporates and leaves the column as a vapor without burning the resins. At the top of the distillation column, there is a thermometer, which should always indicate a temperature very close to the boiling point of the solvent to ensure that the vapor is pure solvent. Solvent as a vapor continues its path across the condenser. Cold water is used to condense the vapor into a pure liquid solvent, which is collected in the solution container. The same distillation process is followed for the aromatics leaving the silica gel column. Once the columns are completely cleaned with no residues of resins or aromatics, the distillation is stopped, and both resins and aromatics are dried in an oven and are then weighed. Figure 2.6 shows the crude oil and the SARA fractions separated from the crude oil by the automated chromatographic columns system (Rezaee et al. 2017a).

2.2   ASPHALTENES

2.2.1   COMPOSITION AND CHEMICAL STRUCTURE

As described in Chapter 1, asphaltenes consist primarily of carbon, hydrogen, nitrogen, oxygen, and sulphur, as well as trace amounts of vanadium, nickel, and other metals. It is practically undisputed that a molecule of asphaltene consists of a number of polyaromatic clusters with side aliphatic chains and other functional groups as previously shown in Figure 1.2. ^{13}C NMR studies suggest that the clusters are formed by pericondensed aromatic groups instead of catacondensed groups (Mullins 2008).

It has been found by X-Ray absorption near-edge structures (XANES) that nitrogen is mainly present in the aromatic structure as pyrrolic nitrogen followed by pyridinic nitrogen (Mitra-Kirtley et al. 1993). Oxygen has been chemically identified as aliphatic hydroxyl groups but also in the form of aliphatic ketones, quinones, ethers, esters, carboxylic acids, and in combination with sulfur (Moschopedis and Speight 1976). By combustion and XANES tests, it has been determined that sulfur is most commonly found as part of thiophenes, sulfides, and sulfoxides (Mitra-Kirtley et al.

1999; Pomerantz et al. 2013). Nickel and vanadium are the most abundant metals in crude oil with concentrations reaching up to 340 ppm of nickel and 1580 ppm of vanadium. They can be found as porphyrins (organometallic complexes) or as high molecular weight complexes associated with asphaltenes (Barwise 1990).

In a larger scale, asphaltenes are defined according to their solubility in selected solvents, that is, n-alkanes and aromatic solvents such as toluene; in this way, n-C_{5+} (or simply C_{5+}) asphaltenes are those that were precipitated with n-pentane, and n-C_{7+} are those precipitated with n-heptane, etc.

Even if asphaltenes have a similar solubility, their chemical characteristics can be very different (Luo et al. 2010). It is believed that asphaltenes contain up to 10^5 different types of molecules without repeating a molecular unit (Wiehe and Liang 1996).

A great variety of analytical techniques has been used to investigate the structure of asphaltenes. In this section, we will describe some of the most practical methods to obtain the elemental and chemical composition and the thermal decomposition pattern.

2.2.1.1 Determination of the Elemental Composition

To have a better understating of the chemical structure of crude oil and its fractions, elemental analysis experiments are performed. Asphaltenes mainly consist of hydrocarbons along with small amounts of heteroatoms such as oxygen, nitrogen, sulfur, and traces of metals including, nickel, vanadium, and iron. However, crude oils from different origins have different asphaltene content and elemental compositions. Ancheyta et al. (2002) carried out the elemental compositions along with asphaltene content of three different crude oils including Maya, Isthmus, and Olmeca. The results are tabulated in Table 2.2

As demonstrated, the content of heteroatoms in asphaltene fractions is higher than the crude oil itself, which suggests that the heteroatoms are likely to concentrate in asphaltene fraction of crude oil. Moreover, the C_{7+} fractions have lower hydrogen-to-carbon (H/C) ratio and higher heteroatoms and metals content than the C_{5+} fraction. This observation is in agreement with asphaltenes' solubility behavior, which indicates that heptane insoluble asphaltenes (C_{7+} fraction) are less paraffinic, heavier and more polar than C_{5+} asphaltenes.

In a similar study, Gaweł et al. (2014) reported the compositional analysis of nine different crude oils alongside with their subsequent SARA fractions including, saturates, aromatics, resins, and asphaltenes. As expected, saturates have the highest H/C ratio followed by aromatic, resins, and asphaltenes. As it will be discussed later in this chapter, the H/C ratio can be used as an indication of the extent of ring condensation and a correlation between this ratio and the aromaticity factor can be derived. In general, the overall heteroatoms content of asphaltenes is higher than other fractions of crude oil. The results show that the distribution of sulfur is approximately even among the aromatic, resin, and asphaltene fractions. There is no sign of sulfur in saturates, meaning most of the sulfur exists within the polycyclic aromatic compounds or in alkyl sulfoxide groups attached to a polycyclic aromatic core (Sharma et al. 2002). Moreover, the oxygen and nitrogen contents of asphaltenes and resins are significantly higher than other fractions, which explain

TABLE 2.2
Elemental Analysis of Three Crude Oils and Their Asphaltene Fractions

Origin	Maya			Isthmus			Olmeca		
Elemental Analysis (wt%)	Crude Oil	C_{5+} Asphaltene	C_{7+} Asphaltene	Crude Oil	C_{5+} Asphaltene	C_{7+} Asphaltene	Crude Oil	C_{5+} Asphaltene	C_{7+} Asphaltene
Carbon	83.96	81.23	81.62	85.4	83.9	83.99	85.91	86.94	87.16
Hydrogen	11.8	8.11	7.26	12.68	8	7.3	12.8	7.91	7.38
Oxygen	0.35	0.97	1.02	0.33	0.71	0.79	0.23	0.62	0.64
Nitrogen	0.32	1.32	1.46	0.14	1.33	1.35	0.07	1.33	1.34
Sulfur	3.57	8.25	8.46	1.45	6.06	6.48	0.99	3.2	3.48
H/C Atomic Ratio	1.687	1.198	1.067	1.782	1.144	1.043	1.788	1.092	1.016
Metals (wppm)									
Nickel	53.4	269	320	10.2	155	180	1.6	82	158
Vanadium	298.1	1217	1509	52.7	710	747	8	501	704

Source: Ancheyta, J. et al., *Energy Fuels*, 16, 1121–1127, 2002.

the higher polarity of asphaltenes and resins. Additionally, the metal content of asphaltenes is much greater than other fractions. Metals in asphaltenes are present in mainly two forms. First, metals can exist in the form of metalloporphyrins, which are defined as heterocyclic macrocycle organic compounds in which metals can bind to interior ligands to form complexes. Second, metals can also be present directly in the aromatic core of asphaltenes through defects (Mullins and Sheu 1999).

The metal content of asphaltenes is usually measured by atomic absorption or mass spectroscopy. To do so, the sample needs to be free of organic components. The organic part of the sample is removed by a heat treatment (calcination) method followed by acidic digestion using concentrated (70 wt%) nitric acid. The sample is diluted with DI water to reach a stock solution of approximately 2 wt% nitric acid. A blank sample is also prepared to account for all the impurities and trace amount of metal ions present or entered into the solution during the digestion and dilution processes. Inductively coupled plasma mass spectroscopy (ICP-MS) is commonly used for detection and quantification of metals in solutions with a detection limit of as low as one part in 10^{15}. During the analysis, the solutions are pumped to the nebulizer one by one. A spray of the sample enters the plasma torch section of the instrument. The highly energetic plasma, made of argon ions, is responsible for the ionization of the elements present in the sample. The generated ions are separated based on their mass-to-charge ratio by electric and magnetic fields. Finally, the separated ions are detected and converted into electrical signals, which are processed by the software (Thomas 2016).

X-ray photoelectron spectroscopy (XPS) is a powerful surface characterization tool, mainly used in this study for identification and quantification of elemental composition and chemical state of the elements present at the surface of samples. It is relatively easy to use and requires minimal sample preparation. XPS is able to detect and quantify the composition of elements from the 10 nm outer layer with an atomic number of greater than three. In XPS, the sample is subjected to irradiation by a focused beam of X-rays under high vacuum. The interaction between the X-rays with the molecules of the sample results in the ejection of electrons from the first layers (10 nm). The escaped electrons are further detected by the detector and their kinetic energy is measured using an electron energy analyzer. Because the energy of the electrons that are bound to an atom is discrete and unique depending on the electron configuration of the atom, the binding energy of the ejected electron is a characteristic of the electron and its respective element. The number of ejected electrons and their binding energy values can be used for identification and quantification of the elements, respectively.

Enayat et al. (2017a) performed XPS analysis on C_{5+} asphaltenes from crude oil A, which results are shown in Figure 2.7. This figure includes a survey spectrum along with high-resolution spectra of the detected elements in the sample. The chemical state of each atom present in the sample is investigated by deconvolution of peaks into subpeaks using curve fitting. All the curve fittings are performed by the PHI Multipack software using a Gaussian-Lorentzian function. The atomic concentrations accompanied by relative binding energies of each peak and chemical assignments are provided in Table 2.3.

In Figure 2.7b, the carbon 1s spectrum is composed of three subpeaks assigned to aliphatic or aromatic carbon bonds (C–C/C–H), carbons single bonded to oxygen

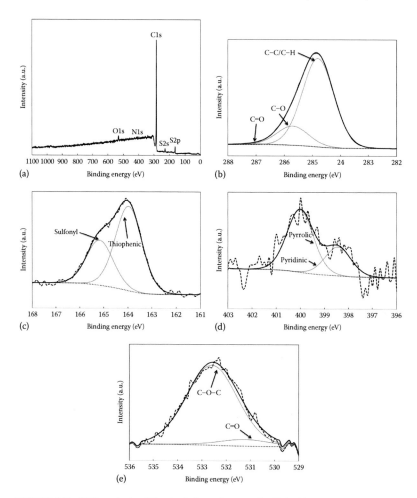

FIGURE 2.7 XPS analysis of C_{5+} Asphaltene from crude oil A: (a) survey spectrum High resolution spectra with fitting of (b) C 1s, (c) S 2p, (d) N 1s, and (e) O 1s. (Enayat, S. et al., Development of asphaltene coated melamine sponges for water-oil separation, In Preparation, 2017a.)

(C–O), and carbons in carbonyl functional groups (C=O) with the binding energies of 284.76, 285.68, and 286.93 (eV), respectively. The deconvolution of the sulfur 2p spectrum, Figure 2.7c, reveals two peaks positioned at 163.95 and 165.21 (eV), which are attributed to thiophenic and sulfonyl groups, respectively. Similarly, two peaks are chosen to fit the nitrogen 1s spectrum; the first peak at 398.44 eV is assigned to pyridinic moieties, whereas the second one (400.01 eV) corresponds to pyrrolic groups. Finally, the oxygen 1s peak is divided into two subpeaks, corresponding to ether/hydroxyl (532.56 eV) and carbonyl functional groups (531.18 eV) (Enayat et al. 2017a).

TABLE 2.3
Summary of Atomic Concentration and Spectral Features of Elements in C$_{5+}$ Asphaltene from Crude Oil A

Element	Atomic Concentration (%)	Binding Energy (eV)	FWHM (eV)	Area	Area Percentage	Functional Group
Carbon	93.18	284.76	1.34	15314	81.45	C—C/C—H
		285.68	1.34	3273	17.41	C—O
		286.93	1.34	215	1.15	C=O
Oxygen	2.98	531.18	2.49	117	8.14	C=O
		532.56	2.49	1316	91.86	C—O
Nitrogen	0.97	398.44	1.29	68	32.42	Pyridinic
		400.01	1.29	142	67.58	Pyrrolic
Sulfur	2.86	163.95	1.34	882	64.43	Thiophenic
		165.21	1.34	487	35.57	Sulfonyl

Source: Enayat, S. et al., Development of asphaltene coated melamine sponges for water-oil separation, In Preparation, 2017a.

2.2.1.2 Determination of the Chemical Structure

2.2.1.2.1 *Fourier Transform Infrared and Raman Spectroscopy*

Fourier transform infrared (FTIR) spectroscopy is an analytical technique used to identify organic and some inorganic materials. It is based on the absorption of mid-infrared radiation, located between the wavelengths of 2.5 μm (4000 cm^{-1}) and 25 μm (400 cm^{-1}) of the light spectrum. When exposed to infrared radiation, the different bonds in the molecule selectively absorb radiation of specific wavelengths, which changes their dipole moment and, therefore, the excited vibrational state. The wavelength of light absorbed by a specific bond is a function of the energy difference between the basal and excited vibrational states, therefore, the wavelengths absorbed by the sample are a characteristic of its molecular structure. The result, the FTIR spectrum, is a plot of absorbance (or transmittance) of radiation versus wavenumber in cm^{-1}, where wavenumber is the reciprocal of the wavelength (Silverstein et al. 2014). Table 2.4 shows the main adsorption peaks for the components of crude oil.

Figure 2.8a shows an example of the FTIR spectrum of C$_{5+}$ asphaltenes from a light oil. The spectrum was recorded in a Nicolete FTIR Infrared Microscope with 64 scans and a 4 cm^{-1} resolution at Rice University. Not all absorption bands are visible because of the small proportion of some functional groups in the asphaltene molecule.

The fact that the functional groups can be determined and their absorbance of radiation can be recorded has been used to determine some characteristics of the crude oil and asphaltenes, that is, the amount of asphaltenes in an oil (Wilt et al. 1998), the paraffinicity and aromaticity of asphaltenes (Castro 2006; Riley et al. 2016), the existence and relative proportion of heteroatomical functional groups versus aliphatic compounds, and the degree of condensation in the polyaromatic compounds (Asemani and Rabbani 2016; Permanyer et al. 2007). These parameters have been used to characterize the geochemical evolution and the correlation of oils

TABLE 2.4
Absorption Bands (FTIR) for the Different Components of Crude Oil

Functional Group	Chemical Structure	Absorption Bands, FTIR (cm^{-1})
Alkanes and cycloalkanes	$CH_3-, -CH_2-$	2950–2850
		1470–1350
		725–720
Aromatics		3080–3030
		1625–1575
		1525–1450
Pyrrolic Nitrogen		3500–3400
		1590–1560
		1540–1500
Pyridinic Nitrogen		3450–3200
		1610–1360
Alcohols	R–OH	3550–3200
		1250–970
Ketones		1750–1680
Quinones		1650–1630
Ethers		1150–1085
Esters		1750–1735
Carboxylic acids		3550–2500
		1780–1710
Sulfoxides		1225–980

(Continued)

TABLE 2.4 (*Continued*)
Absorption Bands (FTIR) for the Different Components of
Crude Oil

Functional Group	Chemical Structure	Absorption Bands, FTIR (cm^{-1})
Thiophenes		710–570
Sulfides	R—S—R$_1$	710–570

Source: Data from Silverstein, R. M. et al., *Spectrometric Identification of Organic Compounds,* John Wiley & Sons, Hoboken, NJ, 2014.

from different locations as well as to model the behavior of asphaltenes in crude oil (Asemani and Rabbani 2016; Permanyer et al. 2005, 2007).

Infrared active molecules are not Raman active. Hence, Raman spectroscopy is a good complement to FTIR in providing additional information to clarify the molecular structure. Raman spectroscopy is not an absorption-based technique but a technique based on light scattering. A monochromatic light, usually from a laser in the visible, near infrared or near ultraviolet range, is irradiated onto a sample. Part of this light is absorbed, and the other part interacts with the polarizability of the sample by inducing a dipole moment. The radiation is emitted by the induced dipole moment at a different frequency and it is called *Raman scattering* (Lewis and Edwards 2001; Person and Zerbi 1982; Simanzhenkov and Idem 2003). The Raman spectrum is a plot of the intensity of Raman scattered radiation as a function of the frequency difference from the incident radiation (in units of cm^{-1}). This difference is called the *Raman shift.* Figure 2.8b shows an example of the Raman Spectrum of asphaltenes.

Raman spectrometry is widely used in the study of commercial graphite-like or diamond-like polycyclic aromatic hydrocarbon (PAH) compounds, and therefore is suitable to be applied to the study of the solid structure of asphaltenes (Bouhadda et al. 2007). As seen in Figure 2.8b, the two frequencies of the Raman bands for asphaltenes are located at around 1600 and 1350 cm^{-1}. The first band (band G) is as a result of the stretching vibration of sp^2 carbons in an ordered polycondensed aromatic structure. The band at 1350 cm^{-1}, or band D1, represents the disorder in the aromatic structure induced by the in-plane defects and the heteroatoms located at the periphery of the microcrystalline structure of the asphaltene molecules (Riedeman et al. 2016). Based on these two bands, several researchers have developed correlations to calculate the size of the fused-ring system and the number of rings contained on it.

Abdallah and Yang (2012) determined the average aromatic sheet size of asphaltenes to be between 1.52 and 1.88 nm of size and to contain seven to eight

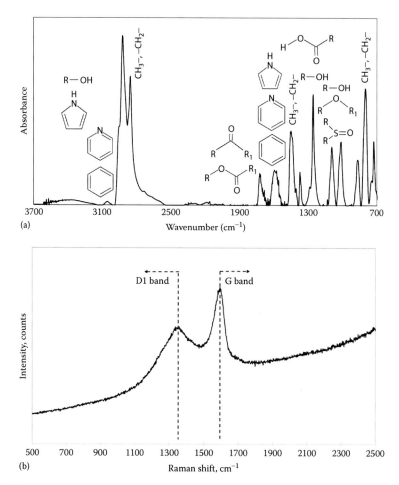

FIGURE 2.8 Typical (a) FTIR and (b) Raman spectra of asphaltene molecules. (Courtesy of Rezaee and Doherty, Rice University.)

aromatic rings. Riedeman et al. (2016) determined that the average aromatic size sheet is 21Å by following the Raman bands at 1350, 1580, and 1600 cm^{-1}, respectively; Bouhadda et al. (2007), determined the aromatic sheet diameter to be 11–17Å by using a combination of Raman spectroscopy and X-Ray diffraction. Wu and Kessler (2015) calculated that approximately six to seven aromatic rings are fused together during the formation of the polyaromatic core of the asphaltene molecules.

2.2.1.2.2 *Nuclear Magnetic Resonance Spectroscopy*

Nuclear magnetic resonance (NMR) is a property of the nucleus of an atom, related to what is known as nuclear spin (the nucleus acting like a miniature bar magnet). When the atomic nucleus spins, it generates a magnetic field, and if an external magnetic field is present, the nucleus aligns either with or against the field of the external magnet. The difference between the energy of the aligned and the misaligned nuclei depends on the applied magnetic field; the greater the strength of the magnetic field, the larger the energy difference. If radio waves are applied, the nuclei in the lower energy state absorb the energy and jump to the higher energy state and then, they undergo relaxation and return to the original energy state. When relaxing, electromagnetic signals are emitted at determined frequencies that depend on the difference in energy. These signals are recorded on a graph of intensity versus signal frequency (or chemical shift, in ppm). The most common nuclei analyzed by NMR are 1H, 13C, 15N, and 31P; however, only 1H and 13C are used in asphaltene characterization. Figure 2.9 shows the NMR spectra of C_{5+} asphaltenes of crude oil S6.

NMR is mainly used in the crude oil industry to calculate the aromatic and aliphatic carbon fractions, the alkyl-substituted, and the unsubstituted aromatic carbons of asphaltenes (Silva et al. 2004; Östlund et al. 2004; Castro 2006). A better discussion on the methodology to perform these calculations can be found in Section 2.2.2.1.

2.2.1.3 Determination of Thermal Decomposition Patterns

The shortage of light and easy accessible crude oil has led to an increase in the refining and conversion of bitumen or heavy crude oil, which is done by different processes such as coking, hydroconversion, and catalytic cracking. Asphaltene, as one of the main components of bitumen or heavy crude oil, has a high tendency to coke formation at high temperatures. If the coking process takes place without the presence of oxygen it is called *pyrolysis*. The asphaltene sample goes into several changes during the coke formation. The results after carbonization indicate that H/C ratio as well as the ratio of aliphatic to aromatic carbons decreases (Enayat et al. 2017b; Barneto et al. 2016; Savage et al. 1988). Based on the results, several reactions have been proposed for asphaltene pyrolysis. According to Savage et al. (1988), the high temperature causes some weaker bonds between aliphatic chains to break down. This leads to the formation of some relatively light hydrocarbons, along with H_2S, leaving the reactor in the form of gas. Moreover, the generated radicals can attach to each other and create aromatic functional groups. Ultimately the naphthenic cycles lose their hydrogen and along with other aromatic rings form bigger aromatic cores in the molecular structure of asphaltenes (Zhao et al. 2010).

Studying the thermal behavior of different asphaltene fractions in the presence of both oxygen and an inert gas such as argon or nitrogen is necessary to have a better understanding of the thermal stability of asphaltenes (Enayat et al. 2017b). In a general thermogravimetric analysis (TGA), a known amount (2–5 mg) of a solid sample is placed in an alumina or a platinum pan, depending on the maximum temperature of the analysis. Then, the pan is placed in a small cylinder that acts as a furnace. The weight of the sample is constantly measured as it gets heated with an average rate of 10 K–20 K/min. Different carrier gasses can be used in the experiment. In the case of an inert gas like nitrogen or argon, the material undergoes thermal degradation or pyrolysis. However, in the presence of oxygen, oxidation cracking is the dominant phenomenon (Enayat et al. 2017b).

As shown in Figure 2.10 pyrolysis of asphaltene mainly consists of two periods: before 673 K, in which, only light and volatile components of asphaltene start to

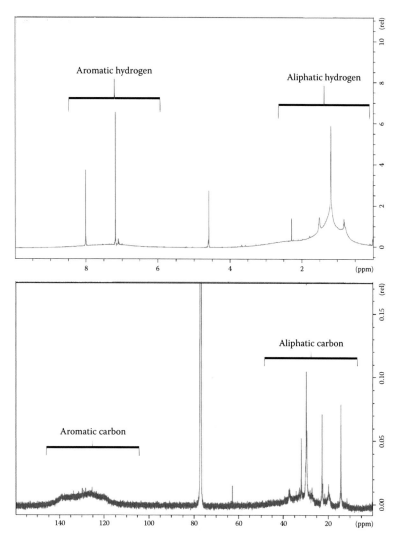

FIGURE 2.9 Typical ¹H and ¹³C of asphaltenes. (From Rezaee, S. et al., Asphaltene characterization, fractionation and chemical properties, In Preparation, 2017b.)

come out. This loss only accounts for less than 5% of the initial mass. At temperatures higher than 673 K, depending on the type of asphaltene and fraction, the sample undergoes thermal degradation and loses about 40%–80% of its initial mass. Most of the mass loss happens between 673 K–773 K and once the temperature reaches 773 K, the weight loss curve becomes a plateau. In the case of having oxygen in the system, the same procedure happens until around 763 K–773 K, at which the combustion of

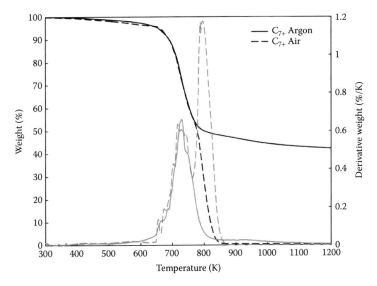

FIGURE 2.10 TGA of C_{7+} asphaltenes extracted from a bitumen in the presence of argon (dash lines) and air (solid lines) with the heating rate of 20 K/min. The peaks (grey lines) show the weight derivative of TGA analysis for both samples. (From Enayat, S. et al., Development of water soluble and photoluminescent carbon based nanoparticles from asphaltenes, In Preparation, 2017b.)

organic matters (mostly char and coke) in the sample starts. At the end, only a very small fraction of the sample's initial weight is left, which is mostly composed of ash and some inorganics such as metal oxides. This residue can be dissolved in an aqueous solution for metal content measurement of the initial sample using different methods such as inductively coupled plasma mass spectroscopy or inductively coupled optical emission spectroscopy (ICP-OES) (Enayat et al. 2017b).

Figure 2.11 shows a thermal behavior comparison between C_{7+} and C_{5-7} fractions of asphaltenes extracted from a bitumen in the presence of argon as the carrier gas. There is no significant change in the weight of the C_{7+} sample up to 673 K. However, above this temperature, the thermal degradation of asphaltene structure starts and the weight of the sample decreases dramatically, which reaches to its maximum rate at 729 K. During the decomposition, most of the relatively weak bonded aliphatic chains are removed, which result in a lower H/C atomic ratio compared to the initial sample. The pyrolysis reaction finishes at 813 K, and the weight reduction curve reaches a plateau. The C_{5-7} fraction follows the same behavior until the temperature around 538 K at which the trend starts to deviate from the C_{7+} thermal behavior. As it is demonstrated, the thermal degradation of the lighter C_{5-7} fraction starts at a lower temperature compared to the heavier C_{7+} fraction. Moreover, the amount of the carbonized char left from the C_{5-7} fraction is almost half of the C_{7+} fraction (21.5% vs. 42%). Therefore, this TGA comparison between the two fractions can demonstrate that the lighter C_{5-7} fraction has more aliphatic chains compared to the heavier C_{7+} fraction, which ultimately explains the higher solubility of the C_{5-7} fraction (Enayat et al. 2017b).

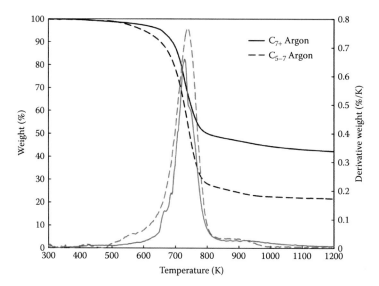

FIGURE 2.11 TGA of C_{7+} asphaltenes extracted from a bitumen (dash line) compared with C_{5-7} fraction (solid line) in the presence of argon. The peaks (grey lines) show the weight derivative of TGA analysis for both samples. (From Enayat, S. et al., Development of water soluble and photoluminescent carbon based nanoparticles from asphaltenes, In Preparation, 2017b.)

Another application of the pyrolysis process for the characterization of asphaltenes comes up when the gases proceeding from the pyrolysis chamber are fed to a gas chromatograph coupled to a mass spectrometer. Because of the large size of asphaltene molecules, their analysis by analytical methods such as FTIR, NMR, and elemental composition, among others, leads us to obtain *bulk* average rather than molecular composition (Behar and Pelet 1984). On the other hand, this large size makes impracticable the use of chromatographic techniques by themselves. However, the addition of a rapid pyrolysis process makes the components of asphaltenes of a suitable size for their gas chromatographic analysis and, by adding a mass spectrometer at the end, the different moieties can be identified and quantified. The resultant technique is known as *pyrolysis-gas chromatography-mass spectroscopy* (py-gc-ms) (Speight 2014)

Speight and Pancirov (1984) showed that a determined fraction of asphaltenes can be divided into subfractions. All of these subfractions contained small polynuclear aromatic systems (one to four rings); however, the distribution of these species was not constant throughout the different fractions of asphaltenes. Another example of the use of pyrolysis in the chromatographic characterization of asphaltenes is the study carried by Behar and Pelet (1984). They found that a low amount of the aliphatic chains of asphaltenes contained more than 30 carbons. With ^{13}C NMR the mean carbon number determined is 4–5, however, it is important to remember that ^{13}C NMR can only report average amounts, contrary to what py-gc-mass spectroscopy is capable to detect.

2.2.2 ASPHALTENE PROPERTIES

2.2.2.1 Aromaticity

The study of asphaltenes structure and aromaticity is necessary to understand their self-association and aggregation behavior. Also, aromaticity is one of the important parameters to investigate the stability of the crude oil. Asphaltene aromaticity range is reported from 0.4 to 0.7 (Bouhadda et al. 2010). If the asphaltene aromaticity is high, there is more tendency for the asphaltenes to aggregate and precipitate. The aggregation of the asphaltene molecules with higher aromaticity happens in lower asphaltene concentration in comparison to the asphaltene with lower aromaticity (León et al. 2000). Besides, the aromaticity of the maltene fraction affects the stability of the crude oil; if the maltene aromaticity is high, it can be a good solvent for the asphaltene and makes the crude oil more stable (Murgich et al. 1996).

Temperature and molecular weight changes influence the asphaltene aromaticity. By increasing the temperature and molecular weight of the sample the aromaticity increases (Speight and Moschopedis 1981). At high temperature, the rheology of asphaltene can be changed (Sharma and Yen 1994); some of the long alkyl functional groups in asphaltene structure are separated from the molecule. Also, because of the polymerization and condensation, some products such as carbon–carboids and coke will form which cause the higher degree of aromaticity (Kayukova et al. 2016).

There are different ways to calculate the aromaticity index; based on elemental composition and by FTIR and NMR spectroscopy. Based on elemental analysis, the aromaticity index (relative aromaticity) is equal to the ratio of C/H (McLean and Kilpatrick 1997).

Aromatic compounds have a higher C/H ratios than naphthenes, which in turn have higher C/H ratios than paraffin. The heavier (denser) the crude oil, the higher its C/H ratio.

In a more recent approach, the aromaticity is defined as the ratio of unbonded carbon/total carbon or unbonded carbon/total hydrogen. The unbonded carbons are the carbons in the aromatic rings, and it is assumed that the aromatic carbons do not have any bonding with hydrogen. So, the unbonded carbons are calculated based on the Equation 2.21 (Rezaee et al. 2017b).

$$UC = TC - \frac{TH}{2} \qquad (2.21)$$

where:
  UC is the unbonded carbon
  TC is the total carbon
  TH is the total hydrogen

Another method is based on NMR spectroscopy; this method can provide accurate values (McLean and Kilpatrick 1997). In this method, the aromaticity is defined as the number of aromatic carbon to the number of total carbon based on the ^{13}C NMR spectrum. The peaks in the 100–170 ppm region are related to aromatic bands and the ones in 10–70 ppm region are related to aliphatic bands (Davarpanah et al. 2015; Calemma et al. 1995). The Aromaticity Index is calculated by the number of aromatic carbons divided by the amount of aromatics plus the amount of aliphatic carbons (Maki et al. 2001; Rogel et al. 2003; Liu and Li 2015). ^{13}CNMR shows that the asphaltene molecule contains 40% aromatic carbon and 90% of hydrogen is in methylene and methyl groups (Groenzin and Mullins 1999).

Aromaticity can be defined based on the FTIR spectrum as well (Craddock et al. 2015). It is defined as the ratio of absorbance at the 1600 cm^{-1} band (which corresponds to the aromatic C=C bonds) to the absorbance at 1400 cm^{-1} band (which corresponds to aliphatic C–H bonds; Petrova et al. 2013). In addition, there is an indirect aromaticity index, which is defined based on the ratio of absorbance at the 1600 cm^{-1} band to the absorbance at 2900 cm^{-1} band (which corresponds to aliphatic C–H bonds; Rogel et al. 2015).

Figure 2.12 shows the correlation between the result of aromaticity of C$_{5+}$ asphaltene fractions for three crude oils (P1, P7, and P60) by FTIR and elemental analysis methods. The aromaticity calculated by elemental analysis is the ratio of UC/TC, UC/TC, and C/H. The result shows that there is a good correlation between the aromaticity calculated by the three methods so they can be used as an index to compare the asphaltene fractions structure. Based on the obtained aromaticity factors, one can conclude that the polycyclic aromatic core for the asphaltenes in the crude oil P1 contains less aliphatic chains and more aromatic rings in comparison to the asphaltene in the crude oil P7. Also, asphaltenes of the crude oil P60 have the lowest aromaticity (Rezaee et al. 2017b).

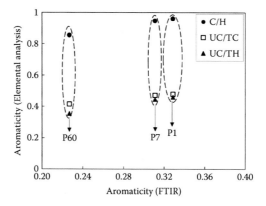

FIGURE 2.12 Aromaticity based on elemental analysis and FTIR spectroscopy. (From Rezaee, S. et al., Asphaltene characterization, fractionation and chemical properties, In Preparation, 2017b.)

In addition to the methods mentioned previously, aromaticity index (AI) can be calculated based on the Equation 2.22 (Koch and Dittmar 2006):

$$AI = \frac{DBE_{AI}}{C_{AI}} = \frac{1+C-O-S-0.5H}{C-O-S-N-P} \qquad (2.22)$$

The Double Bond Equivalent (*DBE*) is equal to the minimum numbers of C–C double bonds plus rings in a structure of a molecule, which has heteroatoms.

$$DBE_{AI} = 1 + \frac{1}{2}\left(2\left(C-O-N-S-P\right)-\left(H-N-P\right)+N+P\right) \qquad (2.23)$$

C_{AI} is equal to the number of carbons which will be reduced by the number of heteroatoms (Koch and Dittmar 2006).

2.2.2.2 Molecular Weight

Molecular weight is one of the important parameters used to develop correlations for the physical properties of asphaltenes (León and Parra 2010). Since asphaltenes are a polydisperse distribution of a broad range of molecules with different sizes and molecular characteristics so a distribution of molar masses is expected for its molecular weight (Yarranton 2005). The average monomer asphaltene molecular weight goes from 500 to 1000 g/mol depending on the source of the sample (Molina et al. 2017). The average molecular weight of the aggregates varies from 3000 to 10000 g/mol and some of the aggregate sizes can reach as high as 50000 g/mol (Yarranton 2005).

Asphaltene molecules have the tendency to self-associate into nano-aggregates, which causes some trouble in the molecular weight measurement (Powers et al. 2016). One of the key concepts in asphaltene self-association modeling, developed by Agrawala and Yarranton (2001), is that asphaltene molecules may contain multiple active sites that can act as propagators or single active sites that can act as terminators. In a mixture of asphaltenes and resins, asphaltenes are considered propagators, which serve as links for other similar molecules and form a chain. Resins serve as the terminators and end the chains. Molecular dynamic simulations have shown that the major driving force for the association is the interaction between the aromatic cores of the asphaltene molecules. In addition, the heteroatoms attached to the aromatic cores have more influence on the association in comparison to the ones attached to the aliphatic chain. Also, the length and number of chains are not as effective on the aggregation of asphaltenes as the aromatic cores; this last one will affect the aggregation size of asphaltenes (Sedghi et al. 2013).

The asphaltene molar mass depends on the solvent, temperature, and concentration. By increasing the concentration and decreasing the temperature and aromaticity of the solvent, the asphaltene molar mass increases (Sztukowski 2005). Yarranton (2005) showed that asphaltene aggregation and self-association decrease in a good solvent (such as tetrahydrofuran [THF]) and high temperature.

There are three most common methods for molecular weight (M_w) measurement; gel permeation chromatography (GPC), vapor pressure osmometry (VPO), and MS.

GPC (also called size exclusion chromatography [SEC]) separates molecules by their size, similarly to a molecular sieve process. The separation is strictly based on the size of the sample in solution. There should be no interaction with the column packing (e.g., adsorption, partition, etc.) as in conventional HPLC. The mode of separation is based on the size of the material (usually a polymer). For a correct GPC analysis, the sample must be dissolved in a suitable solvent (Lambert 1971).

GPC consists of three PSS-SDV (Styrene-divinylbenzene) columns with 100, 1000, and 10000 Å pore sizes. These pores can vary from small to quite large and act as the molecular filters. The larger size molecules will not fit into the smaller pores. Conversely, the smaller molecules will fit into most of the pores and will be retained longer (Lambert 1971).

Figure 2.13 illustrates how the sample is injected into the mobile phase and the path that the sample takes to the detector. The mobile phase normally being used in a GPC method is THF, because of its low refractive index and viscosity, and the calibration standard is based on polystyrene (Hendrickson and Moore 1966).

Inside the GPC, the dissolved sample is injected into a continually flowing stream of solvent (mobile phase). The mobile phase (THF) flows through millions of highly porous, rigid particles (stationary phase) tightly packed together in a column. Data-acquisition accessories control the test automatically, record the results, and calculate the average molecular weight. The most widely used detector today for GPC analysis is the differential refractometer. It is a concentration sensitive detector that simply measures the difference in refractive index between the eluent in the reference side and the sample + eluent on the sample side. Because the refractive index of polymers is usually constant above molecular weights of about 1,000 g/mol, the detector response is directly proportional to the sample concentration (Hendrickson and Moore 1966).

GPC can determine several important parameters including number average molecular weight (Mn), weight average molecular weight (M_w), and polydispersity, which is the ratio of M_w/Mn and is the most fundamental characteristic of a polydisperse system. Mn is affected mainly by the low molecular weights of nonassociated molecules in the sample.

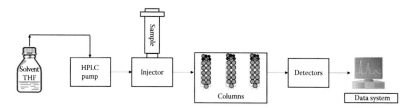

FIGURE 2.13 Schematic of a basic gel permeation chromatography. (Adapted from Walker, J. M. and Rapley, R., *Molecular Biomethods Handbook*, 2nd ed., Humana Press, Totowa, NJ, 2008.)

M_W is affected mainly by the weight of the aggregates and the molecular weights of the biggest monomers, and it will influence physical properties of asphaltenes. M_W is always greater than Mn unless the asphaltene aggregates are completely monodisperse. The ratio of M_W to Mn is used to calculate the polydispersity index (PDI) of a polymer, which provides an indication of aggregation tendency of the asphaltene molecule (Buenrostro-Gonzalez et al. 2002). The broader the molecular weight distribution, the larger the PDI. If the PDI value is close to 1.0, the distribution is narrow while larger values mean a broader distribution (Lambert 1971).

Figure 2.14a shows the Mn distribution for C_{5+} asphaltene fractions of crude oil S9 by GPC. As noticed, the distribution changes dramatically by the asphaltene concentration. Based on this result the distribution is broader for higher asphaltene concentration and as the concentration decreases, the curves become narrower. The same observation was reported by Agrawala and Yarranton (2001). By increasing

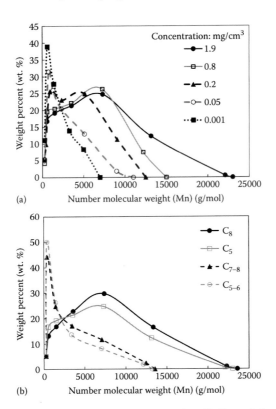

FIGURE 2.14 Asphaltene molecular weight distribution: (a) C_{5+} asphaltene from crude oil S9. (b) Different asphaltene fractions from a light crude oil with the concentration 0.16 wt%. (From Rezaee, S. et al., Asphaltene characterization, fractionation and chemical properties, In Preparation, 2017b.)

the concentration, the molecular weight increases because of the aggregation of the asphaltene molecules, and therefore, the molecular weight shown in this plot is related to the molecular weight of small asphaltene particles and aggregates. Groenzin and Mullins (1999) reported a similar conclusion.

Figure 2.14b shows the Mn distribution for different asphaltene fractions (the procedure of fractionation is described in Section 2.2.3). As expected, the molecular weight of the C_{8+} fraction is the highest, and the distribution is the broadest. The distribution of the molecular weight for C_{5-6} asphaltene is the narrowest among the asphaltene fractions since the structure of C_{5-6} asphaltene fraction is more resinous, and its aromaticity is less than a C_{8+} fraction (Rezaee et al. 2017b).

All of the distributions in Figures 2.14a and b are bimodal, which shows that, at least, there are two different species in the solution. The number of low molecular-weight molecules increases in the first peak for the lowest concentration of asphaltene in Figure 2.14a and the most soluble fraction of asphaltene C_{5-6} in Figure 2.14b (Rezaee et al. 2017b). Also, Mansoori et al. (2007) showed the same trend for the molecular weight changes for C_{5+} and C_{7+} asphaltene.

In addition to the distribution of molecular weight, it is possible to find the average molecular weight of each distribution. The average Mn and PDI measured at 1.9 mg/cm^3 concentration for the C_{5+} asphaltene from the crude oil in this example are 1443 g/mol and 3.4, respectively, whereas the average Mn and PDI measured at 0.001 mg/cm^3 concentration are 710 g/mol and 2.1. The Mn reported for the C_{5+} asphaltene at 1.9 mg/cm^3 is an aggregate weight but not the actual Mn for an asphaltene molecule because of the asphaltene self-association (Badre et al. 2006; Rezaee et al. 2017b).

The GPC methodology has some limitations; the most important one is using polystyrene as the standard. According to Sato et al. (2005), hydrocarbons with a pericondensed polyaromatic structure tend to elute later than linear polymers with similar M_W; therefore, the number-averaged molecular weight determined for asphaltenes is lower than expected. It is considered that the M_W of asphaltenes determined by this method is not the actual value, but rather a polystyrene equivalent M_W. Another limitation is that the sample needs to be filtered through a 0.2 µm filter paper to remove some of the aggregates. As a result, it is not possible to measure asphaltene M_W in different concentrations by GPC.

Another method to measure asphaltene molecular weight is VPO. It is a common method for the determination of asphaltene molecular weight, though it cannot measure an accurate molar mass of asphaltene in a solvent because of the self-association of the asphaltene molecules, which is considered as its major drawback. The main advantage in comparison to the GPC method is that the molar mass of the asphaltene samples with different concentration can be measured by VPO (Yarranton et al. 2000).

The repeatability of the asphaltene molecular-weight measurements based on the VPO method is approximately ±15%. This method is based on the different vapor pressure created by adding a small amount of solute to a solvent. In this method, two separate thermistors contained a pure solvent, and a drop of solute-solvent are placed in a chamber. Each thermistor shows different temperature because of the different vapor pressure caused by the various composition of the solute. The difference between the temperatures causes the resistance or voltage to change in proportional relation to the molecular weight of the solute (Barrera et al. 2013).

$$\frac{\Delta V}{c_s} = K\left(\frac{1}{M_W} + A_1 c_s + A_2 c_s^2 + \ldots\right) \tag{2.24}$$

where:
ΔV is the voltage difference between the thermistors
c_s is the solute concentration
K is a proportionality constant
A_1 and A_2 are coefficients arising from the nonideal behavior of the solution
M_W is the molecular weight of the sample

The equation at low concentration becomes:

$$\frac{\Delta V}{c_s} = K\left(\frac{1}{M_W} + A_1 c_s\right) \tag{2.25}$$

For an ideal system, the second term is zero, and $\Delta V/c_s$ is constant (Peramanu et al. 1999).

VPO shows higher molecular weight in comparison to GPC result because GPC is calibrated based on polystyrene, which does not have the same retention time as crude oil. So, the VPO results are more reliable for lower molecular weights (Peramanu et al. 1999).

Barrera et al. 2013 compared the molecular weight of two asphaltene fractions by VPO: a light cut and a heavy cut. They defined the light cut of asphaltene as the soluble part in a Heptol (n-heptane and toluene) solution, and a heavy cut as the precipitated part in the Heptol solution. They showed that the molecular weight of lighter cuts increases less than the heavier cuts by increasing the asphaltene concentration. This is caused by the fact that the lighter cuts contain less self-association, whereas the heaviest cuts have more (Barrera et al. 2013).

In addition to VPO and GPC methods, mass spectroscopy is also one of the most commonly used methods for measuring molecular weight. This method is mainly based on the volatilization/ionization of the sample. This is a powerful method to measure the mass-to-charge ratio (m/z) of ions to identify and quantify molecules in simple and complex mixtures. In this method, the sample (asphaltene) is vaporized and then ionized by an ion source, which creates molecular ions. These ions will be deflected by electric and magnetic fields. Then, the deflected ions will hit a detector of ions. Based on the strength of the magnetic field, different ions (different m/z) will be detected by the ion detector at different times. The computer connected to the mass spectrometer analyzes the data from the detector and produces a plot of m/z (on the x-axis) against the relative abundance (on the y-axis) of the ions. Assuming that the charge of the ions is +1 (one electron is lost), it is possible to find the weight percentage of each of the ion fractions based on the relative abundance on the y-axis (Coelho and Franco 2013).

Mass spectroscopy results can be misleading in two conditions; incomplete volatilization because of a low laser power or fragmentation because of an excessive power. Controlling the laser energies can solve the problem of fragmentation. It is possible to control the laser energy by the use of a matrix-assisted laser desorption ionization-time-of-flight method (MALDITOF-MS), which is mainly used to analyze biopolymers and synthetic polymers (Acevedo et al. 2005). In this method, because asphaltenes are dissolved in a diluted solution, it is not necessary to measure the distribution of molecular weight (Hortal et al. 2006; Mullins et al. 2008).

Shinya Sato et al. (2005) compared the data of M_W obtained by mass spectroscopy (MS) and GPC. They reported that the values of M_W (MS) is almost the same as M_W by GPC (standard was polystyrene) for M_W greater than 900 amu. For the molecular weight less than 900 amu, the difference between the M_W (MS) and M_W (GPC) increases (Sato et al. 2005).

There is still a debate on the accuracy of the mass spectroscopy method, although most of the studies proof that the matrix-assisted laser desorption ionization measurements, with the asphaltene diluted in a matrix, give the more reliable results.

2.2.2.3   Wetting Properties

Wettability is defined as the tendency of a fluid to wet a surface, which, talking about oil reservoirs, explains how oil occupies the rock pores. Wetting properties of the rock reservoir have a strong influence on the displacement behavior and the crude oil recovery. The two most widely used parameters to study the wettability behavior are the interfacial tension and the contact angle (Hjelmeland and Larrondo 1986).

Surface tension is one of the major issues in the oil industry because it is needed to predict the capillary pressure of the oil in a porous solid. Also, it influences on the relative gas-to-liquid phase permeability. Surface tension is defined as the attractive force between two phases (a liquid and a solid or a liquid and a gas). When this attractive force is between the two immiscible liquids, it is known as *interfacial tension*. The common units for surface tension are dynes/cm or mN/m, which are equivalent. The surface tension is affected by the temperature and the molecular weight (Speight 2014).

Firoozabadi and Ramey (1988) proposed a correlation for estimating oil–water surface tension as a function of density and reduced temperature. The limitation of using this correlation is that the effect of salt concentration is not considered in the equation.

$$\sigma_{hw} = \left(\frac{1.58\left(\rho_w - \rho_h\right) + 1.76}{T_r^{0.3125}} \right)^4 \tag{2.26}$$

where:
   σ_{hw} is the hydrocarbon/water surface tension in dynes/cm
   ρ_w is the density of water in g/cm^3
   ρ_h is the density of the hydrocarbon in g/cm^3
   T_r is the reduced temperature

The contact angle can be measured by Young's equation. This equation considers the equilibrium between the force factors at the oil–water–rock system contact line (Xu 2005):

$$\sigma_{so} = \sigma_{sw} + \sigma_{wo} \cos\theta \qquad (2.27)$$

θ is the angle at which the fluid–fluid interface meets the surface, and it is defined as the direct measure of the surface wettability. The value of θ goes from $0°$ to $180°$. For values between $0°$ and $75°$, it is considered that water is wetting the surface. At θ between $75°$ and $105°$, a neutral wettability or an intermediately wet surface happens. When θ is greater than $105°$, oil is wetting the surface (Rajayi and Kantzas 2011; Speight 2014). σ is the interfacial energy between solid–oil, solid–water, and water–oil (Xu 2005). There are several different methods to measure the contact angle (Yuan and Lee 2013), that is, the telescope goniometer (Bigelow et al. 1946), the captive bubble method (Adanson and Gast 1997; Brandon et al. 2003), the dual-drop dual-crystal technique (Rao and Girard 1996), the tilting plate method, the Wilhelmy balance method, the capillary rise at a vertical plate, the capillary tube, the capillary penetration method for powders and granules and the capillary bridge method. In the oil industry, the most widely used methods for measuring the contact angle are the captive bubble and dual-drop dual-crystal techniques (Xu 2005; Rajayi and Kantzas 2011).

In the sessile drop method, a single solid plate is placed in the test cell, which is filled with fluid as its environment, whereas in the dual-drop dual-crystal method, two parallel plates are used in a cuvette that contains the fluid environment. The solid plate is usually made out of quartz or calcite crystals to simulate the reservoir rock surface (Anderson 1986a; Morrow 1990).

The first step for the measurement of the contact angle by sessile drop method is to place the oil droplet on the solid surface by injecting the oil with a U-shaped needle (Figure 2.15). During the process of oil expansion on the solid plate, the measured angle is called the *receding contact angle* (RCA) with respect to water, and during the withdrawal of the oil, the contact angle is defined as the *advancing contact angle* (ACA) with respect to water.

Before measuring ACA, one must wait enough time to reach the equilibrium of the receding contact angle. Six days are needed to reach the equilibrium contact angle for a sample of 0.1 wt% C_{5+} asphaltene in toluene (Rezaee et al. 2017b). Also, Anderson reported about 1000 h to reach the equilibrated contact angle (Anderson 1986b).

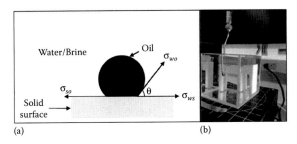

(a)   (b)

FIGURE 2.15   (a and b) Contact angle at oil–water–solid interface. (From Rezaee, S. et al., Asphaltene characterization, fractionation and chemical properties, In Preparation, 2017b.)

Because of the long time needed to reach equilibrium, Buckley et al. (1997) proposed to age the plate before measuring the contact angle at a temperature higher than 333 K. However, Hopkins et al. (2017) reported that both the aged and non-aged chalk cores were mixed wet in the presence of crude oil.

After the aging process, the bulk oil should be washed by the appropriate solvent. Figure 2.16 shows how washing or soaking with different solvent affects the contact angle measurement.

Decane acts as the precipitant, and based on the contact angle when using decane, the plate is considered completely oil wet. Cyclohexane is neither a solvent nor a precipitant for asphaltenes, but based on the asphaltene type, it may act as a pure solvent. Toluene is a good solvent for asphaltenes, but rinsing with toluene can just remove the bulk oil from the surface of calcite, while soaking with toluene may dissolve some the asphaltenes adsorbed to the surface (Buckley et al. 1997).

There are some parameters that affect the wetting properties of crude oil such as physical and chemical properties of the solid surface, crude oil fractions (acidic and basic groups), brine composition (ionic strength, pH) and ambient conditions (e.g., temperature, pressure and dissolved carbon dioxide) (Buckley 1998).

The effect of temperature and pressure on wettability alteration depends on the condition of the test which is under study. Wang and Gupta (1995) showed that pressure does not have a significant effect on contact angle of oil–brine on two substrates of quartz and calcite plates. They also found that by increasing the temperature, calcite becomes more water wet, while quartz becomes more oil wet. Rajayi and Kantzas (2011) demonstrated that the contact angle of bitumen–water decreases (more water wet) as the pressure increased. Najafi-Marghmaleki et al. (2016) investigated the wettability alteration of four carbonate reservoir by temperature. They concluded that temperature and pressure had a negligible effect on wettability alteration of the reservoir with the lowest asphaltene content.

Brine composition has a significant impact on wetting properties of rock reservoirs. Previous studies have shown that potential determining ions, such as Mg^{2+}, Ca^{2+},

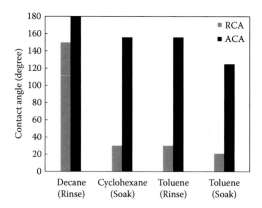

FIGURE 2.16   Effect of washing on RCA and ACA measurement. (From Buckley, J. S. et al., *SPE J*, 2, 107–119, 1997.)

CO_3^{2-}, and SO_4^{2-} present in the brine, increase the water wetness of the chalk rocks (Puntervold et al. 2007). Puntervold et al. (2007) demonstrated that the water wetness of the chalk reservoir increases with temperature because of the SO_4^{2-} affinity to the chalk surface in the presence of Ca^{2+} and Mg^{2+}. Generosi et al. (2017) studied the effect of Mg^{2+} on calcite wettability alteration. Based on their result, the surface becomes more water wet by replacing NaCl brine with a solution containing $MgSO_4$ (The organic attraction to the surface of calcite is weaker in the presence of Mg^{2+}). Their result is in agreement with other studies available in the literature. For example, Prabhakar and Melnik (2017) found that Mg^{2+} and SO_4^{2-} ions make the surface of calcite more water wet in case of having an excess of Mg^{2+} because Mg^{2+} forms Mg acetate in the presence of oil in the system. Mg acetate is less sticky on the surface of calcite, and therefore, causes the more water wetness condition.

Based on the study by Andersson et al. (2016), calcite surface is more hydrophilic in the presence of divalent transition metal ions (Mn, Fe, Co, Ni, Cu, and Zn). They proposed to use divalent ion Mn for the low temperature and Fe for the high temperature carbonate reservoir in order to enhance oil recovery (Andersson et al. 2016). In addition, different studies have shown the importance of low salinity water injection in enhancing oil recovery (Tang and Morrow 1997; Al Shalabi et al. 2013). However, there are still debates on understanding the mechanism of increasing oil recovery by low salinity water flood (Yousef et al. 2010; Lee et al. 2010; Mahani et al. 2017).

Crude oil composition has a significant effect on wettability alteration, which is mainly because of the presence of resins and asphaltenes as the most polar fractions in the crude oil (Anderson 1986a,b). In case of carbonate rocks, the carboxylic acid group is responsible for the adsorption to the positively charged calcite surface. Therefore, it is important to measure the acid number (AN) of the test oil, which shows the amount of acidic compounds (Hopkins et al. 2017). Different studies available in the literature have presented a correlation between acid number and wettability alteration. They also investigated the effect of asphaltene on changing the wettability of reservoir rocks (Morrow 1990; Pedroza et al. 1996; Buckley et al. 1997; Drummond and Israelachvili 2004).

The interactions between the solid plate and oil can be ionic and surface precipitation. Ionic interaction happens when acids and bases ionize at the interface of oil–water or solid–water. Surface precipitation interaction depends on the solvent properties (crude oil media) of asphaltene (Al-Maamari and Buckley 2003). The adsorption of asphaltenes on the solid surfaces depends mostly on the stability of the crude oil (Akhlaq et al. 1996).

Figure 2.17 shows the results of RCA and ACA for C_{5+} asphaltene in toluene. Based on the results RCA at time zero is approximately $20°$ and increases until it reaches equilibrium at $38°$. It is shown that during the measurement of ACA, a rigid film is formed and remains on the surface of calcite. This film formation around the droplet is not related to the concentration of asphaltene since it was is seen at 0.01 wt% C_{5+} asphaltene in toluene (Rezaee et al. 2017b). Hjelmeland and Larrondo (1986) showed that a thick rigid film at the brine–oil interface would form, especially at low temperature. The rigid film is formed because the surface active components of asphaltene are adsorbed at the interface of brine–model oil (Buckley et al. 1997). Rezaee et al. (2017b) showed that the rigid film is formed even in the presence of the lightest fraction of asphaltene (C_{5-6} asphaltene) (refer to Section 2.2.3).

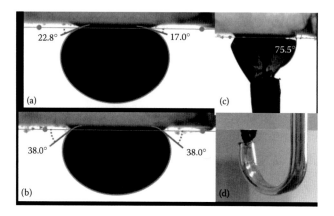

FIGURE 2.17 RCA and ACA measurement for 0.1 wt% C_{5-6} asphaltene in toluene from crude oil S6 at room temperature. (From Rezaee, S. et al., Asphaltene characterization, fractionation and chemical properties, In Preparation, 2017b.) (a) RCA measurement at time zero. (b) Equilibrium RCA after 6 days. (c) ACA measurement, rigid film formation. (d) The rigid film remained after reducing the drop volume.

Hjelmeland and Larrondo (1986) showed that the rigid film formation is temperature dependent. They observed that the rigid film disappears above 333 K after a short time.

Regarding the comparison of the surface activity of different asphaltene fractions, the only difference was that the rigid film is thicker for the heavier fractions of asphaltene (C_{8+} asphaltene in comparison to C_{5-6} asphaltene) (Rezaee et al. 2017b).

Rao and Girard (1996) proposed a dual-drop dual-crystal technique to remove the problem of skin formation. In this method, the RCA and ACA are measured by shifting the lower calcite plate. A drop of oil is placed between the two plates, and the RCA and ACA are measured by moving the plates toward each other. In this case, because the volume of the droplet is not changing, there is no rigid film (no asphaltene accumulation in the interface of water–oil) during the advancing contact angle measurement. Also, the results reported by this method are more reproducible in comparison to the other methods (Rao and Girard 1996).

2.2.3 ASPHALTENE FRACTIONATION

A common strategy to study the asphaltene characterization is the fractionation based on their solubility. As advantages, this strategy reduces the complexity of the material to study and provides a distribution of properties for the each of the asphaltene fractions. Rogel et al. (2015) fractionated asphaltene into four parts: fraction 1 was extracted with 15/85 CH_2Cl_2/n-heptane at room temperature; fraction 2 was extracted with 30/70 CH_2Cl_2/n-heptane at room temperature; fraction 3 was extracted with 100% CH_2Cl_2 at room temperature; and fraction 4 was extracted with 90/10 CH_2Cl_2/methanol. They showed that there is a decrease in the H/C molar ratio from fraction 1 to 3, which is inconsistent with a lower solubility. There was an increase in the H/C ratio in fraction 4. This is caused by the low solubility of this fraction and might be partially as a result of the presence of polar functionalities. Also, infrared spectroscopy shows that there is an increase in the carbonyl signal for fraction 4. It also shows that a low amount of hydrogen would cause an increase in aromaticity and the size of the polyaromatic rings in the molecules.

Davarpanah et al. (2015) fractionated the asphaltene into two C_{5+} and C_{7+} asphaltenes based on their solubility in n-alkanes. They showed that the aromaticity (defined as the number of aromatic carbon to the total carbon using ^{13}C NMR) of the C_{7+} fraction is higher than the one for C_{5+}. Also, the number of carbons for the C_{5+} fraction is higher than the C_{7+} fraction based on ^1H NMR results. The number of carbons per alkyl substituent based on ^1H NMR is calculated by the equation:

$$n_{carbon} = \frac{\left[H_\alpha + \left(H_\beta + H_\gamma \right) \right]}{H_\alpha} \tag{2.28}$$

where H_α, H_β, and H_γ refer to the α-hydrogen-to-aromatic ring (2 to 5 ppm region), β-hydrogen-to-aromatic ring (-0.5 to 2 ppm region), and γ-hydrogen-to-aromatic ring (-0.5 to 2 ppm region), respectively.

The aromatic hydrogen band refers to the aromatic hydrogens (5 to 10 ppm region). Strausz et al. (2002) separated Athabasca asphaltenes into five fractions using GPC fractionation by separating them with different ratio of solvent (benzene) to sample. The GPC result showed that, by diluting the sample the retention time of the asphaltene in GPC response increases and M_W decreases from 17000 to below 12000. Also, they showed that the aromaticity decreases from 48% to 35% by increasing the ratio of solvent. The nitrogen, oxygen, and sulfur content is the same in different fractions.

Östlund et al. (2004) fractionated asphaltenes based on their different polarities using binary mixtures of DCM and n-pentane. In the first step the asphaltene is dissolved in DCM, and it is mixed completely. In the second step, n-pentane is added and mixed for 30 minutes. Then the mixture is centrifuged, and the undissolved asphaltene is separated. This process is repeated three more times, and the asphaltene of each step is separated. They showed that the asphaltene separated in the first step has the highest tendency to flocculate and the highest molecular weight. The heteroatoms content was approximately the same for the four asphaltenes fractions.

Tojima et al. (1998) developed a new method of fractionation by using heptane-toluene and separating the light and heavy C_{7+} asphaltene fraction. In their method the C_{7+} asphaltene is separated first, then a toluene-to-heptane a ratio of 65:35 is added to the C_{7+} asphaltene. In this step, the undissolved asphaltene is separated and named the lowest soluble portion of C_{7+} asphaltene. As the next step, the soluble asphaltenes remained in toluene solution are recovered by evaporation of the solvent. To separate the most soluble fraction of C_{7+}, the process will be repeated two more times with a toluene-to-heptane ratio of 25:75 and 18:82. Finally, the most soluble fraction of C_{7+} asphaltene is separated by evaporation of toluene. They showed that the aromaticity of heavier asphaltenes is higher than for the lighter asphaltenes, and it has the most highly condensed polynuclear aromatics. As a result, they concluded that the heavy asphaltenes are considered as peptized material and the light ones are peptizing material.

Rezaee et al. (2017b) proposed a new method for asphaltene fractionation based on the solubility difference of four n-alkanes. The asphaltenes separation is based

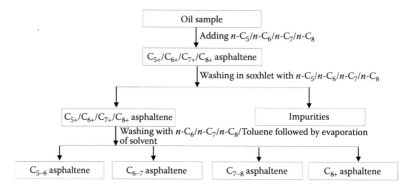

FIGURE 2.18 Asphaltene fractionation; aging time was 48 hours. (From Rezaee, S. et al., Asphaltene characterization, fractionation and chemical properties, In Preparation, 2017b.)

on the IP-143 procedure (ASTM D6560 2005). Figure 2.18 shows the different steps of the asphaltene fractionation.

In this method, the chosen precipitant (n-pentane, n-hexane, n-heptane, and n-octane), is added to the crude oil at a 40:1 volume-to-weight ratio (40 cm³ of precipitant per 1 g of oil). The oil is mixed with the precipitant in a glass vial and the vial is sealed to prevent the evaporation of the solvent. The mixture is sonicated once a day for 30 minutes to enhance the mixing of the oil and the precipitant, and the subsequent precipitation of asphaltenes. The aging process of the asphaltene aggregates occurs at ambient temperature. After two days, the aging process is completed and the solution is filtered using a 0.2 μm Nylon filter.

In the next step, the filtrate, containing the maltenes fraction, is analyzed by the automated chromatographic columns system described in Section 2.1.2.3.

The filter cake, containing the asphaltenes fraction, is placed inside a Soxhlet apparatus (Figure 2.19) using a paper thimble and refluxed with the corresponding asphaltene precipitant (n-C_5, n-C_6, n-C_7, and n-C_8) for two days to purify the asphaltene fraction and remove any co-precipitated material. The solvent plus the impurities are then added to the maltene fraction. The heating rate, condenser cooling rate, and fluid level of the solvent in the Soxhlet apparatus have to be adjusted to maintain the appropriate level of solvent in the Soxhlet and flask. To increase the rate of asphaltenes purification, a motor is installed on top of the condenser column to vertically move the thimble containing the filtered asphaltenes. Using this technique, the diffusion rate of impurities into the washing solvent increases, and therefore, the efficiency of extraction improves.

In the next step, the filter papers containing the purified asphaltenes are washed with toluene to dissolve all asphaltenes. The end point of this step is determined by the color of the toluene in contact with the thimble; the process will stop when toluene is colorless.

In the final step, the solutions of toluene and asphaltenes are transferred to evaporation plates to allow the complete evaporation of toluene at a temperature of 378 K–393 K. The samples are taken to constant weight and the weights of C_{5+}, C_{6+}, C_{7+}, and C_{8+} asphaltenes are calculated.

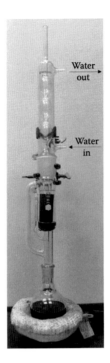

FIGURE 2.19 Asphaltene separation by Soxhlet extraction. (Courtesy of Vargas Research Laboratory, Rice University, Houston, Texas.)

Finally, by subtracting the C_{6+} asphaltene content from the C_{5+} asphaltene content, the amount of C_{5-6} asphaltene is quantified. With the same methodology, the amounts of C_{6-7} and C_{7-8} asphaltenes are obtained.

In order to separate C_{5-6}, C_{6-7}, C_{7-8}, and C_{8+} asphaltene fractions for further analyses, it is needed to wash the C_{5+}, C_{6+}, C_{7+}, and C_{8+} fractions with n-C_6, n-C_7, n-C_8, and toluene, respectively. The final step is to remove the added solvent by evaporation. The residue is the corresponding asphaltene fraction.

Table 2.5 shows the results of the asphaltene fractionation of two light crude oils. Based on the results, the asphaltene distribution is not the same for the two crude oils. Different asphaltene fractions have different chemical compositions and structures (Section 2.2.1.1). Not only asphaltene structure varies from one crude oil to the other, but also the asphaltene subfractions look different.

Although it is not applicable for all asphaltene fractions from different crude oils, some of the subfractions may also look different. In a few cases, the C_{5-6} asphaltenes are viscous liquid, whereas the C_{5+} asphaltenes are powdery solid (Figure 2.20).

Based on the results described in this chapter, C_{5-6} asphaltene has the lowest aromaticity and molecular weight, whereas the C_{8+} has the highest molecular weight and aromaticity (Rezaee et al. 2017b).

TABLE 2.5
Asphaltene Fractionation of Crude Oil S9 and S6

Asphaltene Fraction	Crude Oil S6 Asphaltene Content (wt%)	Crude Oil S9 Asphaltene Content (wt%)
C_{5+} (Total)	2.1	1.0
C_{6+}	1.8	0.8
C_{7+}	1.7	0.6
C_{8+}	1.5	0.5
C_{5-6}	0.3	0.2
C_{6-7}	0.1	0.2
C_{7-8}	0.2	0.1

Source: Rezaee, S. et al., Asphaltene characterization, fractionation and chemical properties, In Preparation, 2017b.

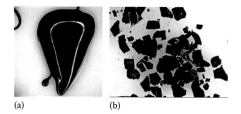

(a)   (b)

FIGURE 2.20 (a) C_{5-6} and (b) C_{5+} asphaltenes separated from the crude oil A. (Courtesy of Vargas Research Laboratory, Rice University, Houston, Texas.)

2.3   FINAL REMARKS

In this chapter, the methods and techniques currently used to determine the physical and chemical properties and characteristics of both crude oil and asphaltenes are discussed. Most of the methodologies for crude oil analysis presented here are based on ASTM standards. In other cases, the property is calculated from another property, which was obtained experimentally, under the same ASTM conditions, as in the case of the solubility parameter. Despite the reproducibility and reliability of these tests, the conditions of temperature and pressure at which these tests take place are not representative of the actual conditions of the oil wells. One solution to this problem is the use of more sophisticated equipment, such as the Anton Paar DMA HPM HPHT apparatus to obtain the density at conditions of high temperature and pressure. However, this type of equipment is expensive and not commonly found in laboratories. To provide a solution, mathematical correlations have been proposed. Section 2.1.1.7 describes several equations that can be used to predict properties, such as the solubility parameter, the viscosity, and the molecular weight.

Next, the main methods to characterize crude oils by fractionation are described. Knowing that crude oil is composed of many different types of hydrocarbons and that some of these molecules share similar characteristics, fractionation techniques have been divided into three categories: based on the boiling point of the different groups, based on similar chemical structures, and based on similar polarity. All three methods have pros and cons; the differences are mainly in the use of expensive versus affordable equipment, large versus small amounts of sample and solvent, the need for asphaltene removal, and the time needed to complete the tests. Vargas and coworkers are currently developing a new methodology, based on polarity, which optimizes all these factors mentioned.

The last part of this chapter explains in detail the different methodologies and techniques to determine the properties and chemical composition of asphaltenes. All these properties will be used in further chapters.

REFERENCES

Abdallah, W. A., and Y. Yang. 2012. Raman spectrum of asphaltene. *Energy & Fuels* 26 (11): 6888–6896. doi:10.1021/ef301247n.

Acevedo, S., L. B. Gutierrez, G. Negrin, J. C. Pereira, B. Mendez, F. Delolme, G. Dessalces, and D. Broseta. 2005. Molecular weight of petroleum asphaltenes: A comparison between mass spectrometry and vapor pressure osmometry. *Energy & Fuels* 19 (4): 1548–1560. doi:10.1021/ef040071+.

Adamson, A. W., and A. P. Gast. 1997. *Physical Chemistry of Surfaces*, 6th ed. New York: Wiley. doi:10.1126/science.160.3824.179.

Agrawala, M., and H. W. Yarranton. 2001. An asphaltene association model analogous to linear polymerization. *Industrial & Engineering Chemistry Research* 40 (21): 4664–4672. doi:10.1021/ie0103963.

Aitani, A. M. 2004. Oil refining and products. In *Encyclopedia of Energy*, 715–729. doi:10.1016/B0-12-176480-X/00259-X.

Akhlaq, M. S., D. Kessel, and W. Dornow. 1996. Separation and chemical characterization of wetting crude oil compounds. *Journal of Colloid and Interface Science* 180 (2): 309–314. doi:10.1006/jcis.1996.0308.

Al-Maamari, R. S. H., and J. S. Buckley. 2003. Asphaltene precipitation and alteration of wetting: The potential for wettability changes during oil production. *SPE Reservoir Evaluation & Engineering* 6 (04): 210–214. doi:10.2118/84938-PA.

Alomair, O., M. Jumaa, A. Alkoriem, and M. Hamed. 2016. Heavy oil viscosity and density prediction at normal and elevated temperatures. *Journal of Petroleum Exploration and Production Technology* 6 (2): 253–263. doi:10.1007/s13202-015-0184-8.

Al-Rawahi, N., G. Vakili-Nezhaad, I. Ashour, and A. Fatemi. 2012. A new correlation for prediction of viscosities of Omani Fahud-field crude oils. In *Advances in Modeling of Fluid Dynamics*, C. Liu (Ed.). InTech.

AlShaikh, M., and J. Mahadevan. 2014. Impact of brine composition on carbonate wettability: A sensitivity study. *Society of Petroleum Engineers*. doi:10.2118/172187-MS.

Al Shalabi, E., Delshad, M. and Sepehrnoori, K., 2013. Does the double layer expansion mechanism contribute to the LSWI effect on hydrocarbon recovery from carbonate rocks? In *Society of Petroleum Engineers*. doi:10.2118/165974-MS.

Ancheyta, J., G. Centeno, F. Trejo, G. Marroquín, J. A. García, E. Tenorio, and A. Torres. 2002. Extraction and characterization of asphaltenes from different crude oils and solvents. *Energy & Fuels* 16 (5): 1121–1127. doi:10.1021/ef010300h.

Ancheyta, J., and J. G. Speight. 2007. *Hydroprocessing of Heavy Oils and Residua*. Boca Raton, FL: CRC Press.

Andersson, M. P., K. Dideriksen, H. Sakuma, and S. L. S. Stipp. 2016. Modelling how incorporation of divalent cations affects calcite wettability–implications for biomineralisation and oil recovery. *Scientific Reports* 6 (June): 28854. doi:10.1038/srep28854.

Anderson, W. G. 1986a. Wettability literature survey—Part 2: Wettability measurement. *Journal of Petroleum Technology* 38 (11): 1–246.

Anderson, W. G. 1986b. Wettability literature survey—Part 1: Rock/oil/brine interactions and the effects of core handling on wettability. *Journal of Petroleum Technology* 38 (10): 1125–1144. doi:10.2118/13932-PA.

Asemani, M., and A. R. Rabbani. 2016. Oil-oil correlation by FTIR spectroscopy of asphaltene samples. *Geosciences Journal* 20 (2): 273–283. doi:10.1007/s12303-015-0042-1.

Aske, N. 2002. Characterisation of crude oil components, asphaltene aggregation and emulsion stability by means of near infrared spectroscopy and multivariate analysis.

ASTM D86. 2016. *Standard Test Method for Distillation of Petroleum Products and Liquid Fuels at Atmospheric Pressure.* ASTM D86-16. West Conshohocken, PA: American Society for Testing and Materials.

ASTM D445. 2017. *Standard Test Method for Kinematic Viscosity of Transparent and Opaque Liquids (and Calculation of Dynamic Viscosity).* ASTM D445-17. West Conshohocken, PA: American Society for Testing and Materials.

ASTM D1160. 2015. *Standard Test Method for Distillation of Petroleum Products at Reduced Pressure.* ASTM D1160-15. West Conshohocken, PA: American Society for Testing and Materials.

ASTM D1218. 2002. *Standard Test Method for Refractive Index and Refractive Dispersion of Hydrocarbon Liquids.* ASTM D1218-02. West Conshohocken, PA: American Society for Testing and Materials.

ASTM D1747. 2004. *Standard Test Method for Refractive Index of Viscous Materials.* ASTM D1747-99. West Conshohocken, PA: American Society for Testing and Materials.

ASTM D2007. 2007. *Standard Test Method for Characteristic Groups in Rubber Extender and Processing Oils and Other Petroleum–Derived Oils by the Clay–Gel Absorption Chromatographic Method.* ASTM D 2007. West Conshohocken, PA: American Society for Testing and Materials.

ASTM D2502. 2014. *Standard Test Method for Estimation of Mean Relative Molecular Mass of Petroleum Oils from Viscosity Measurements.* ASTM D2502-14. West Conshohocken, PA: American Society for Testing and Materials.

ASTM D2503. 1992. *Standard Test Method for Relative Molecular Mass (Molecular Weight) of Hydrocarbons by Thermoelectric Measurement of Vapor Pressure.* ASTM D2503. West Conshohocken, PA: American Society for Testing and Materials.

ASTM D2709. 1996. *Standard Test Method for Water and Sediment in Middle Distillate Fuels by Centrifuge.* ASTM D2709-96. West Conshohocken, PA: American Society for Testing and Materials.

ASTM D2892. 2016. *16 Standard Test Method for Distillation of Crude Petroleum (15-Theoretical Plate Column).* ASTM D2892-16. West Conshohocken, PA: American Society for Testing and Materials.

ASTM D4006. 2005. *Standard Test Method for Water in Crude Oil by Distillation.* ASTM D4006-81. West Conshohocken, PA: American Society for Testing and Materials.

ASTM D4007. 2002. *Standard Test Method for Water and Sediment in Crude Oil by the Centrifuge Method.* ASTM D4007. West Conshohocken, PA: American Society for Testing and Materials.

ASTM D4287. 2014. *Standard Test Method for High-Shear Viscosity Using a Cone/Plate Viscometer.* ASTM D4287-14. West Conshohocken, PA: American Society for Testing and Materials.

ASTM D4928. 2000. *Standard Test Methods for Determination of Water in Crude Oils by Coulometric Karl Fischer Titration.* ASTM D4928. West Conshohocken, PA: American Society for Testing and Materials.

ASTM D5481. 2013. *Standard Test Method for Measuring Apparent Viscosity at High-Temperature and High-Shear Rate by Multicell Capillary Viscometer.* ASTM D5481-13. West Conshohocken, PA: American Society for Testing and Materials.

ASTM D6304. 2007. *Standard Test Method for Determination of Water in Petroleum Products, Lubricating Oils, and Additives by Coulometric Karl Fischer Titration.* ASTM D6304-07. West Conshohocken, PA: American Society for Testing and Materials.

ASTM D6560. 2005. *Standard Test Method for Determination of Asphaltenes (Heptane Insolubles) in Crude Petroleum and Petroleum Products.* ASTM D6560-05. West Conshohocken, PA: American Society for Testing and Materials.

ASTM D7042. 2016. *Standard Test Method for Dynamic Viscosity and Density of Liquids by Stabinger Viscometer (and the Calculation of Kinematic Viscosity).* ASTM D7042-16. West Conshohocken, PA: American Society for Testing and Materials.

ASTM D7152. 2011. *Standard Practice for Calculating Viscosity of a Blend of Petroleum Products.* ASTM D7152-11(2016). West Conshohocken, PA: American Society for Testing and Materials.

ASTM D7395. 2007. *Standard Test Method for Cone/Plate Viscosity at a 500 S-1 Shear Rate.* ASTM D7395-07(2012). West Conshohocken, PA: American Society for Testing and Materials.

ASTM D7996. 2015. *Standard Test Method for Measuring Visible Spectrum of Asphaltenes in Heavy Fuel Oils and Crude Oils by Spectroscopy in a Microfluidic Platform.* ASTM D7996-15. West Conshohocken, PA: American Society for Testing and Materials.

Badre, S., C. C. Goncalves, K. Norinaga, G. Gustavson, and O. C. Mullins. 2006. Molecular size and weight of asphaltene and asphaltene solubility fractions from coals, crude oils and bitumen. *Fuel* 85 (1): 1–11. doi:10.1016/j.fuel.2005.05.021.

Barneto, A. G., J. A. Carmona, and M. J. F. Garrido. 2016. Thermogravimetric assessment of thermal degradation in asphaltenes. *Thermochimica Acta* 627–629: 1–8. doi:10.1016/j.tca.2016.02.004.

Barrera, D. M., D. P. Ortiz, and H. W. Yarranton. 2013. Molecular weight and density distributions of asphaltenes from crude oils. *Energy & Fuels* 27 (5): 2474–2487. doi:10.1021/ef400142v.

Barton, A. F. M. 1991. *CRC Handbook of Solubility Parameters and Other Cohesion Parameters*, 2nd ed. Boca Raton, FL: CRC Press.

Barwise, A. J. G. 1990. Role of nickel and vanadium in petroleum classification. *Energy & Fuels* 4 (6): 647–652. doi:10.1021/ef00024a005.

Beggs, H. D., and J. R. Robinson. 1975. Estimating the viscosity of crude oil systems. *SPE: Journal of Petroleum Technology* 27 (9). doi:10.2118/5434-PA.

Behar, F., and R. Pelet. 1984. Characterization of asphaltenes by pyrolysis and chromatography. *Journal of Analytical and Applied Pyrolysis* 7 (1–2): 121–135. doi:10.1016/0165-2370(84)80045-5.

Bigelow, W. C., D. L. Pickett, and W. A. Zisman. 1946. Oleophobic monolayers: I. Films adsorbed from solution in non-polar liquids. *Journal of Colloid Science* 1 (6): 513–538. doi:10.1016/0095-8522(46)90059-1.

Biltz, H. 1899. *Practical Methods for Determining Molecular Weights.* Easton, PA: Chemical Publishing Company.

Bird, R. B., W. E. Stewart, and E. N. Lightfoot. 2002. *Transport Phenomena*, 2nd ed. New York: John Wiley & Sons.

Bouhadda, Y., D. Bormann, E. Sheu, D. Bendedouch, A. Krallafa, and M. Daaou. 2007. Characterization of Algerian Hassi-Messaoud asphaltene structure using Raman spectrometry and X-ray diffraction. *Fuel* 86 (12–13): 1855–1864. doi:10.1016/j.fuel.2006.12.006.

Bouhadda, Y., P. Florian, D. Bendedouch, T. Fergoug, and D. Bormann. 2010. Determination of Algerian Hassi-Messaoud asphaltene aromaticity with different solid-state NMR sequences. *Fuel* 89 (2): 522–526. doi:10.1016/j.fuel.2009.09.018.

Brandon, S., N. Haimovich, E. Yeger, and A. Marmur. 2003. Partial wetting of chemically patterned surfaces: The effect of drop size. *Journal of Colloid and Interface Science* 263 (1): 237–243. doi:10.1016/S0021-9797(03)00285-6.

Buckley, J. S. 1998. Wetting alteration of solid surfaces by crude oils and their asphaltenes. *Oil & Gas Science and Technology* 53 (3): 303–312. doi:10.2516/ogst:1998026.

Buckley, J. S. 1999. Predicting the onset of asphaltene precipitation from refractive index measurements. *Energy Fuels* 13 (2): 328–332. doi:10.1021/ef980201c.

Buckley, J. S., G. J. Hirasaki, Y. Liu, S. Von Drasek, J.-X. Wang, and B. S. Gill. 1998. Asphaltene precipitation and solvent properties of crude oils. *Petroleum Science and Technology* 16 (3–4): 251–285. doi:10.1080/10916469808949783.

Buckley, J. S., X. Xie, Y. Liu, and N. R. Morrow. 1997. Asphaltenes and crude oil wetting-the effect of oil composition. *SPE Journal* 2 (2): 107–119. doi:10.2118/35366-PA.

Buenrostro-Gonzalez, E., S. I. Andersen, J. A. Garcia-Martinez, and C. Lira-Galeana. 2002. Solubility/molecular structure relationships of asphaltenes in polar and nonpolar media. *Energy & Fuels* 16 (3): 732–741. doi:10.1021/ef0102317.

Burke, J. 1984. Solubility parameters: Theory and application. Text. Article. X-Unrefereed. Retrieved from http://cool.conservation-us.org/coolaic/sg/bpg/annual/v03/bp03-04.html.

Buryakovsky, L., N. A. Eremenko, M. V. Gorfunkel, and G. V. Chilingarian. 2005. *Geology and Geochemistry of Oil and Gas*. Amsterdam, the Netherlands: Elsevier.

Calemma, V., P. Iwanski, M. Nali, R. Scotti, and L. Montanari. 1995. Structural characterization of asphaltenes of different origins. *Energy & Fuels* 9 (2): 225–230. doi:10.1021/ef00050a004.

Castro, A. T. 2006. NMR and FTIR characterization of petroleum residues: Structural parameters and correlations. *Journal of the Brazilian Chemical Society* 17 (6): 1181–1185. doi:10.1590/S0103-50532006000600016.

Ceric, E. 2012. *Crude Oil, Processes and Products*. Sarajevo: IBC.

Chew, J. N., and C. A. Connally Jr. 1959. A viscosity correlation for gas-saturated crude oils, January. Retrieved from https://www.onepetro.org/general/SPE-1092-G.

Coelho, A. V., and C. Franco. 2013. *Tandem Mass Spectrometry—Molecular Characterization*. London, UK: InTech. doi:10.5772/56703.

Correra, S., M. Merlini, A. Di Lullo, and D. Merino-Garcia. 2005. Estimation of the solvent power of crude oil from density and viscosity measurements. *Industrial & Engineering Chemistry Research* 44 (24): 9307–9315. doi:10.1021/ie0507272.

Craddock, P. R., T. V. Le Doan, K. Bake, M. Polyakov, A. M. Charsky, and A. E. Pomerantz. 2015. Evolution of kerogen and bitumen during thermal maturation via semi-open pyrolysis investigated by infrared spectroscopy. *Energy & Fuels* 29 (4): 2197–2210. doi:10.1021/ef5027532.

Davarpanah, L., F. Vahabzadeh, and A. Dermanaki. 2015. Structural study of sphaltenes from Iranian heavy crude oil. *Oil & Gas Science and Technology—Revue d'IFP Energies Nouvelles* 70 (6): 1035–1049. doi:10.2516/ogst/2012066.

Donaldson, E. C., G. V. Chilingarian, and T. F. Yen. 1985. *Enhanced Oil Recovery, I: Fundamentals and Analyses*, 1st ed. Amsterdam, the Netherlands: Elsevier.

Drummond, C., and J. Israelachvili. 2004. Fundamental studies of crude oil–surface water interactions and its relationship to reservoir wettability. *Journal of Petroleum Science and Engineering* 45 (1): 61–81. doi:10.1016/j.petrol.2004.04.007.

Egbogah, E. O., and J. T. Ng. 1990. An improved temperature-viscosity correlation for crude oil systems. *Journal of Petroleum Science and Engineering* 4 (3): 197–200. doi:10.1016/0920-4105(90)90009-R.

El Ghandoor, H., E. Hegazi, I. Nasser, and G. M. Behery. 2003. Measuring the refractive index of crude oil using a capillary tube interferometer. *Optics & Laser Technology* 35 (5): 361–367. doi:10.1016/S0030-3992(03)00029-X.

Enayat, S., F. Lejarza, M. Tavakkoli, and F. M. Vargas. 2017a. Development of asphaltene coated melamine sponges for water-oil separation. In Preparation.

Enayat, S., M. Tavakkoli, and F. M. Vargas. 2017b. Development of water soluble and photoluminescent carbon based nanoparticles from asphaltenes. In Preparation.

Ese, M.-H., J. Sjöblom, J. Djuve, and R. Pugh. 2000. An atomic force microscopy study of asphaltenes on mica surfaces. Influence of added resins and demulsifiers. *Colloid and Polymer Science* 278 (6): 532–538. doi:10.1007/s003960050551.

Fan, T., and J. S. Buckley. 2002. Rapid and accurate SARA analysis of medium gravity crude oils. *Energy & Fuels* 16 (6): 1571–1575. doi:10.1021/ef0201228.

Fan, T., J. Wang, and J. S. Buckley. 2002. Evaluating crude oils by SARA analysis. In *Society of Petroleum Engineers Journal*, 1–6. Tulsa, OK. doi:10.2118/75228-MS.

Fedors, R. F. 1974. A method for estimating both the solubility parameters and molar volumes of liquids. *Polymer Engineering & Science* 14 (2): 147–154. doi:10.1002/pen.760140211.

Firoozabadi, A., and H. J. Ramey Jr. 1988. Surface tension of water-hydrocarbon systems at reservoir conditions. *Journal of Canadian Petroleum Technology* 27 (03). doi:10.2118/88-03-03.

Gaweł, B., M. Eftekhardadkhah, and G. Øye. 2014. Elemental composition and fourier transform infrared spectroscopy analysis of crude oils and their fractions. *Energy & Fuels* 28 (2): 997–1003. doi:10.1021/ef402286y.

Generosi, J., M. Ceccato, M. P. Andersson, T. Hassenkam, S. Dobberschütz, N. Bovet, and S. L. S. Stipp. 2017. Calcite wettability in the presence of dissolved Mg^{2+} and SO_4^{2-}. *Energy & Fuels* 31 (1): 1005–1014. doi:10.1021/acs.energyfuels.6b02029.

Glaso, O. 1980. Generalized pressure-volume-temperature correlations. *Journal of Petroleum Technology* 32 (05): 785–795. doi:10.2118/8016-PA.

González, G., and M. B. C. Moreira. 1991. The wettability of mineral surfaces containing adsorbed asphaltene. *Colloids and Surfaces* 58 (3): 293–302. doi:10.1016/0166-6622(91)80229-H.

Groenzin, H., and O. C. Mullins. 1999. Asphaltene molecular size and structure. *The Journal of Physical Chemistry A* 103 (50): 11237–11245. doi:10.1021/jp992609w.

Groysman, A. 2014. *Corrosion in Systems for Storage and Transportation of Petroleum Products and Biofuels: Identification, Monitoring and Solutions*. Dordrecht, the Netherlands: Springer Science & Business Media.

Hansen, C. M. 1967. The three dimensional solubility parameter. *Danish Technical: Copenhagen*, 14.

Hansen, C. M. 2007. *Hansen Solubility Parameters: A User's Handbook*. 2nd ed. Boca Raton, FL: CRC Press.

Hay, G., H. Loria, and M. A. Satyro. 2013. Thermodynamic modeling and process simulation through PIONA characterization. *Energy & Fuels* 27 (6): 3578–3584. doi:10.1021/ef400286m.

Hendrickson, J. G., and J. C. Moore. 1966. Gel permeation chromatography. III. Molecular shape versus elution. *Journal of Polymer Science Part A-1: Polymer Chemistry* 4 (1): 167–188. doi:10.1002/pol.1966.150040111.

Hjelmeland, O. S., and L. E. Larrondo. 1986. Experimental investigation of the effects of temperature, pressure, and crude oil composition on interfacial properties. *SPE Reservoir Engineering* 1 (04): 321–328. doi:10.2118/12124-PA.

Hopkins, P. A., I. Omland, F. Layti, S. Strand, T. Puntervold, and Tor Austad. 2017. Crude oil quantity and its effect on chalk surface wetting. *Energy & Fuels* 31 (5): 4663–4669. doi:10.1021/acs.energyfuels.6b02914.

Hortal, A. R., B. Martínez-Haya, M. D. Lobato, J. M. Pedrosa, and S. Lago. 2006. On the determination of molecular weight distributions of asphaltenes and their asggregates in laser desorption ionization experiments. *Journal of Mass Spectrometry* 41 (7): 960–968. doi:10.1002/jms.1056.

Huc, A. Y. 2010. *Heavy Crude Oils: From Geology to Upgrading: An Overview*. Paris: Editions TECHNIP.

International Energy Statistics. 2017. Accessed April 21. http://tinyurl.com/ybmoofhe.

Jafarinejad, S. 2016. *Petroleum Waste Treatment and Pollution Control*. Cambridge, MA: Butterworth-Heinemann.

Jewell, D. M., J. H. Weber, J. W. Bunger, H. Plancher, and D. R. Latham. 1972. Ion-exchange, coordination, and adsorption chromatographic separation of heavy-end petroleum distillates. *Analytical Chemistry* 44 (8): 1391–1395. doi:10.1021/ac60316a003.

Jiang, C., S. R. Larter, K. J. Noke, and L. R. Snowdon. 2008. TLC–FID (Iatroscan) analysis of heavy oil and tar sand samples. *Organic Geochemistry* 39 (8): 1210–1214. doi:10.1016/j.orggeochem.2008.01.013.

Karger, B. L., L. R. Snyder, and C. Eon. 1978. Expanded solubility parameter treatment for classification and use of chromatographic solvents and adsorbents. *Analytical Chemistry* 50 (14): 2126–2136. doi:10.1021/ac50036a044.

Kayukova, G. P., A. T. Gubaidullin, S. M. Petrov, G. V. Romanov, N. N. Petrukhina, and A. V. Vakhin. 2016. Changes of asphaltenes' structural phase characteristics in the process of conversion of heavy oil in the hydrothermal catalytic system. *Energy & Fuels* 30 (2): 773–783. doi:10.1021/acs.energyfuels.5b01328.

Kitak, T., A. Dumičić, O. Planinšek, R. Šibanc, and S. Srčič. 2015. Determination of solubility parameters of Ibuprofen and Ibuprofen lysinate. *Molecules* 20 (12): 21549–21568. doi:10.3390/molecules201219777.

Koch, B. P., and T. Dittmar. 2006. From mass to structure: An aromaticity index for high-resolution mass data of natural organic matter. *Rapid Communications in Mass Spectrometry* 20 (5): 926–932.

Krevelen, D. W., and K. Te Nijenhuis. 2009. *Properties of Polymers: Their Correlation with Chemical Structure; Their Numerical Estimation and Prediction from Additive Group Contributions*. Amsterdam, the Netherlands: Elsevier.

Lambert, A. 1971. Review of gel permeation chromatography. *British Polymer Journal* 3 (1): 13–23.

Lee, S. Y., K. J. Webb, I. Collins, A. Lager, S. Clarke, M. Sullivan, A. Routh, and X. Wang. 2010. Low salinity oil recovery: Increasing understanding of the underlying mechanisms. SPE Improved Oil Recovery Symposium, April 24–28, Tulsa, Oklahoma, Society of Petroleum Engineers. doi:10.2118/129722-MS.

León, A. Y., and M. J. Parra. 2010. Determination of molecular weight of vacuum residue and their SARA fractions. *CT&F—Ciencia, Tecnología Y Futuro* 4 (2): 101–112.

León, O., E. Rogel, J. Espidel, and G. Torres. 2000. Asphaltenes: Structural characterization, self-association, and stability behavior. *Energy & Fuels* 14 (1): 6–10.

Lewis, I. R., and H. Edwards. 2001. *Handbook of Raman Spectroscopy: From the Research Laboratory to the Process Line*. New York: CRC Press.

Liu, Y. J., and Z. F. Li. 2015. Structural characterisation of asphaltenes during residue hydrotreatment with light cycle oil as an additive. *Journal of Chemistry* 2015: e580950.

Lundanes, E., and T. Greibrokk. 1994. Separation of fuels, heavy fractions, and crude oils into compound classes: A review. *Journal of High Resolution Chromatography* 17 (4): 197–202.

Luo, P., X. Wang, and Y. Gu. 2010. Characterization of asphaltenes precipitated with three light alkanes under different experimental conditions. *Fluid Phase Equilibria* 291 (2): 103–110.

Mahani, H., A. L. Keya, S. Berg, and R. Nasralla. 2017. Electrokinetics of carbonate/brine interface in low-salinity waterflooding: Effect of brine salinity, composition, rock type, and pH on ζ-potential and a surface-complexation model. *SPE Journal* 22 (01): 53–68. doi:10.2118/181745-PA.

Maki, H., T. Sasaki, and S. Harayama. 2001. Photo-oxidation of biodegraded crude oil and toxicity of the photo-oxidized products. *Chemosphere* 44 (5): 1145–1151.

Mansoori, G. A., D. Vazquez, and M. Shariaty-Niassar. 2007. Polydispersity of heavy organics in crude oils and their role in oil well fouling. *Journal of Petroleum Science and Engineering* 58 (3): 375–390.

McLean, J. D., and P. K. Kilpatrick. 1997. Comparison of precipitation and extrography in the fractionation of crude oil residua. *Energy & Fuels* 11 (3): 570–585.

Microfluidic SARA Analysis | Schlumberger. 2017. Coupling microfluidics and spectroscopy for precise SARA measurements. Accessed April 22. http://www.slb.com/services/characterization/reservoir/core_pvt_lab/fluid_lab_services/microfluidic-analysis.aspx.

Mitra-Kirtley, S., O. C. Mullins, J. F. Branthaver, and S. P. Cramer. 1993. Nitrogen chemistry of kerogens and bitumens from X-ray absorption near-edge structure spectroscopy. *Energy & Fuels* 7 (6): 1128–1134.

Mitra-Kirtley, S., O. C. Mullins, C. Y. Ralston, and C. Pareis. 1999. Sulfur characterization in asphaltene, resin and oils fractions of two crude oils. *ACS: New Orleans.* http://web.anl.gov/PCS/acsfuel/preprint%20archive/Files/44_4_NEW%20ORLEANS_08-99_0763.pdf.

Molina V., D., E. Ariza, and J. C. Poveda. 2017. Structural differences among the asphaltenes in colombian light crudes from the colorado oil field. *Energy & Fuels* 31 (1): 133–139.

Morrow, N. R. 1990. Wettability and its effect on oil recovery. *Journal of Petroleum Technology* 42 (12): 1–476.

Moschopedis, S. E., and J. G. Speight. 1976. Oxygen functions in asphaltenes. *Fuel* 55 (4): 334–336.

Mullins, O. C. 2008. Review of the molecular structure and aggregation of asphaltenes and petroleomics. *SPE Journal* 13 (01): 48–57.

Mullins, O. C., B. Martínez-Haya, and A. G. Marshall. 2008. Contrasting perspective on asphaltene molecular weight. This comment vs the overview of A. A. Herod, K. D. Bartle, and R. Kandiyoti. *Energy & Fuels* 22 (3): 1765–1773.

Mullins, O. C., and E. Y. Sheu. 1999. *Structures and Dynamics of Asphaltenes.* New York: Springer Science & Business Media.

Murgich, J., J. Rodríguez, and Y. Aray. 1996. Molecular recognition and molecular mechanics of micelles of some model asphaltenes and resins. *Energy & Fuels* 10 (1): 68–76.

Nadkarni, R. A. 2000. *Guide to ASTM Test Methods for the Analysis of Petroleum Products and Lubricants.* ASTM Manual Series 44. West Conshohocken, PA: ASTM.

Najafi-Marghmaleki, A., A. Barati-Harooni, A. Soleymanzadeh, S. J. Samadi, B. Roshani, and A. Yari. 2016. Experimental investigation of effect of temperature and pressure on contact angle of four Iranian carbonate oil reservoirs. *Journal of Petroleum Science and Engineering* 142: 77–84.

Östlund, J. A., P. Wattana, M. Nydén, and H. S. Fogler. 2004. Characterization of fractionated asphaltenes by UV–vis and NMR self-diffusion spectroscopy. *Journal of Colloid and Interface Science* 271 (2): 372–380.

Otremba, Z. 2000. The impact on the reflectance in VIS of a type of crude oil film floating on the water surface. *Optics Express* 7 (3): 129–134.

Pearson, C. D., and S. G. Gharfeh. 1986. Automated high-performance liquid chromatography determination of hydrocarbon types in crude oil residues using a flame ionization detector. *Analytical Chemistry* 58 (2): 307–311.

Pedroza, T. M. de, G. Calderon, and A. Rico. 1996. Impact of asphaltene presence in some rock properties. *SPE Advanced Technology Series* 4 (1): 185–191. doi:10.2118/27069-PA.

Peramanu, S., B. B. Pruden, and P. Rahimi. 1999. Molecular weight and specific gravity distributions for athabasca and cold lake bitumens and their saturate, aromatic, resin, and asphaltene fractions. *Industrial & Engineering Chemistry Research* 38 (8): 3121–3130.

Permanyer, A., L. Douifi, N. Dupuy, A. Lahcini, and J. Kister. 2005. FTIR and SUVF spectroscopy as an alternative method in reservoir studies. Application to Western Mediterranean oils. *Fuel* 84 (2–3): 159–168.

Permanyer, A., C. Rebufa, and J. Kister. 2007. Reservoir compartmentalization assessment by using FTIR spectroscopy. *Journal of Petroleum Science and Engineering* 58 (3–4): 464–471.

Person, W. B., and G. Zerbi. 1982. *Vibrational Intensities in Infrared and Raman Spectroscopy.* Amsterdam, the Netherlands: Elsevier.

Petrova, L. M., N. A. Abbakumova, I. M. Zaidullin, and D. N. Borisov. 2013. Polar-solvent fractionation of asphaltenes from heavy oil and their characterization. *Petroleum Chemistry* 53 (2): 81–86.

Pomerantz, A. E., D. J. Seifert, K. D. Bake, P. R. Craddock, O. C. Mullins, B. G. Kodalen, S. Mitra-Kirtley, and T. B. Bolin. 2013. Sulfur chemistry of asphaltenes from a highly compositionally graded oil column. *Energy & Fuels* 27 (8): 4604–4608.

Powers, D. P. 2014. Characterization and asphaltene precipitation modeling of native and reacted crude oils. University of Calgary. Retrieved from http://theses.ucalgary.ca/handle/11023/1873.

Powers, D. P., H. Sadeghi, H. W. Yarranton, and F. G. A. van den Berg. 2016. Regular solution based approach to modeling asphaltene precipitation from native and reacted oils: Part 1, molecular weight, density, and solubility parameter distributions of asphaltenes. *Fuel* 178: 218–233.

Prabhakar, S., and R. Melnik. 2017. Wettability alteration of calcite oil wells: Influence of smart water ions. *Scientific Reports* 7 (1): 17365. doi:10.1038/s41598-017-17547-z.

Puntervold, T., S. Strand, and T. Austad. 2007. New method to prepare outcrop chalk cores for wettability and oil recovery studies at low initial water saturation. *Energy & Fuels* 21 (6): 3425–3430. doi:10.1021/ef700323c.

Rajayi, M., and A. Kantzas. 2011. Effect of temperature and pressure on contact angle and interfacial tension of quartz/water/bitumen systems. *Journal of Canadian Petroleum Technology* 50 (06): 61–67.

Rao, D. N., and M. G. Girard. 1996. A new technique for reservoir wettability characterization. *Journal of Canadian Petroleum Technology* 35 (01). https://www.onepetro.org/journal-paper/PETSOC-96-01-05.

Rezaee, S., R. H. Doherty, M. Tavakkoli, and F. M. Vargas. 2017a. Crude oil characterization, fractionation and SARA analysis. In Preparation.

Rezaee, S., M. Tavakkoli, R. H. Doherty, and F. M. Vargas. 2017b. Asphaltene characterization, fractionation and chemical properties. In Preparation.

Riazi, M. R. 2005. *Characterization and Properties of Petroleum Fractions.* West Conshohocken, PA: ASTM International.

Riazi, M. R., and G. N. Al-Otaibi. 2001. Estimation of viscosity of liquid hydrocarbon systems. *Fuel* 80 (1): 27–32.

Riedeman, J. S., N. Reddy Kadasala, A. Wei, and H. I. Kenttämaa. 2016. Characterization of asphaltene deposits by using mass spectrometry and Raman spectroscopy. *Energy & Fuels* 30 (2): 805–809.

Riley, B. J., C. Lennard, S. Fuller, and V. Spikmans. 2016. An FTIR method for the analysis of crude and heavy fuel oil asphaltenes to assist in oil fingerprinting. *Forensic Science International* 266: 555–564.

Rogel, E., O. León, E. Contreras, L. Carbognani, G. Torres, J. Espidel, and A. Zambrano. 2003. Assessment of asphaltene stability in crude oils using conventional techniques. *Energy & Fuels* 17 (6): 1583–1590.

Rogel, E., M. Roye, J. Vien, and T. Miao. 2015. Characterization of asphaltene fractions: Distribution, chemical characteristics, and solubility behavior. *Energy & Fuels* 29 (4): 2143–2152.

Sato, S., T. Takanohashi, and R. Tanaka. 2005. Molecular weight calibration of asphaltenes using gel permeation chromatography/mass spectrometry. *Energy & Fuels* 19 (5): 1991–1994.

Savage, P. E., M. T. Klein, and S. G. Kukes. 1988. Asphaltene reaction pathways. 3. Effect of reaction environment. *Energy & Fuels* 2 (5): 619–628.

Sedghi, M., L. Goual, W. Welch, and J. Kubelka. 2013. Effect of asphaltene structure on association and aggregation using molecular dynamics. *The Journal of Physical Chemistry B* 117 (18): 5765–5776.

Sharma, A., H. Groenzin, A. Tomita, and O. C. Mullins. 2002. Probing order in asphaltenes and aromatic ring systems by HRTEM. *Energy and Fuels* 16 (2): 490–496.

Sharma, M. K., and T. F. Yen. 1994. *Asphaltene Particles in Fossilfuel Exploration, Recovery, Refining, and Production Processes*. New York: Springer.

Shimadzu excellence in science. 2017. PONA Analysis (GC): SHIMADZU (Shimadzu Corporation). *SHIMADZU (Shimadzu Corporation)*. Accessed April 22. http://www.shimadzu.com/an/industry/machineryautomotive/fuel_battery0403020.htm.

Sieben, V. J., A. J. Stickel, C. Obiosa-Maife, J. Rowbotham, A. Memon, N. Hamed, J. Ratulowski, and F. Mostowfi. 2017. Optical measurement of saturates, aromatics, resins, and asphaltenes in crude oil. *Energy & Fuels* 31 (4): 3684–3697. doi:10.1021/acs.energyfuels.6b03274.

Silva, R. C., P. R. Seidl, S. M. C. Menezes, and M. A. G. Teixeira. 2004. 1H and 13C NMR for determining average molecular parameters of asphaltenes from vacuum residue distillation. *Annals of Magnetic Resonance* 3: 63–67.

Silverstein, R. M., F. X. Webster, D. J. Kiemle, and D. L. Bryce. 2014. *Spectrometric Identification of Organic Compounds*. Hoboken, NJ: John Wiley & Sons.

Simanzhenkov, V., and R. Idem. 2003. *Crude Oil Chemistry*. Boca Raton, FL: CRC Press.

Speight, J. G. 2014. *The Chemistry and Technology of Petroleum*, 5th ed. Boca Raton, FL: CRC Press.

Speight, J. G., and S. E. Moschopedis. 1981. On the molecular nature of petroleum asphaltenes. In *Chemistry of Asphaltenes*, 1–15. Advances in Chemistry 195. Washington, DC: American Chemical Society. doi:10.1021/ba-1981-0195.ch001.

Speight, J. G., and R. J. Pancirov. 1984. Structural types in petroleum asphaltenes as deduced from pyrolysis/gas chromatography/mass spectroscopy. *Liquid Fuels Technology* 2 (3): 287–305.

Stefanis, E., and C. Panayiotou. 2008. Prediction of Hansen solubility parameters with a new group-contribution method. *International Journal of Thermophysics* 29 (2): 568–585.

Strausz, O. P., P. Peng, and J. Murgich. 2002. About the colloidal nature of asphaltenes and the MW of covalent monomeric units. *Energy & Fuels* 16 (4): 809–822.

Sztukowski, D. M. 2005. Asphaltene and solids-stabilized water-in-oil emulsions. Retrieved from http://adsabs.harvard.edu/abs/2005PhDT.......171S.

Tang, G. Q., and N. R. Morrow. 1997. Salinity, temperature, oil composition, and oil recovery by waterflooding. *SPE Reservoir Engineering* 12 (04): 269–276.

The World Factbook. Central Intelligence Agency. 2017. www.cia.gov/library/publications/the-world-factbook. Accessed April 21, 2017.

Thomas, R. 2016. *Practical Guide to ICP-MS: A Tutorial for Beginners*, 3rd ed. Boca Raton, FL: CRC Press.

Tojima, M., S. Suhara, M. Imamura, and A. Furuta. 1998. Effect of heavy asphaltene on stability of residual oil. *Catalysis Today* 43 (3–4): 347–351.

U.S. Energy Information Administration. 2017. U.S. Energy Facts. Accessed April 19. Retrieved from https://www.eia.gov/energyexplained/?page=us_energy_home.

Vargas, F. M., and W. G. Chapman. 2010. Application of the one-third rule in hydrocarbon and crude oil systems. *Fluid Phase Equilibria* 290 (1–2): 103–108.

Vazquez, M., and H. D. Beggs. 1980. Correlations for fluid physical property prediction. *Journal of Petroleum Technology* 32 (06): 968–970.

Walker, J. M., and R. Rapley. 2008. *Molecular Biomethods Handbook*, 2nd ed. Totowa, NJ: Humana Press.

Wang, F., T. J. Threatt, and F. M. Vargas. 2016. Determination of solubility parameters from density measurements for non-polar hydrocarbons at temperatures from (298–433) K and pressures up to 137 MPa. *Fluid Phase Equilibria* 430: 19–32.

Wang, J., and J. S. Buckley. 2001. A two-component solubility model of the onset of asphaltene flocculation in crude oils. *Energy & Fuels* 15 (5): 1004–1012.

Wang, W., and A. Gupta. 1995. Investigation of the effect of temperature and pressure on wettability using modified pendant drop method. In *Society of Petroleum Engineers*. doi:10.2118/30544-MS.

Wiehe, I. A., and K. S. Liang. 1996. Asphaltenes, resins, and other petroleum macromolecules. *Fluid Phase Equilibria* 117 (1): 201–210.

Wilt, B. K., W. T. Welch, and J. G. Rankin. 1998. Determination of asphaltenes in petroleum crude oils by fourier transform infrared spectroscopy. *Energy & Fuels* 12 (5): 1008–1012.

Woods, J., J. Kung, D. Kingston, L. Kotlyar, B. Sparks, and T. McCracken. 2008. Canadian crudes: A comparative study of SARA fractions from a modified HPLC separation technique. *Oil & Gas Science and Technology—Revue de l'IFP* 63 (1): 151–163.

The World Factbook—Central Intelligence Agency. 2017. Accessed April 21. Retrieved from https://www.cia.gov/library/publications/the-world-factbook/rankorder/2244rank.html.

Wu, H., and M. R. Kessler. 2015. Asphaltene: Structural characterization, molecular functionalization, and application as a low-cost filler in epoxy composites. *RSC Advances* 5 (31): 24264–24273.

Xu, W. 2005. Experimental Investigation of Dynamic Interfacial Interactions at Reservoir Conditions. Louisiana State University. http://etd.lsu.edu/docs/available/etd-04112005-141253/.

Yarranton, H. W. 2005. Asphaltene self-asssociation. *Journal of Dispersion Science and Technology* 26 (1): 5–8.

Yarranton, H. W., H. Alboudwarej, and R. Jakher. 2000. Investigation of asphaltene association with vapor pressure osmometry and interfacial tension measurements. *Industrial & Engineering Chemistry Research* 39 (8): 2916–2924.

YingShi, H., Y. HongQing, Z. XiaoPing, J. Guitarte, X. ChengGang, L. WeiMin, C. Xiang, and G. HongZhi. 2012. Cased hole formation testing in challenging operational conditions reveals reservoir fluids distribution: South China sea case study. In *IPTC 2012: International Petroleum Technology Conference*.

Yousef, A. A., S. Al-Saleh, A. U. Al-Kaabi, and M. S. Al-Jawfi. 2010. Laboratory investigation of novel oil recovery method for carbonate reservoirs. Canadian Unconventional Resources and International Petroleum Conference, October 19–21, Calgary, Alberta, Canada, Society of Petroleum Engineers. doi:10.2118/137634-MS.

Yuan, Y., and T. R. Lee. 2013. Contact angle and wetting properties. In *Surface Science Techniques*, G. Bracco and B. Holst (Eds.), 51:3–34. Berlin, Germany: Springer.

Zhao, Y., F. Wei, and Y. Yu. 2010. Effects of reaction time and temperature on carbonization in asphaltene pyrolysis. *Journal of Petroleum Science and Engineering* 74 (1–2): 20–25.

3 Experimental Investigation of Asphaltene Precipitation

A. T. Khaleel, F. Wang, E. Song,
M. Tavakkoli, and F. M. Vargas

CONTENTS

3.1 ASPHALTENE PRECIPITATION, AGGREGATION, AND AGING

Asphaltenes in solution tend to form nano-aggregates by the π-stacking of its polyaromatic hydrocarbon backbone (Indo et al. 2009). The precipitation of asphaltenes is caused by the instability of these nano-aggregates as a result of changes in pressure, temperature, and composition. Asphaltene precipitation is a phase separation process that was defined as a liquid-liquid separation by Hirschberg et al. (1984). The reversibility of asphaltene precipitation has been a controversy for many years because the phenomenon by which precipitation occurs is not clearly understood. Different researchers have varying opinions in this matter. For example, Pfeiffer and Saal (1940) have experimentally proved that asphaltene precipitation is irreversible. On the other hand, Hirschberg et al. (1984) concluded that asphaltene precipitation is reversible from experimental evidence and their developed thermodynamic model. Kokal et al. (1992) have directly shown that precipitated asphaltenes from two different crude oils could be redissolved by adding heavy oil. Even without the addition of an aromatic solvent. Hammami et al. (2000) experimentally proved the reversibility of asphaltene precipitation by changing pressure. Precipitated asphaltenes were successfully dissolved by altering the decompression and recompression steps in a pressure-volume-temperature (PVT) cell. A similar experiment was conducted by Pina et al. (2006), in which they measured the change in the amount of dissolved asphaltenes during a pressure depletion test. The crude oil samples were depressurized from high pressure to bubble point pressure and then repressurized. Asphaltene precipitation reversibility was evaluated by comparing the content of dissolved asphaltenes initially and after recompression. Pina et al. (2006) observed opposite results in two different crude oils, where in one crude they were able to return to the initial state after recompression, but this was not the case for the other crude oil despite the long time that was allowed to reach equilibrium. Peramanu et al. (2001) investigated the reversibility of asphaltene precipitation as a result of changes in composition by the addition of a precipitant and because of changes in temperature. Asphaltenes precipitated with n-heptane were found to be fully reversible. On the other hand, the temperature reversibility experiments with n-dodecane showed only partial reversibility. Recently, Chaisoontornyotin et al. (2017) concluded that asphaltene precipitation is fully reversible with changes in temperature through temperature cycling experiments. They also pointed out, that this was not observed in previous studies because a precipitant was used to help in precipitating asphaltenes along with changing temperature. This represents two competing asphaltene precipitation mechanism, and thus the precipitant-induced aggregation needs to be isolated (Chaisoontornyotin et al. 2017).

Vargas et al. (2014) performed a more comprehensive study on asphaltene precipitation reversibility. In their study, asphaltenes were forced to precipitate by the addition of iso-octane; this was carried out on a watch glass and observed under an optical microscope. As soon as iso-octane comes in contact with the oil, asphaltene aggregates are formed, as shown in Figure 3.1a. By the evaporation of iso-octane, asphaltenes were seen to fully dissolve back into the oil (Figure 3.1b and c).

The microstructure of the asphaltene aggregate was studied using scanning electron microscope (SEM) (Vargas et al. 2014). The aggregate appears to possess a

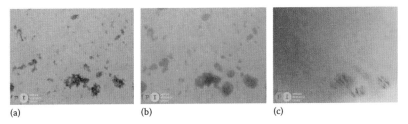

(a)   (b)   (c)

FIGURE 3.1 Microscope images that demonstrate the reversibility of asphaltene precipitation: (a) Asphaltene precipitation upon addition of iso-octane; (b) asphaltene precipitate upon evaporation of iso-octane; and (c) asphaltene precipitate redissolution into the same oil it was precipitated from. (Reprinted from Vargas, F. M. et al., Offshore Technology Conference, doi:10.4043/25294-MS, 2014.)

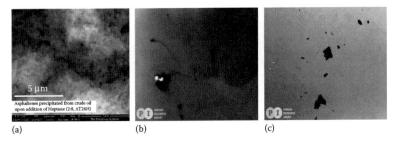

(a)   (b)   (c)

FIGURE 3.2 (a) Scanning electron microsocope image of asphaltene precipitate; (b) asphaltene precipitant dried at ambient temperature, then put into contact with the original oil; and (c) asphaltene precipitant dried at 393 K, then brought into contact with the original oil. (Reprinted from Vargas, F. M. et al., Offshore Technology Conference. doi:10.4043/25294-MS, 2014.)

porous structure that was formed by the agglomeration of particles that are 300–500 nm in size (Figure 3.2a). Vargas et al. (2014) referred to these particles as *primary particles*. The spherical shape of these primary particles indicates that asphaltene precipitation is a liquid-liquid separation, where the viscous liquid forms spheres to minimize surface energy. Additionally, Vargas et al. (2014) investigated the aging of asphaltenes. They separated and dried asphaltenes at two different conditions, ambient and 393 K, until no change in the sample's mass was observed. The dried asphaltenes were placed into the same crude oil they were separated from and then observed under an optical microscope. The microscope images are shown in Figure 3.2b and c; some redissolution can be observed for the sample that was dried at ambient conditions unlike the sample dried at 393 K. Thus, the authors concluded that the precipitation of asphaltenes, that is, the formation of the porous aggregates, is a fully reversible process. During the aging process, asphaltenes' microstructure becomes more compact (less porous) making it more difficult to redissolve. Thus, the aging process is irreversible (Vargas et al. 2014). Putting these experimental observations together, Vargas et al. (2014) proposed a conceptual mechanism that can explain asphaltene precipitation, aggregation, and aging, which is presented in Figure 1.6.

3.2   EXPERIMENTAL METHODS AT HIGH-PRESSURE AND HIGH-TEMPERATURE

During crude oil extraction process, the reservoir fluid is depressurized from high-pressure conditions in the reservoir to a much lower pressure at the wellhead. These changes might induce asphaltenes precipitation, at an intermediate pressure that is lower than the reservoir pressure but higher than the bubble pressure (BP). The pressure at which asphaltenes start to precipitate is known as the upper asphaltene onset pressure (AOP). The amount of asphaltenes that precipitates keeps increasing until it reaches a maximum at the BP. Below the BP, the light components of the crude oil start to evaporate, making the remaining liquid phase heavier and more aromatic and thus a better solvent for asphaltenes. At pressures below the BP, the precipitated asphaltenes start to redissolve because the light components of the crude oil, which are strong asphaltene precipitants, are no longer in the oil. This defines a lower boundary for AOP, that is, lower AOP below which all asphaltenes are in solution. This boundary is difficult to identify experimentally because of the slow kinetics of the pressure-induced asphaltene redissolution.

Various techniques have been developed over the years to study asphaltene precipitation at high pressure and high temperature trying to mimic reservoir conditions. This chapter will cover the basics of each technique, its advantages and disadvantages, and comparison between the results from the different methods.

3.2.1   GRAVIMETRIC TECHNIQUE

The gravimetric technique uses a conventional PVT cell, where the cell is depressurized to induce asphaltene precipitation. At each pressure step, the content of asphaltenes in solution is measured using saturate, aromatic, resin, and asphaltene (SARA) fractionation. At pressures, below the upper AOP, asphaltenes start to precipitate, and because of the effect of gravity, they will separate out and settle at the bottom of the PVT cell. The drop in the amount of asphaltene measured in the supernatant fluid defines the asphaltene precipitation boundary. If the cell is depressurized further below the upper AOP, the BP and the lower AOP can also be detected (Jamaluddin et al. 2001; Akbarzadeh et al. 2007).

In this technique, a sample of the reservoir fluid is charged into the PVT cell at a constant pressure that is higher or equal to reservoir pressure to maintain a single-phase fluid. The system is then kept for 24 hours to establish thermal equilibrium at the temperature of interest, usually the reservoir temperature. The system is then depressurized in carefully chosen pressure steps and allowed to stabilize for 24 hours at each pressure. At each pressure interval, a sample of the fluid is flashed to atmospheric conditions to analyze the amount of asphaltenes. The pressure steps used in this experiment are crucial to obtain reliable results and determine the actual AOP. At the BP, the gas bubbles are purged from the PVT cell. Then, the PVT cell is agitated for 6 hours and allowed to stabilize for 24 hours before analyzing asphaltene content from the supernatant fluid as done previously (Jamaluddin et al. 2001).

Jamaluddin et al. (2001) used the gravimetric method to detect the upper AOP, BP, and lower AOP for oil from the Middle East. The result from this experiment is shown in Figure 3.3. The experiment was performed at reservoir temperature, 389 K and a pressure step of 5 MPa. The amount of asphaltene precipitated was quantified as n-pentane and n-heptane insoluble asphaltenes. The pressure at which the asphaltene content in the supernatant fluid starts to decrease is defined as the upper AOP (at 42.75 MPa, from Figure 3.3). The amount of asphaltene keeps decreasing at a constant rate until it reaches a pressure where there is a sudden increase. This pressure corresponds to the BP, and Jamaluddin et al. (2001) validated the value using a constant mass expansion experiment. After the inflection point, the amount of asphaltene stabilizes around the original amount of asphaltenes in the reservoir fluid. This shows asphaltenes redissolution as a result of the evaporation of the light components after the BP. Thus, one of the advantages of the gravimetric technique is that it can detect the lower boundary of AOP, which is around 13.5 MPa in this example. The plot presented in Figure 3.3 also validates that the minimum solubility of asphaltenes is at the BP (Jamaluddin et al. 2001).

The accuracy of this method depends on the size of the pressure step used and the precision of the method used to measure the asphaltene content in the supernatant fluid. Better accuracy can be achieved by using smaller pressure steps, but this might be time-consuming and will require a large volume of the reservoir fluid. On the other hand, if the pressure steps are too large, the onset point can be misrepresented (Jamaluddin et al. 2001; Akbarzadeh et al. 2007). Thus, an optimum pressure step size needs to be defined without sacrificing the accuracy.

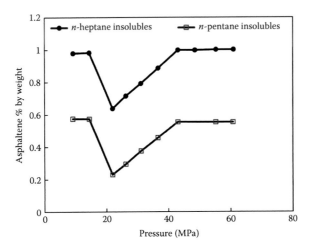

FIGURE 3.3 The amount of asphaltene precipitated using the gravimetric technique as a function of pressure. (Reprinted from Jamaluddin, A.K.M. et al., In SPE-72154-MS, Society Petroleum Engineers, Kuala Lumpur, Malaysia, doi:10.2118/72154-MS, 2001.)

3.2.2 ACOUSTIC RESONANCE TECHNIQUE

Acoustic resonance technique (ART) is based on measuring the response of a fluid, contained in a cylindrical cavity, to applied acoustic stimulation. This nonoptical technology can be used to get an insight on fluid phase behavior or phase transition by studying the state and time evolution of the resonance response of the fluid with changes in pressure, volume, or temperature. It applies to any single-phase fluid. Thus, it applies to crude oils of any type and color (Sivaraman et al. 1998; Kabir et al. 2002). This technique has shown success in measuring the upper AOP onset by some researchers (Sivaraman et al. 1998; Jamaluddin et al. 2001; Kabir et al. 2002; Akbarzadeh et al. 2007). ART uses a PVT cell that is thermally insulated using an air bath. The PVT cell is equipped with an acoustic transducer that transmits sound waves through the fluid. On the other end of the PVT cell, an acoustic receiver detects the generated resonance, where the layout of these standing waves depends on the nature of the fluid and its state. Thus, on changing the pressure, temperature or composition of the system, to induce asphaltene precipitation, the variations in the resonance detected will indicate the onset point (Sivaraman et al. 1998; Jamaluddin et al. 2001).

PVT control elements in the system allow precise control of the experimental conditions. In ART, the PVT cell is first heated to the experimental temperature and pressurized to the reservoir pressure. A 10 cm^3 sample of the reservoir fluid is then injected isobarically into the PVT cell and allowed to stabilize. The system is then depressurized at a decreasing depressurizing rate to induce asphaltene precipitation. Throughout the run, raw time domain amplitude data are collected and processed to amplitude versus frequency plots. The changes in frequencies are then plotted as a function of the independent variables (i.e., pressure, temperature, or volume) to identify different phase change phenomena that might be taking place (Kabir et al. 2002).

Jamaluddin et al. (2001) used the ART to measure the onset of asphaltene precipitation for the Middle East crude oil that was tested using the gravimetric technique, as highlighted previously. Figure 3.4 presents a typical acoustic response, where the y-axis is the normalized values of the sonic frequency in Hertz. The two sharp changes illustrated in Figure 3.4 are results of the upper AOP and the BP. The lower boundary of asphaltene precipitation could not be detected using ART because the gradual change (redissolution of asphaltenes) had no effect on the resonance characteristics. This set of experiments was performed at 389 K and depressurized at an initial rate of 0.28 MPa/min. From Figure 3.4, there is a sharp decrease in the acoustic response around 55 MPa, this is followed by a sudden increase around 43 MPa. The point at which the acoustic response starts to increase is defined as the upper AOP (Jamaluddin et al. 2001). The speed of sound waves in a denser phase is faster than in a lighter phase. Thus, the presence of the precipitated solid asphaltenes results in an increase in acoustic response (Sivaraman et al. 1998). Further depressurization of the system below the upper AOP results in another sharp change in the acoustic resonance response. This drop in acoustic resonance response is associated with the appearance of bubbles, and thus, this pressure is defined as the BP, around 23 MPa (Jamaluddin et al. 2001).

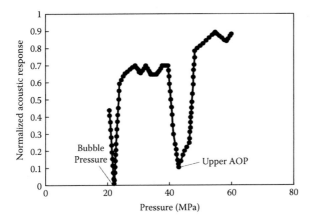

FIGURE 3.4   Acoustic response for isothermal depressurization of crude oil obtained using acoustic resonance technique. AOP, asphaltene onset pressure. (Reprinted from Akbarzadeh, K. et al., *Oilfield Rev.*, 19, 22–43, 2007.)

The upper AOP and BP obtained using ART agree with the values obtained using the gravimetric technique (Kabir et al. 2002).

This technique is accurate to ± 0.70 MPa and is not able to detect the lower onset of asphaltene precipitation. The main advantage of this method is that it needs a small amount of sample to perform the test and is relatively fast compared to the gravimetric technique (Akbarzadeh et al. 2007). Sivaraman et al. (1998) have shown that the onset point obtained by ART is in excellent agreement with that obtained using a light scattering technique. However, there is no mixing in the system, which risks inaccurate onset measurements that can be caused by the nonhomogenous distribution of asphaltenes. Additionally, the changes in the resonance response can be caused by the presence of any secondary phase, not only by asphaltene aggregates (Akbarzadeh et al. 2007).

3.2.3   LIGHT SCATTERING TECHNIQUE

The light scattering technique (LST) is based on measuring the transmittance of near-infrared (NIR) light through a sample of reservoir fluid undergoing temperature, pressure, or composition changes. NIR light with 800 to 2200 nm wavelength is specifically used because of the dark color of crude oil (Jamaluddin et al. 2001). The experimental setup is composed of a PVT cell, made of transparent Pyrex glass to allow the transmittance of NIR light. It also contains a magnetic impeller mixer to ensure the sample homogeneity. On one side of the PVT cell, an NIR light source is mounted across the window to generate light at a specific transmittance power. The system conditions (temperature, pressure, or fluid composition) are controlled by automated equipment and process variables (transmitted light power level and time) are detected by a fiber-optic sensor on the opposite side of the NIR source (Jamaluddin et al. 2001; Akbarzadeh et al. 2007).

In a typical LST experiment, a 30 cm³ volume of a single-phase reservoir fluid is charged into the PVT cell at or above the reservoir pressure. The content of the cell is mixed at a maximum speed of 1,400 rpm (revolutions per minute) for 30 minutes. Once the system is ready, a light transmittance scan is performed to obtain a reference baseline. If an isothermal depressurization experiment is performed, the light transmittance is simultaneously measured while the system is being depressurized. In this technique, a depressurization rate lower than 0.30 MPa is usually used. A plot of the average transmitted light power as a function of pressure can be constructed. Below the AOP, the detected transmittance drops as a result of light scattering induced by the precipitated asphaltene particles. Once the BP is reached, the experiment is performed in steps. At each pressure step, the system is first allowed to equilibrate, then the gas is purged out of the PVT cell, and the light transmittance is recorded. The experiment is ended once an experimental pressure of 3.4 MPa is reached, but the start pressure and end pressure usually depends on the crude oil being tested (Jamaluddin et al. 2001).

Figure 3.5 presents the NIR response of crude oil from Gulf of Mexico. An isothermal depressurization experiment was performed to measure the AOP. From the transmittance plot, it can be seen that the NIR transmittance increases as pressure decreases. This is because of the decrease in density of the single-phase fluid as the light components in the mixture expand above the BP. In this case, a drop in light transmittance is observed at 36.5 MPa, indicating the formation of asphaltene particles. As the system is further depressurized, more light is scattered as asphaltene aggregates grow, which decreases the transmittance value to zero. The BP is identified by the appearance of gas bubbles, at 29.4 MPa. By further depressurization, the light transmittance value starts to increase around 26 MPa. This increase is because of the redissolution of asphaltene particles between the

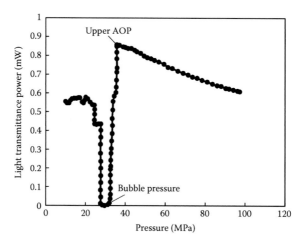

FIGURE 3.5 Near-infrared response during isothermal depressurization of crude oil. (Reprinted from Akbarzadeh, K. et al., *Oilfield Rev.*, 19, 22–43, 2007.)

bubble pressure and the lower onset point. Any further depressurization will have no effect on the light transmittance value.

This technique has gained more interest compared to other techniques because it is relatively fast and requires a small amount of sample. Light scattering coupled with a high-pressure microscope has become the industry standard to measure AOP under high-pressure, high-temperature (HPHT) conditions. However, one of the drawbacks of this technique is that the measured onset point depends on the rate of depressurization used (Akbarzadeh et al. 2007). A higher depressurization rate will result in a lower onset pressure. This is probably because LST can only detect particles that are larger than 1 μm.

3.2.4 HIGH-PRESSURE MICROSCOPY TECHNIQUE

High-pressure microscopy (HPM) consists of two major units: PVT cell and a high-magnification microscope. This is a direct technique that allows the visualization of multiple phases at HPHT. At the onset of asphaltene precipitation, the appearance of asphaltene particles can be imaged using this technique. HPM is usually used in conjunction with LST to validate the onset point determined by LST.

Jamaluddin et al. (2002), measured the asphaltene precipitation onset and the BP of the same crude oil used in LST and compared the results. Initially, the reservoir fluid was allowed to stabilize at the experimental pressure and temperature. It was then injected into the HPM assembly and flushed several times to ensure sample homogeneity. The fluid was then depressurized, and any visual or phase changes were recorded by the HPM. The results were recorded in the form of images. At the initial pressure, few solid particles were observed. These can be solid impurities such as sand particles because the sample was not pretreated before the experiment. So the amount and size of solid particles at the starting pressure can be used as a baseline reference to detect the onset of precipitation. Around 37 MPa, the number of solid particles slightly increased, indicating that the upper onset of precipitation was reached. This validates the onset obtained from LST. Upon further depressurization, the amount and size of solid particles increased even more. The BP was detected visually by observing the formation of the bubble (Jamaluddin et al. 2002). HPM can only provide qualitative information about the amount and size of the precipitated particle. However, the HPM images can be analyzed by particle size analysis software to quantify the amount and size of particles. These values can also be used to define asphaltene precipitation onset pressure if the visual comparison is not clear (Akbarzadeh et al. 2007). In another example, Karan et al. (2003) used HPM to test the effectiveness of different inhibitors. The inhibitors were evaluated based on the pressure at which asphaltene particles start to precipitate and the size of the particles.

3.2.5 HIGH PRESSURE, HIGH TEMPERATURE FILTRATION TECHNIQUE

The filtration technique uses a conventional PVT cell, where the cell is depressurized to induce asphaltene precipitation. The content of the PVT cell is mixed continuously with a magnetic stirrer throughout the experiment. At each pressure step, a small amount of the fluid is extracted through a filter while maintaining the experimental pressure and temperature. SARA analysis is performed on the material trapped by the filter. As the upper onset of asphaltene precipitation is reached, the amount of asphaltenes in the material trapped by the filter increases and a maximum value is reached at the bubble pressure. The lower onset point can also be detected using this technique (Jamaluddin et al. 2001).

A sample of crude oil from the Gulf of Mexico was tested using HPHT filtration to compare this technique with LST (Jamaluddin et al. 2001). A 60 cm³ volume of reservoir fluid was injected into the PVT cell and kept at reservoir pressure and temperature. At each pressure interval, a 10 cm³ sample was passed through a 0.45 μm filter, and the trapped material was analyzed. The amount of asphaltene at each pressure interval was plotted as a function of pressure in Figure 3.6. The NIR response from the LST was also plotted for comparison.

At 69 MPa, the content of asphaltenes on the material collected by the filter is equal to the amount of asphaltenes in the stock tank liquid. This indicates that all asphaltenes are in solution and the onset pressure is not reached yet. At 36.5 MPa, the NIR response shows a drop in transmittance as a result of asphaltene precipitation. However, the amount of asphaltenes in the filtered sample does not show an increase or an indication of asphaltenes precipitation. Yet, this is probably because of the size limitation associated with this technique. The precipitated asphaltenes can be smaller than 0.45 μm and thus was not trapped by the filter. As the pressure is further reduced, the amount of asphaltenes in the trapped sample increases

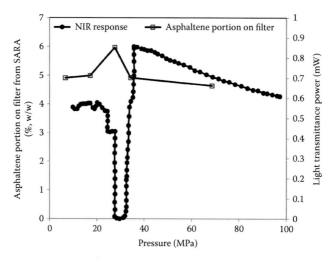

FIGURE 3.6 Comparison between the amounts of asphaltenes collected in the filter during high-pressure and high-temperature filtration method to the near-infrared response obtained using light scattering techniques for crude oil from the Gulf of Mexico. (Reprinted from Jamaluddin, A.K.M. et al., In SPE-72154-MS, Society Petroleum Engineers, Kuala Lumpur, Malaysia, doi:10.2118/72154-MS, 2001.)

until it reaches a maximum at a pressure lower than the BP detected by LST. Below the bubble pressure, the content of asphaltenes in the material trapped on the filter starts to decrease because of the redissolution of asphaltenes until it stabilizes around the initial amount of asphaltene in the stock tank liquid (Jamaluddin et al. 2001).

The HPHT filtration technique is less accurate in comparison to the LST. This is because when asphaltenes first precipitate they are liquid like. Thus, they cannot be effectively trapped by the filter paper, and the onset point detected using this method is not accurate. It is also time-consuming and requires more time to run than other techniques, such as ART and LST. Nevertheless, HPHT filtration technique allows the quantification of the amount of asphaltenes precipitated. Another advantage is that asphaltenes are physically extracted and can be characterized by electron microscopy techniques, mass spectrometry, molecular diffusion studies, etc (Akbarzadeh et al. 2007).

3.2.6   QUARTZ CRYSTAL RESONATOR TECHNIQUE

Techniques based on quartz crystal resonator (QCR) have shown a great potential to study asphaltene adsorption, precipitation, and deposition. Using this technique to detect asphaltene was first suggested by Abudu and Goual (2008) after they have noticed the adsorption of asphaltenes on QCR even in toluene solutions. Then, Daridon et al. (2013) developed a sensitive indirect technique to detect AOP at ambient conditions using QCR under titration experiments. More recently, Cardoso et al. (2014) developed a way to perform the QCR technique at HPHT where asphaltene precipitation is induced by depressurization. The QCR technique is based on the resonant frequency of the quartz crystal used. The precipitation of asphaltene induces a shift in the resonance frequency and the half band–half width. This is because asphaltenes tend to deposit on the QCR, and the resonant frequency is proportional to the dimension of the quartz crystal.

Cardoso et al. (2014) performed a set of experiments to show the potential of using QCR in measuring asphaltene onset and its advantages over optical methods such as HPHT filtration. The experiments were performed on a recombined fluid containing 40.7 mol% Brazilian dead oil, and 38.3 mol% methane, and 31.0 mol% CO_2. Isobaric filtration was used to test the presence of precipitated asphaltene to make sure that the recombined oil was a single phase. The resonance frequency and bandwidth of the QCR were collected while in contact with the recombined oil during depressurization. The crude oil was assumed to behave as a Newtonian fluid. Thus, the product of fluid density and viscosity depends linearly on the pressure. The Kanazawa-Gordon equation shows that the shift in the resonance frequency is directly proportional to the product of fluid density and viscosity. As a result, for Newtonian fluids the change in resonance is expected to increase linearly with pressure and any deviation would be an indication of asphaltene precipitation. Similarly, the bandwidth is expected to increase linearly with pressure for Newtonian fluids. As per the conclusion reported by Daridon et al. (2013), deviations in the bandwidth provide a more accurate detection of the asphaltene precipitation onset. This is because the shift in resonance frequency results from an increase in solution viscosity and adsorption, whereas the deviations in the bandwidth are only related to the change in solution viscosity. Thus, the pressure at which a deviation from linearity was observed on the bandwidth reading was defined as the AOP by Cardoso et al. (2014).

Using QCR, at 331.85 K an AOP of 42.0 MPa for the recombined oil was detected. However, for the same crude oil mixture and experimental conditions, the filtration method failed to detect an onset point. This was justified by the size limitation associated with the sieve size used in the filtration experiment, where only particles of 0.45 μm or larger can be detected. This indicates that QCR is more sensitive and can detect particles that are smaller than 0.45 μm. However, this technique is more expensive than the previously mentioned techniques. It was also found that onset pressure obtained using QCR depends on the depressurization rate used. It is claimed that this finding confirms that asphaltene precipitation is a time-dependent process (Cardoso et al. 2014).

3.3 EXPERIMENTAL METHODS AT AMBIENT CONDITIONS

The experimental determination of AOP at HPHT is relatively expensive. In addition to that, bottom-hole samples are not always readily available. Thus, experiments at ambient conditions were used along with a thermodynamic model to predict the asphaltene behavior at reservoir conditions. The thermodynamic models will be discussed in Chapter 4. Asphaltene precipitation onset at ambient condition is usually defined as the minimum amount of precipitant, mainly *n*-alkanes, required to destabilize asphaltene from crude oil. A number of techniques have been developed to detect the onset of asphaltene precipitation by detecting the first precipitated asphaltene particles with the addition of precipitant to the crude oil. The measured onset varies not only among different crude oils but also with the precipitant used, the time needed to reach equilibrium, and the detection technique. This section provides an overview of the popular experimental methods used to determine the onset of asphaltene precipitation at ambient conditions. The advantages and limitations of each technique will be covered.

3.3.1 FILTRATION METHOD

Filtration is a simple and straightforward way to measure asphaltene precipitation onset at ambient conditions. By adding various amounts of precipitant to the crude oil, the well-mixed crude oil and precipitant mixtures are filtered through a filter paper. The measured onset is defined as the minimum amount of precipitant at which asphaltene particles are detected on the filter paper after drying. This method is developed from titration experiments implemented by various researchers studying the asphaltene precipitation process and the maximum precipitation amount (Chung et al. 1991; Rassamdana et al. 1996).

The filtration method is well presented by Buenrostro-Gonzalez et al. (2004) with two Mexican crude oils and four *n*-alkanes. The crude oils were prefiltrated using a 0.45 μm Teflon filter paper to remove any suspended particles. Then different volumes of *n*-alkane were mixed with 5 g of crude oil sample in a flask, respectively, using ultrasonic mixing for 15 minutes. The oil/*n*-alkane mixtures were left overnight in the sealed container and then filtered using another 0.45 μm Teflon filter paper (known weight) under vacuum. The flask and the filter paper were rinsed with a small amount of the used *n*-alkane to remove any residual oil. Then the filter paper together with the precipitated asphaltene particles was dried in a vacuum oven at 333 K and 0.1 bar (gauge pressure) for more than 6 hours and then measured the weight again. The difference between the final weight and the initial weight of the filter paper was the mass of the precipitated asphaltene.

Filtration method requires only basic lab instruments and supplies. It is the easiest way to determine the asphaltene precipitation onset. The asphaltene precipitation amount is also of keen interest to researchers and can be determined by filtration method (Kokal et al. 1992; Spiecker et al. 2003; Trejo et al. 2004). However, the measured asphaltene onset is not reliable compared to other techniques. The accuracy of filtration method is hampered by many factors, such as the incremental amounts of precipitant added, the precision of the balance, the mesh size of the filter paper, and so on. The 0.45 μm mesh size is larger than the size of first precipitated asphaltene particles that the measured onset is shifted to higher precipitant region.

3.3.2 Microscopy Method

Microscopy is a common technique to measure asphaltene precipitation onset at ambient conditions as well as at HPHT. Precipitant/crude oil mixtures with different ratios are prepared and viewed under an optical microscope. The onset is determined by the micrograph that first shows the first asphaltene particles. Angle et al. (2006) measured the asphaltene precipitation onset of three toluene-diluted heavy crude oils (API ≤ 11) by adding *n*-heptane using a Nikon microscope. The microscopic images showed that the first precipitated asphaltene particles were small dots (~1 μm observation limit) and then grew into larger particles over time.

The time dependence of the asphaltene precipitation from crude oils was also studied using optical microscopy (Maqbool et al. 2009). Figure 3.7 presents the micrographs of a crude/heptane mixture made with an equal ratio of crude oil and heptane that is sealed for a different amount of time. There was some haze observed after waiting for 0.9 hours. The particle size was around 0.2–0.3 μm, which was smaller than the resolution limit of the used microscope. The precipitation onset time was defined as 1.4 hours with clearly observed asphaltene particles with a size of about 0.5 μm.

Evaporation of the volatile precipitant leads to unreliable observations under the microscope, making it more difficult to determine the asphaltene precipitation onset from the micrographs. Buckley (2012) used a capillary glass tube (50 mm in length by 200 μm across by 20 μm deep) with closed ends to test the asphaltene precipitation onset of three different crude oils with *n*-heptane.

The precipitation onset determined by microscopy technique is highly dependent on the resolution limit of the used instrument. With the development of more advanced microscopes, researchers introduced asphaltene onset measurements using a high-resolution transmission electron microscope (HRTEM; Goual et al. 2014). The observed asphaltene particles were of the size of ~100 nm. Besides the

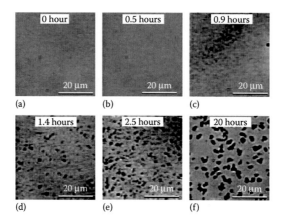

FIGURE 3.7   (a–f) Micrographs showing the time dependence of asphaltene precipitation for a crude/heptane mixture containing 50 vol % heptane. (Reprinted from Maqbool, T. et al., *Energy Fuels* 23, 3681–3686, 2009.)

resolution limit of the instrument, the measured onset is often likely to change with the visual acuity of the observer. Automatic measurements are difficult to achieve with microscopy technique, and the evaporation of precipitant during observation adds error to the result.

3.3.3   SPECTROSCOPY METHOD

Spectroscopy is one of the most popular methods to determine the onset of asphaltene precipitation based on changes in transmittance intensity by the addition of a precipitant. The absorbance first decreases because of the dilution effect as a result of adding the precipitant. When the onset is reached, the precipitated asphaltene particles cause scattering of the transmitted light, which leads to an increment of the absorbance. The point at which the trend deviates from linearity is the measured onset. UV-visible spectrophotometer has been used by various researchers to measure the asphaltene precipitation onset from dilute oils and model oils (asphaltene dissolved in a solvent) (Fuhr et al. 1986; Cimino et al. 1995; Andersen 1999). In one of the experiments, the absorbance at 800 nm was monitored over time by continuously pumping precipitant into the flow cell filled with dilute bitumen sample (Fuhr et al. 1986). The amount of precipitant where the trend deviates from linearity was defined as the onset point. Andersen (1999) chose a wavelength of 740 nm to study the effect of titration flow rate on the precipitation onset of a dilute toluene solution of asphaltene or crude oil. The results showed that the onset appeared early under a high flow rate because of severe local precipitation.

It is difficult to detect the asphaltene precipitation onset from original oils by UV spectroscopy because of the dark color of oils. NIR spectroscopy is more practical to measure the onset directly from crude oils demonstrated by previous researchers (Fuhr et al. 1991; Oh and Deo 2002; Oh et al. 2004; Tharanivasan et al. 2009).

Oh and Deo (2002) investigated the effect of the carbon number of the used n-alkanes on the precipitation onset using NIR spectrometer with a fiber optic detector. In their experiment, a clean glass flask was filled with a certain amount of crude oil, and the precipitant of interest was pumped into the flask continuously. The oil/alkane mixture was well mixed with the help of magnetic stirring. The NIR probe was placed inside the mixture to obtain the absorbance. The absorbance of the crude oil was measured at a wavelength of 1600 nm with the addition of heptane up to 240 cm^3. The flow rate of heptane was 1 cm^3/min. The wavelength of 1600 nm was used because of its low background absorbance. The onset was defined as the amount of heptane that results in a deviation from the linear trend.

The high repeatability of the spectroscopy technique makes it a good candidate for asphaltene onset measurements. LST is also a powerful tool to study the asphaltene aggregation process for a better understanding of the mechanism (Yudin and Anisimov 2007). Burya et al. (2001) observed two different aggregation regimes, diffusion-limited and reaction-limited, and a crossover regime between those two with the help of dynamic light scattering.

3.3.4 CAPILLARY VISCOSIMETRY METHOD

The asphaltene precipitation onset can be detected by viscosity measurements because of the non-Newtonian rheological behavior of crude oil-asphaltene-precipitant suspensions. Before the asphaltene precipitation onset, the viscosity of the crude oil/precipitant mixture decreases with the addition of precipitant. At the onset, the precipitated asphaltene particles are suspended in the fluid, which increases the apparent viscosity. The viscometric method was proposed to determine the onset of asphaltene precipitation from a crude oil with three different n-alkane precipitants (Escobedo and Mansoori 1995, 1997). More than 30 crude oil/n-alkane samples were prepared to cover the entire range of precipitant concentration varying from 0 to 100 vol%. Glass capillary viscometers with different sizes were used to measure the viscosity of the dark crude oil/n-alkane samples at the entire viscosity range. Figure 3.8 presents the viscosity trend of the crude oil/n-alkane samples with the addition of n-heptane and n-nonane, respectively. The onset of the asphaltene precipitation was determined graphically and enhanced by comparison with a reference system consisting of nonprecipitating solvents, such as toluene, tetrahydrofuran, and their mixtures. The measured onset of Maya crude oil was 32.9 wt% for n-heptane and 35.8 wt% for n-nonane.

Viscosimetry method can also be used to study the asphaltene aggregation and the amount of asphaltene precipitated out of solution by calculating the effective volume fraction of suspended asphaltene particles (Escobedo and Mansoori 1997). The onset of micellization of asphaltenes in aromatics and polar solvents have been investigated by viscosity technique (Mousavi-Dehghani et al. 2004; Priyanto et al. 2001). The expression of relative viscosity, the ratio of the viscosity of the suspension over that of the pure solvent, was used to determine both the upper and lower limits of asphaltene micellization. Asphaltene micellization is one of the different theories used to explain asphaltene stabilization, details on the different theories will be discussed in Section 4.2.1.

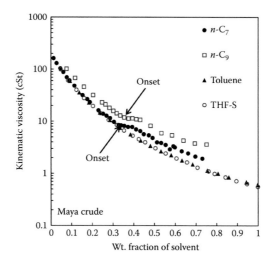

FIGURE 3.8 Kinematic viscosity versus weight fraction of solvent in solutions of Maya crude with various solvents. (Reprinted from Escobedo, J., and Mansoori, G.A., *SPE Prod. Facil.*, 12, 116–122, 1997.)

Glass capillary viscometer is an inexpensive equipment with acceptable accuracy to determine asphaltene precipitation onset compared to other instruments. It works for both light and heavy crude oils by using capillary viscometers of different sizes. The measured onset depends on the sensitivity of the suspension viscosity to the size of the asphaltene particles. In other words, the asphaltene particles need to grow to a certain size to change the characteristic property of the host fluid. The main limitation of viscosity technique is equipment clogging due to severe asphaltene precipitation in the narrow capillary.

3.3.5 CONDUCTIVITY METHOD

Asphaltene is the heaviest and most polarizable fraction of crude oil. Functional groups with heterogeneous elements, such as nitrogen, sulfur, oxygen, vanadium, and nickel, have been found in asphaltene fractions that can be ionized and carry charge in an electric field. Therefore, the asphaltene precipitation onset can be determined from electric conductivity measurements of crude oil by continuously adding a precipitant. A sudden change of the measured conductivity indicates the onset of asphaltene precipitation.

Fotland et al. (1993) introduced the conductivity method to determine asphaltene precipitation onsets of four crude oils with various asphaltene concentrations. A conductivity cell consisting of thin concentric cylinders was designed. This particular setup showed a high resistance of the empty cell that made it suitable for low conductivity measurements of crude oil/precipitant samples. The electrodes were coated with gold to obtain more stable readings. A Hewlett-Packard impedance

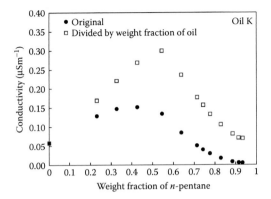

FIGURE 3.9 Conductivity as a function of added *n*-pentane (lower curve) and conductivity divided by weight fraction of oil as a function of added *n*-pentane (upper curve). (Reprinted from Fotland, P. et al., *Fluid Phase Equilib.*, 82, 157–164, 1993.)

meter was used to collect the conductivity and capacitance at 1.1 V and 1500 Hz. Figure 3.9 presents the measured conductivity of crude oil K with the addition of *n*-pentane. The conductivity of the crude oil/*n*-pentane mixture initially increased with increasing concentration of *n*-pentane because of the higher mobility of the conducting components in the mix. After reaching the maximum, the dilution effect of added *n*-pentane became dominant, resulting in decreased conductivity. A steeper drop was observed at the asphaltene precipitation onset as a consequence of losing the conducting species in the mixture. The asphaltene precipitation onset can be determined from the mass fraction normalized conductivity (upper curve in Figure 3.9) more conveniently. The onset corresponded to the maximum of the mass fraction normalized conductivity curve.

The extent of asphaltene precipitation can also be determined from conductivity measurements (Fotland et al. 1993). Compared with gravimetric measurements, the conductivity method shows a late asphaltene precipitation onset because of its capability to distinguish wax formation from asphaltene particles. Wax particles are paraffin molecules with no capacity to carry a charge. Thus, its crystallization has no significant effect on the conductivity of the fluid. The conductivity method outperforms other methods to detect asphaltene precipitation onset at temperatures lower than the wax appearance temperature. There is potential to operate this technique at HPHT for onset measurements. Previous researchers used electrical conductivity measurements to study the molecular interactions between asphaltenes and polar solvents, such as tetrahydrofuran and nitrobenzene (Behar et al. 1998).

Two different types of behaviors were observed for asphaltene in model solutions corresponding to both high and low asphaltene concentrations. The solvent was found to have little effect on asphaltene aggregation in dilute solutions, which was opposite to high asphaltene concentration solutions.

3.3.6 DENSITY METHOD

Asphaltene precipitation onset can be determined from density measurements of the crude oil solvent/precipitant mixtures. The measured density should decrease gradually with the addition of precipitant because of the dilution effect. An abrupt drop in the measured density is expected at the asphaltene precipitation onset. However, the dominant dilution effect makes it difficult to capture this sudden change precisely from the density plot (Ekulu et al. 2004). A reduced density was defined and used to determine the asphaltene precipitation onset instead (Ekulu et al. 2004, 2010). The reduced density was defined as the density difference between the crude oil/solvent/n-alkane mixture and the deasphalted oil/solvent/n-alkane mixture measured at the same temperature, pressure, and concentration. The change in density as a result of asphaltene precipitation was illustrated by the reduced density. The densitometry technique was used to study the asphaltene precipitation onset of seven different crude oils from various sources. The Anton Paar densitometer, model DMA 60, was used for density measurements calibrated by high-purity n-heptane and toluene. Figure 3.10 presents the reduced density profile of a crude oil/toluene/n-heptane system as a function of the ratio between n-heptane concentration and concentration of crude oil. The breakpoint observed in the plot corresponds to the measured asphaltene precipitation onset. The results were compared to those measured by spectroscopy technique with good agreement. The minimum amount of precipitant required to precipitate out all asphaltenes from the crude oil can also be determined using density method. The density method is relatively simple compared to other techniques. The limitation of the technique is potential clogging of the density measuring cell for samples with a large number of precipitated asphaltene particles.

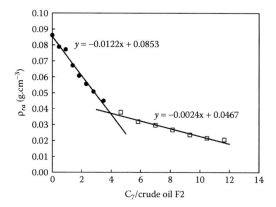

FIGURE 3.10 Reduced density of the system {crude oil F2 (5 g) + toluene (5 g)} + n-heptane. The crude oil F2 contains 5.0% (w/w) of asphaltenes. (Reprinted from Ekulu, G. et al., *J. Dispers. Sci. Technol.*, 25, 321–331, 2004.)

3.3.7 Refractive Index Method

Asphaltene precipitation is dominated by interactions between asphaltene and other components in the crude oil/precipitant mixture, which are mainly London dispersion interactions that can be characterized by the refractive index of the mixture measured at the sodium-D line (Buckley et al. 1998). The capability of the mixture to stabilize asphaltenes in solution, usually characterized by solubility parameters, was directly related to the refractive index. Additionally, asphaltene precipitation was observed at a narrow range of mixture refractive index (Buckley 1996). The method based on refractive index is a powerful tool to evaluate asphaltene instability in different solutions.

Buckley (1999) proposed an experimental method to predict the asphaltene precipitation onset using refractive index measurements. A given crude oil was titrated with heptane to induce asphaltene precipitation. The refractive index of crude oil/n-heptane mixtures with various heptane concentrations was measured to determine the asphaltene precipitation onset. Figure 3.11 presents the measured refractive index of the crude oil/n-heptane mixture as a function of the crude oil volume fraction in the mix. The refractive index first decreases linearly along the line connecting the refractive index of pure crude oil and pure n-heptane because of the dilution effect of n-heptane. The refractive index further decreases as asphaltene precipitates, which causes deviation from the previous line. The first observed deviation is defined as the precipitation onset. The refractive index decreases along the maltene/n-heptane line after all asphaltenes are precipitated out of the oil. Wattana et al. (2003) also used refractive index measurements to study asphaltene precipitation for several crude oils.

The refractive index measurement is straightforward and easy compared to other methods, requiring a minuscule amount of sample. However, there are difficulties in defining the onset from the refractive index plot for some oils without significant

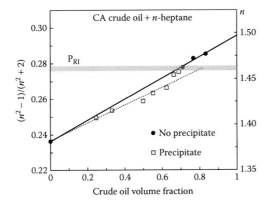

FIGURE 3.11 The refractive index of the crude oil/n-heptane mixtures versus the crude oil volume fraction in the mixture. P_{RI}, refractive index at the onset. (Reprinted from Buckley, J. S., *Energy Fuels*, 13, 328–332, 1999.)

deviation from the first region. The solvent condition at the onset can be quantified by refractive index measurements, which is applied in asphaltene instability trend (ASIST) method to predict asphaltene precipitation at HPHT (Wang and Buckley 2001, 2003). The asphaltene instability trend method is discussed in Section 4.2.2.4.

3.3.8 INTERFACIAL TENSION METHOD

The asphaltene precipitation onset can be detected by measuring the interfacial tension between oil/precipitant mixture and water phases (Kim et al. 1990; Mousavi-Dehghani et al. 2004). The start of asphaltene precipitation induces abrupt changes in the measured interfacial tension. An unstable region is observed for the interfacial tension measurement after asphaltene precipitation because of the migration of precipitated asphaltene particles to the oil and water interface. The asphaltene precipitation onset of three different crude oils with various asphaltene contents (0.8–6.8 wt%) was measured by an interfacial tension (IFT) technique (Mousavi-Dehghani et al. 2004). Crude oil/n-heptane mixtures were prepared with different weight percentages of n-heptane. The IFTs of the mixtures and water systems were measured by the DuNouy ring method at 298 K. The surface tensions of the crude oil/n-heptane mixtures were also measured. Figure 3.12 presents the measured IFT and surface tension as a function of the weight percentage of n-heptane. The continuous line shows the changing trend of IFT with different amount of added n-heptane without asphaltene precipitation. The experimental data indicate an unstable region in the measured IFT with significant deviations from the solid line. The onset is defined as the point at which the slope of the IFT curve first changes. The asphaltene precipitation has little effect on the surface tension of the mixture.

The instability of measured IFT is more significant after the precipitation onset as a consequence of the adhesion or deposition of asphaltene particles at

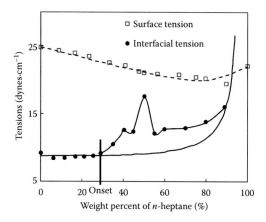

FIGURE 3.12 Onset of asphaltene precipitation of crude oil by adding n-heptane determined from interfacial tension measurements. (Reprinted from Mousavi-Dehghani, S.A. et al., *J. Pet. Sci. Eng.*, 42, 145–156, 2004.)

the interface. The IFT method is claimed to be more sensitive because a small amount of surface-active material at the interface can cause a noticeable change in the IFT. Additionally, the IFT technique is used by various researchers to study the asphaltene aggregation in different organic solvents (Yarranton et al. 2000; Fossen et al. 2007). The IFT method is time-consuming and difficult to perform. It is also challenging to extend to HPHT.

3.3.9 GRAVIMETRIC METHOD

The gravimetric technique can be used to determine the asphaltene precipitation onset because the precipitated asphaltenes are heavier than the rest of crude oil/precipitant mixture. The asphaltene particles are driven to the bottom after centrifugation. The amount of precipitated asphaltene can be determined by removing the supernatant. The precipitation onset is the first sample with observed precipitated asphaltene particles at the bottom. The measured onset is highly dependent on the driving force generated by centrifugation and is not consistent with those measured by other techniques. The gravimetric method is more appropriate for quantifying the amount of asphaltene precipitated. The procedure starts with adding a significant amount of precipitant to the oil. Then the mixture is sonicated for a better mix and left to settle for a period of time. The mixture is then centrifuged, and the supernatant is carefully removed. The precipitate is washed with hot pure precipitant and centrifuged again until the supernatant is colorless. The precipitate is finally dried and weighed.

3.3.10 NUCLEAR MAGNETIC RESONANCE METHOD

Prunelet et al. (2004) studied and demonstrated that the precipitation onset in an asphaltene solution could be measured using nuclear magnetic resonance (NMR). This method uses the wealthy information on molecular motions and large-scale interactions between molecules given by transverse relaxation time. Because NMR relaxation time is related to the sizes of asphaltene particle, it will vary during the precipitation process, and precipitation of asphaltene can be detected because aggregated asphaltene alters the distribution of relaxation time.

In Prunelet et al.'s (2004) study, asphaltene extracted with n-heptane were mixed with toluene in a sealed container for several hours. Samples were prepared by mixing asphaltene solution with a known quantity of n-heptane inside a sealed 10 mm NMR tubes. The transverse magnetization decays were measured at 60 MHz using standard CPMG sequences. They found that asphaltene has a well-defined signature typically at relaxation times in the order of 1 ms and that the asphaltene signature becomes bimodal near the flocculation threshold. An example plot is shown in Figure 3.13, in which the star indicates actual flocculation in solution, as observed visually.

For all the oils tested, the bimodal signature was observed near asphaltene precipitation onset except for one. Thus, this method might be a useful indicator of asphaltene precipitation, but other independent measurement should be performed to verify the result. The sensitivity of this method should also be further improved. For some sets, the bimodal signature starts to show up at a precipitant concentration higher than the concentration in which precipitated asphaltene can be spotted visually.

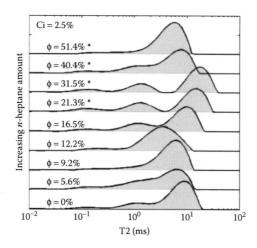

FIGURE 3.13   Relaxation time distributions of the asphaltene aggregates. (Reprinted from Prunelet, A. et al., *Comptes Rendus Chim.*, 7, 283–289, 2004.)

3.3.11   QUARTZ CRYSTAL RESONATOR METHOD

The use of quartz crystal resonator (QCR) to detect asphaltene flocculation in real time was first suggested by Abudu and Goual (2008). This is due to its ability to detect both mass deposition and rheological changes in the surrounding medium. Daridon et al. (2013) proposed the use of QCR to assess the onset of asphaltene precipitation at ambient conditions during titration experiment. They tested the technique using four different samples of dead oil. Toluene was used as the solvent for the oil and *n*-heptane as the precipitant. The measurements were performed by recording the resonant properties of the QCR, during continuous addition of the flocculating agent. The solution was continuously stirred using a magnetic stirrer, during the measurement.

Daridon et al. (2013) showed that both resonant frequency and dissipation are extremely sensitive to asphaltene precipitation. They have concluded that changes in resonant frequency are due to the adsorption of asphaltene on the gold electrode surface, which can occur before asphaltene precipitation. Thus, changes in resonant frequency cannot be used to determine the onset of asphaltene precipitation. On the other hand, changes in dissipation are only related to changes in viscosity. Asphaltene precipitation onset can be detected by plotting the dissipation against precipitant weight percent, as shown in Figure 3.14. The initial drop in dissipation is due to the addition of *n*-heptane, which decreases the viscosity of the solution. As the fraction of *n*-heptane increases, a small addition of *n*-heptane induces a sharp increase in dissipation. This increase in dissipation indicates a significant increase in viscosity, which is caused by the formation of asphaltene aggregates close to the quartz surface. The main advantage of this technique is its sensitivity and its applicability at high pressure and high temperature, as discussed in Section 3.2.6.

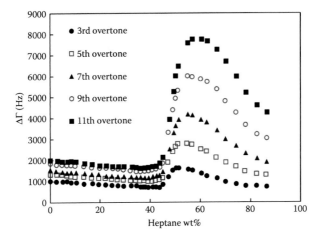

FIGURE 3.14 Change in the dissipation of different overtones as a function of heptane mass percentage. (Reprinted from Daridon, J.L. et al., *Energy Fuels*, 27, 4639–4647, 2013.)

3.4 INDIRECT METHOD

Tavakkoli et al. (2015) proposed a novel technique, referred to as the *indirect method* to determine asphaltene precipitation onset. This technique utilizes the currently available technologies, where it is based on gravimetric and light scattering techniques. This method is unique in a sense that the absence of asphaltene particles is detected unlike the conventional "direct methods," which requires direct observation of the precipitated asphaltene particles. The indirect method was found to surpass the currently available techniques. The indirect method is more sensitive than currently available commercial methods, where it can detect particles as small as 100 nm. Using the indirect method, one can quantify the amount of asphaltene precipitated. Unlike methods based on direct detection of asphaltene precipitation, the indirect method can be used to detect asphaltene precipitation onset for all types of crude oils (high or low asphaltene content).

Unstable crude oil needs to be treated before utilizing the indirect method, in which the free water and sediments need to be removed. In this context, unstable crude oil is defined as the crude oil with a large amount of sediment due to the highly unstable asphaltenes. A separation funnel is used to separate free water from the crude oil, where the crude oil is transferred to the funnel and left to stand for 10 minutes. As for the separation of sediments and precipitated phase, centrifugation is usually used. However, upon centrifugation and removal of sediments, the unstable oil may produce more precipitate to reach equilibrium. Also, the remaining oil composition after centrifugation may not be representative of the composition of the actual oil. Therefore, if the oil is very unstable and the amount of precipitated phase is significant, it is recommended to re-dissolve the precipitated phase back into the crude oil. The precipitated phase can be re-dissolved by heating the sample at a temperature lower than 343 K. A temperature of 343 K or less is recommended to prevent a significant loss of light components in the crude oil. If the precipitated phase is only partially dissolved at 343 K, an excess amount of a good solvent, such as toluene, can be added to the crude oil. The oil/toluene mixture is referred to as *modified oil*. The effect of excess solvent on the measured absorbance of the sample can be removed later in the study (the procedure will be discussed later in this section).

Once the unstable oil is treated, the indirect method can be performed by adding a precipitant (typically *n*-heptane) in different ratios and allowed to stand undisturbed for a specific aging time. The aging time is defined as the time the sample is left undisturbed after sample preparation and before centrifugation. To investigate the effect of water and electrolytes, the procedure is modified as follows. The aqueous solution is mixed with crude oil using a homogenizer before mixing the oil with the precipitant. Also, the samples are stirred using magnetic stir bars during the aging time to prevent separation of oil and aqueous phases. Then, the samples are centrifuged at 11,000 relative centrifuge force (RCF) for 15 minutes, which removes unstable asphaltene particles with sizes of 100 nm and above from the supernatant. This specific speed was chosen because SEM images suggest that asphaltene particles start to precipitate at sizes between 100 and 400 nm (Vargas et al. 2014). Then, a certain amount of supernatant is taken from the centrifuged sample and diluted with enough toluene to avoid absorbance signal saturation. The absorbance values at a wavelength ranging from 1300 to 700 nm are measured, using toluene as a blank. The absorbance value for supernatant at a specific wavelength can be calculated using Equation 3.1.

$$A_{supernatant} = A_{measured} - A_{toluene} \tag{3.1}$$

The absorbance value calculated above does not represent the actual absorbance value of the oil sample because it has been diluted by precipitant and toluene. To remove the dilution effect, the sample is weighed every time a new liquid is added to keep track of the actual volume of crude oil, heptane, and toluene in the mixture. The overall dilution ratio (DL) is calculated using Equation 3.2, and the corrected absorbance value is the multiplication of absorbance for supernatant and the overall dilution ratio.

$$DL = \frac{V_{precipitant}}{V_{precipitant} + V_{oil}} * \frac{V_{supernatant}}{V_{supernatant} + V_{toluene}} \tag{3.2}$$

Next, the absorbance value for supernatant versus the volume fraction of precipitant is plotted (Figure 3.15). The sudden deviation from linear behavior (here between 40 vol% and 50 vol% of *n*-heptane) corresponds to asphaltene precipitation onset. The main idea behind the proposed procedure is that the precipitated asphaltene particles they are removed by centrifugation change the optical density of the remaining liquid. As a result, the absorbance value drops after the precipitation onset, because the remaining liquid becomes less dense, or more transparent. This method is called *indirect method* because it studies the change in optical properties of supernatant instead of directly looking for precipitated particles.

The amount of asphaltene precipitated can be quantified by constructing a calibration curve, which relates the concentration of asphaltene to the absorbance value.

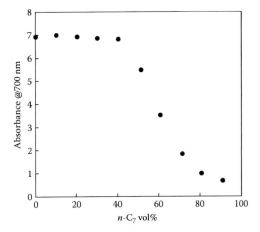

FIGURE 3.15 Results for indirect method after correcting the dilution effect for model oil (asphaltene extracted from crude oil S) with 0.5 wt% asphaltene concentration. The model oil was precipitated with *n*-heptane and aged for 1 day at ambient conditions. (Reprinted from Tavakkoli, M. et al., *Energy Fuels*, 29, 2890–2900, 2015.)

A linear regression is applied to the data, and the regressed slope is used to find the percentage of asphaltene precipitated (Equation 3.3).

$$Asph.\% precipitated_{x vol\% precipitant} = \frac{A_{0 vol\% precipitant} - A_{x vol\% precipitant}}{slope\, of\, regression\, line} * 100\% \quad (3.3)$$

3.4.1 EFFECT OF WATER

The effect of emulsified water on asphaltene stability at ambient condition has been studied using the indirect method by Tavakkoli et al. (2016b). It was found that emulsified water has different effects on different crude oils. For crude oil CN (Figure 3.16a), emulsified water did not cause any changes to the asphaltene onset pressure or the amount of asphaltene precipitated, it only lowered the absorbance value of the supernatant before the onset point. The reduction in absorbance can be due to the adsorption of some asphaltenes at the water–oil interface. These adsorbed asphaltenes are removed during centrifugation, which reduces the optical density of the remaining fluid and thus reduces the absorbance value. For crude oil SE (Figure 3.16b), the results with and without emulsified water are almost identical. According to Table 3.1, crude oil CN is heavier compared to crude oil SE, and therefore, one may conclude that heavier asphaltene particles have higher affinity to the water–oil interface. To verify this hypothesis, Tavakkoli et al. (2016b) also applied the indirect method to bitumen A1, which is a highly viscous oil that contains 22 wt% C_{5+} asphaltene. Two types of asphaltenes were extracted from bitumen A1 using *n*-pentane and *n*-heptane to prepare model oils (0.5 wt% asphaltene in toluene), and the indirect method results are shown in Figures 3.16c and d. These

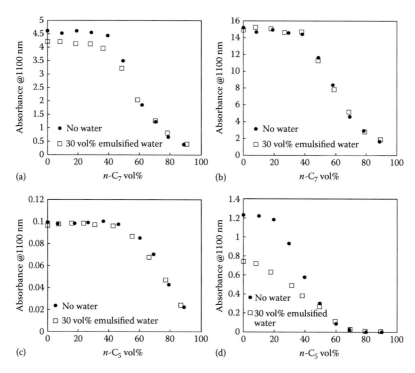

FIGURE 3.16 Effect of emulsified water on model oil prepared with asphaltene extracted from (a) CN, (b) SE, (c) bitumen A1 (C_{5-7} asphaltene), and (d) bitumen A1 (C_{7+} asphaltene). Samples in (a) and (b) were diluted with *n*-heptane and samples in (c) and (d) were diluted with *n*-pentane. All samples were emulsified with 30 vol% water, and aged for 1 day at 296 K and ambient pressure. (Reprinted from Tavakkoli, M. et al., *Energy Fuels*, 30, 3676–3686, 2016b.)

TABLE 3.1
Properties of Crude Oils CN and SE at Ambient Conditions

Property	Crude Oil CN	Crude Oil SE
Density (g/cm³)	0.899	0.893
Molecular weight (g/mol)	278.92	220.41
Viscosity (cP)	55.01	33.79
Water content (wt%)	0.076	0.183
Saturates (wt%)	31.90	55.27
Aromatics (wt%)	25.10	15.25
Resins (wt%)	29.90	19.45
C_{5+} asphaltenes (wt%)	13.10	10.03

results confirm the proposed hypothesis, that is, emulsified water does not alter the behavior of lighter asphaltene particles in solution. Yet, it significantly destabilizes the heavier asphaltene particles in solution. Therefore, the less soluble fraction of asphaltene has a high affinity for the oil–water interface, and only crude oils with heavier asphaltenes are destabilized by the presence of water.

Tavakkoli et al. (2016b) also studied the temperature effect on crude oil CN. The results suggest that without the presence of emulsified water, less amount of asphaltene precipitated at a higher temperature, as oil becomes a better solvent for asphaltene. However, when emulsified water is introduced to the system, more asphaltene particles precipitate compared to those precipitated at room temperature. One possible explanation is that at a higher temperature, both density and viscosity of the oil decrease, and thus mobility of asphaltene in the solution increases. At higher temperature, asphaltene particles can move to oil–water interface more quickly so that more asphaltene particles will precipitate out during a specific aging time.

3.4.2 EFFECT OF ELECTROLYTES

Sung et al. (2016) also studied the effect of electrolytes Al (III), Cr (III), and Na on asphaltene precipitation onset. As previous studies have shown that emulsified water and acidity can affect asphaltene stability, the effects of the two must be removed to investigate the effect of electrolytes only (Nassar et al. 2012; Tavakkoli et al. 2016b). To remove the effect of water, light crude oil was used, such that asphaltenes do not adsorb on the oil–water interfaces as shown by the work of Tavakkoli et al. (2016b). To remove the effect of acidity, all samples were adjusted to the same acidity (pH) before sample aging. Also, the effect of acidity alone was examined to separate its influence from effects of electrolyte on asphaltene precipitation.

Figure 3.17 shows the effects of Al (III), Cr (III), and Na on asphaltene stability for model oil with asphaltene extracted from crude oil P. These ions do not significantly change the absorbance values before precipitation onset, whereas after the onset, absorbance values drop more compared to the *no aqueous phase* set. One proposed explanation is that the acidity and charge of the metal ions both contribute to the drop in absorbance values. Nassar et al. (2012) concluded that the acidity of oil contributes to asphaltene precipitation because asphaltene aggregation is stimulated by the positively charged protons. Also, Schramm (2000) suggested that the positively charged protons may induce the formation of asphaltene aggregates through charge neutralization. For the effect of brine solution, Serrano-Saldaña et al. (2004) and Alotaibi et al. (2011) showed that the IFT of n-C_{12}/brine interface decreases with increasing concentration of brine because ions prefer to localize close to the interface. This might be the reason for the drop of absorbance value after precipitation onset with the presence of brine.

3.4.3 EFFECT OF IRON

Several studies have found that iron ions are released to oil from seawater or pipes and that it has a detrimental effect on asphaltene aggregation and precipitation (Ibrahim and Idem 2004; Ovalles and Rechsteiner 2015). It is found that the formation of

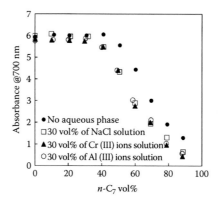

FIGURE 3.17 Effect of electrolytes on asphaltene precipitation for model oil (asphaltene extracted from crude oil P) with 0.5 wt% asphaltene concentration. Model oil was diluted by *n*-heptane, emulsified with 30 vol% ionic solutions, and aged for 1 day at 296 K and ambient pressure. (Reprinted from Sung, C.A. et al., In Offshore Technology Conference, doi:10.4043/27008-MS, 2016.)

asphaltene iron complex is associated with aromaticity, hydrogen-to-carbon (H/C) ratio and the formation of bonds on heteroatoms (Murgich et al. 2001). The specific mechanics behind Fe (III) and asphaltene particles were further studied using the indirect method by Sung et al. (2016).

The effects of emulsified water and acidity of oil were removed or identified using techniques explained in the previous section to isolate the effect of iron. For the effect of acidity, previous research also showed that the extent of asphaltic sludge precipitation in crude oil caused by Fe (III) is more severe than those caused by the acidic protons alone (Jacobs 1989). Sung et al. (2016) also verified using the indirect method that acidic protons did not affect asphaltene precipitation onset for model oil with asphaltene extracted from crude oil P.

As shown in Figure 3.18, the presence of Fe (III) drastically destabilizes the asphaltene particles in the system, in which the precipitation onset shifts to the left and the amount of asphaltene precipitated increases. The absorbance value drops from 0 vol% of *n*-heptane, but there is a significant drop in absorbance value from 30 vol% to 40 vol% of *n*-heptane, which stands for the precipitation onset. Sung et al. (2016) inferred that the change in absorbance value before the onset is caused by the presence of iron-asphaltene complex in the supernatant. After the onset, iron ions might be more attracted to the precipitated and aggregated asphaltene, so the absorbance value drops more drastically compared to the case with no aqueous phase.

To alleviate the negative effect from Fe (III) on asphaltene instability, Sung et al. (2016) proposed the addition of ethylenediaminetetraacetic acid (EDTA). EDTA has been used to form a chelate with metal ions, and thus the addition of EDTA can capture iron ions from the organic phase to prevent the formation of the iron-asphaltene complex. The modified indirect method with the addition of EDTA is a 3-day procedure, in which the mixture of crude oil and Fe (III) solution is stirred for 24 hours, then EDTA is added to the system and aged for 24 hours, and lastly, the precipitant is

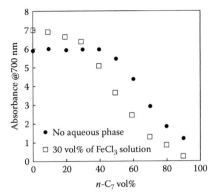

FIGURE 3.18 Effect of Fe (III) on asphaltene precipitation using indirect method for model oil (asphaltene extracted from crude oil P) with 0.5 wt% asphaltene concentration. Model oil was diluted by *n*-heptane, emulsified with 30 vol% FeCl$_3$ solution, and aged for 1 day at 296 K and ambient pressure. (Reprinted from Sung, C.A. et al., In Offshore Technology Conference, doi:10.4043/27008-MS, 2016.)

added and stirred for another 24 hours. As shown in Figure 3.19, the restored asphaltene precipitation onset suggests that the addition of EDTA successfully captures the Fe (III) in the solution. The higher absorbance values after the onset compared to the Fe (III) set also indicate that the amount precipitated out is smaller. However, the absorbance values are still lower than those in the *no aqueous phase* set as a result of acidic pH and charge of metal ions. In conclusion, EDTA can be used to mitigate the detrimental effects from irons on asphaltene instability.

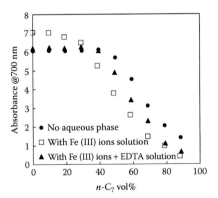

FIGURE 3.19 Effect of EDTA addition in model oil (asphaltene extracted from crude oil P) with 0.5 wt% asphaltene concentration and Fe (III) mixture on asphaltene instability. Model oil was diluted by *n*-heptane and aged for 1 day at 296 K and ambient pressure. (Reprinted from Sung, C.A. et al., In Offshore Technology Conference, doi:10.4043/27008-MS, 2016.)

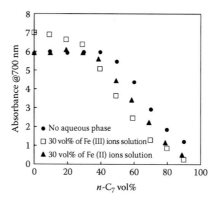

FIGURE 3.20 Comparison of the effect of Fe (II) and Fe (III) on asphaltene precipitation for model oil (asphaltene extracted from crude oil P). Model oil was diluted by *n*-heptane, emulsified with 30 vol% ionic solution, and aged for 1 day at 296 K and ambient pressure. (Reprinted from Sung, C.A. et al., In Offshore Technology Conference, doi:10.4043/27008-MS, 2016.)

The effect of Fe (II) on asphaltene stability ions was also studied by Sung et al. (2016), as shown in Figure 3.20. Compared to Fe (III), Fe (II) does not form a strong iron-asphaltene complex, and its effect on asphaltene stability resembles the effect from other electrolytes such as Al (III) and Cr (III). The larger drop in absorbance value after the onset can be explained using the same reasoning given previously.

3.4.4 Effect of Polydispersity

Asphaltene is a polydisperse mixture of the heaviest and most polarizable component in crude oil. The use of different *n*-alkane precipitants results in precipitates with significant differences in chemical and physical properties.

Tavakkoli et al. (2016a), showed that lighter hydrocarbons are stronger asphaltene precipitants, using the indirect method. As shown in Figure 3.21, the precipitated amount of asphaltene increases when precipitant carbon number decreases, which suggests that lighter hydrocarbons are stronger precipitants (Table 3.2).

It is also known that asphaltene precipitated by heavier precipitant are heavier and more aromatic than those precipitated by lighter hydrocarbons. As shown in Figure 3.22, asphaltenes separated with heavier hydrocarbons require less amount of precipitant to precipitate out of the solution. This plot also shows that the volume fraction of precipitant needed to induce asphaltene precipitation increases with increase in carbon number of precipitant until it reaches a maximum, after which the volume fraction of precipitant decreases with increase in carbon number of precipitant. This indicates that the asphaltene precipitation onset first shifts to the right and

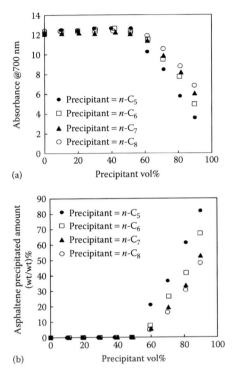

(a)

(b)

FIGURE 3.21 (a) Effect of polydispersity on asphaltene instability using indirect method for modified oil S4 (properties see Table 3.2) and (b) amount of asphaltene precipitated using calibration curve. All sets were aged for 1 day at ambient temperature. (Reprinted from Tavakkoli, M. et al. *Fluid Phase Equilibria*, 416, 120–129, 2016.)

TABLE 3.2

Properties of Crude Oil S4 at 1 atm and 293 K

Property	Crude Oil S4
Density (g/cm³)	0.826
Molecular weight (g/mol)	176
Viscosity (cP)	5.4
Saturates (wt%)	69.60
Aromatics (wt%)	22.02
Resins (wt%)	7.17
C_{5+} asphaltenes (wt%)	1.21
C_{6+} asphaltenes (wt%)	1.03
C_{7+} asphaltenes (wt%)	0.66
C_{8+} asphaltenes (wt%)	0.62

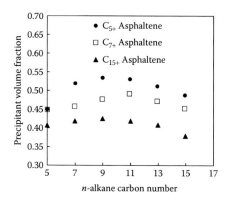

FIGURE 3.22 Precipitant volume fraction for asphaltene precipitation onset versus n-alkane carbon number for Lagrave asphaltenes-toluene/n-alkane mixtures. All experiments were conducted at 293 K and 1 bar. (Reprinted from Tavakkoli, M. et al., *Fuel*, 117, 206–217, 2014.)

then shifts back to the left as the carbon number of precipitant increases. This change in onset volume fraction is related to Gibbs free energy of mixing, which is affected by the solubility parameter difference between n-paraffin and asphaltene, and the ratio of the molar volume of asphaltene to the molar volume of n-paraffin (Wiehe et al. 2005). Before the volume fraction maximum, increase in carbon number of n-paraffin lowers the solubility parameter difference and the molar volume ratio in the entropy of mixing becomes smaller. Both changes decrease Gibbs free energy of mixing and make asphaltene and precipitant more compatible. However, as carbon number further increases, the rate of changes in solubility decreases, whereas the volume ratio increases linearly. Because enthalpy of mixing increases as the molar volume of n-paraffin increases, at one point the molar volume ratio effect outweighs the solubility effect and the net Gibbs free energy of mixing increases as carbon number increases. Thus, less precipitant is needed to reach asphaltene precipitation onset. A detailed explanation can be found in Tavakkoli et al. (2014).

3.5 KINETICS OF ASPHALTENE PRECIPITATION

The onset of asphaltene precipitation measured using the different techniques mentioned previously was obtained either immediately or after waiting for a short period of time (varies from 1 hour to 1 day). The time needed to reach equilibrium has not been standardized in the literature. Maqbool et al. (2009) showed that the time needed to reach equilibrium is dictated by the amount of precipitant added to destabilize asphaltenes, and the time can vary from few minutes to several months. This means that for every precipitant concentration there is an onset of asphaltene precipitation. Their experimental observations were obtained using optical microscopy and centrifugation-based separation. They obtained the solubility of asphaltenes as a function of precipitant concentration from long-term experiments and concluded that the solubility could be overpredicted by short-term experiments. Thus, the true thermodynamic onset needed for thermodynamic models must be obtained from long-term experiments, which is not practical. However, a recent study by Tavakkoli et al. (2016a) showed that the thermodynamic models could be tuned to the amount of asphaltene precipitated instead of the onset point itself. Using the indirect method, they demonstrated that at an aging time of 1 hour and a precipitant concentration of 90 vol%, the system reaches equilibrium and the amount of asphaltene precipitated does not change with time. The effect of temperature was also studied to reduce the time needed to reach equilibrium (Maqbool et al. 2011). They have shown that at higher temperatures, less time is required to reach equilibrium and the solubility of asphaltenes increases. The authors explained this observation by the change in viscosity with temperature. The change in viscosity affects the rate of aggregation which controls the onset time of precipitation because of its very slow kinetics (which is the rate-limiting step). They also considered the expansion of hydrocarbons, oxidation of crude oil and the evaporation of light components in their study (Maqbool et al. 2011).

3.6   FINAL REMARKS

This chapter includes a thorough discussion on a number of experimental techniques used to detect the onset of asphaltene precipitation at reservoir conditions. Among the different available experimental methods, LST, and HPM have become the industry's standard to measure asphaltene precipitation onset. The HPHT conditions used to mimic reservoir conditions are associated with a high cost. Thus, researchers have suggested the detection of asphaltene precipitation onset through experiments at ambient conditions and predicting the onset at reservoir conditions using thermodynamic models. Among the different proposed techniques, the indirect method was found to surpass the other approaches by sensitivity, applicability, and effectiveness. The indirect method is the only available technique that is sensitive to particles that are as small as 100 nm in diameter. It is also applicable to dark crude oil with high asphaltene content, which usually limits the other techniques. It can also be used to quantify the amount of asphaltene precipitated (Tavakkoli et al. 2015). Using the indirect method, the effect of emulsified water, electrolyte, iron ions, and asphaltene polydispersity on asphaltene stability were investigated by Vargas and coworkers. As part of future work, Vargas and coworkers are working on designing a HPHT unit to perform the indirect method at reservoir conditions. This is discussed in more details in Section 9.2.2.

   Maqbool et al. (2009) highlighted the effect of kinetics on the onset of asphaltene precipitation. They concluded that the true thermodynamic onset needed for thermodynamic models must be obtained from long-term experiments. The accuracy of the model predictions will highly depend on the experimental data used to tune the model parameters. However, Tavakkoli et al. (2016a), addressed this issue using the indirect method where the amount of asphaltene precipitated at 90 vol% of precipitant was found to be time independent. Thus, thermodynamic models can be tuned using the amount of asphaltene precipitated rather than the onset point itself (refer to Section 8.1.5).

REFERENCES

Abudu, A., and L. Goual. 2008. Adsorption of crude oil on surfaces using quartz crystal microbalance with dissipation (QCM-D) under flow conditions. *Energy Fuels* 23 (3): 1237–1248.

Akbarzadeh, K., A. Hammami, A. Kharrat, D. Zhang, S. Allenson, J. L. Creek, C. S. Kabir et al. 2007. Asphaltenes—problematic but rich in potential. *Oilfield Rev* 19 (2): 22–43.

Alotaibi, M. B., R. A. Nasralla, and H. A. Nasr-El-Din. 2011. Wettability studies using low-salinity water in sandstone reservoirs. *SPE Reserv Eval Eng* 14 (6): 713–725. doi:10.2118/149942-PA.

Andersen, S. I. 1999. Flocculation onset titration of petroleum asphaltenes. *Energy Fuels* 13 (2): 315–322. doi:10.1021/ef980211d.

Angle, C. W., Y. Long, H. Hamza, and L. Lue. 2006. Precipitation of asphaltenes from solvent-diluted heavy oil and thermodynamic properties of solvent-diluted heavy oil solutions. *Fuel* 85 (4): 492–506. doi:10.1016/j.fuel.2005.08.009.

Behar, E., N. Hasnaoui, C. Achard, and M. Rogalski. 1998. Study of asphaltene solutions by electrical conductivity measurements. *Oil Gas Sci Technol* 53 (1): 41–50.

Buckley, J. S. 1996. Microscopic investigation of the onset of asphaltene precipitation. *Fuel Sci Technol Int* 14 (1–2): 55–74. doi:10.1080/08843759608947562.

Buckley, J. S. 1999. Predicting the onset of asphaltene precipitation from refractive index measurements. *Energy Fuels* 13 (2): 328–332. doi:10.1021/ef980201c.

Buckley, J. S. 2012. Asphaltene deposition. *Energy Fuels* 26 (7): 4086–4090. doi:10.1021/ef300268s.

Buckley, J. S., G. J. Hirasaki, Y. Liu, S. Von Drasek, J-X. Wang, and B. S. Gill. 1998. Asphaltene precipitation and solvent properties of crude oils. *Pet Sci Technol* 16 (3–4): 251–285. doi:10.1080/10916469808949783.

Buenrostro-Gonzalez, E., C. Lira-Galeana, A. Gil-Villegas, and J. Wu. 2004. Asphaltene precipitation in crude oils: Theory and experiments. *AIChE J* 50 (10): 2552–2570. doi:10.1002/aic.10243.

Burya, Y. G., I. K. Yudin, V. A. Dechabo, V. I. Kosov, and M. A. Anisimov. 2001. Light-scattering study of petroleum asphaltene aggregation. *Appl Opt* 40 (24): 4028–4035. doi:10.1364/AO.40.004028.

Cardoso, F. M. R., H. Carrier, J.-L. Daridon, J. Pauly, and P. T. V. Rosa. 2014. CO_2 and temperature effects on the asphaltene phase envelope as determined by a quartz crystal resonator. *Energy Fuels* 28 (11): 6780–6787. doi:10.1021/ef501488d.

Chaisoontornyotin, W., A. W. Bingham, and M. P. Hoepfner. 2017. Reversibility of asphaltene precipitation using temperature-induced aggregation. *Energy Fuels* 31: 3392–3398.

Chung, F., P. Sarathi, and R. Jones. 1991. *Modeling of Asphaltene and Wax Precipitation.* Bartlesville, OK: National Institute for Petroleum and Energy Research. doi:10.2172/6347484.

Cimino, R., S. Correra, P. A. Sacomani, and C. Carniani. 1995. *Thermodynamic Modelling for Prediction of Asphaltene Deposition in Live Oils.* San Antonio, TX: Society of Petroleum Engineers. doi:10.2118/28993-MS.

Daridon, J. L., M. Cassiède, D. Nasri, J. Pauly, and H. Carrier. 2013. Probing asphaltene flocculation by a quartz crystal resonator. *Energy Fuels* 27 (8): 4639–4647. doi:10.1021/ef400910v.

Ekulu, G., P. Magri, and M. Rogalski. 2004. Scanning aggregation phenomena in crude oils with density measurements. *J Dispers Sci Technol* 25 (3): 321–331. doi:10.1081/DIS-120037681.

Ekulu, G., A. Sadiki, and M. Rogalski. 2010. Experimental study of asphaltenes flocculation onset in crude oils using the densitometry measurement technique. *J Dispers Sci Technol* 31 (11): 1495–1503. doi:10.1080/01932690903293867.

Escobedo, J., and G. A. Mansoori. 1997. Viscometric principles of onsets of colloidal asphaltene flocculation in paraffinic oils and asphaltene micellization in aromatics. *SPE Prod Facil* 12 (2): 116–122. doi:10.2118/28729-PA.

Escobedo, J., and G. A. Mansoori. 1995. Viscometric determination of the onset of asphaltene flocculation: A novel method. *SPE Prod Facil* 10 (2): 115–118. doi:10.2118/28018-PA.

Fossen, M., H. Kallevik, K. D. Knudsen, and J. Sjöblom. 2007. Asphaltenes precipitated by a two-step precipitation procedure. 1. Interfacial tension and solvent properties. *Energy Fuels* 21 (2): 1030–1037. doi:10.1021/ef060311g.

Fotland, P., H. Anfindsen, and F. H. Fadnes. 1993. Detection of asphaltene precipitation and amounts precipitated by measurement of electrical conductivity. *Fluid Phase Equilib* 82: 157–164. doi:10.1016/0378-3812(93)87139-R.

Fuhr, B. J., C. Cathrea, L. Coates, H. Kalra, and A. I. Majeed. 1991. Properties of asphaltenes from a waxy crude. *Fuel* 70 (11): 1293–1297. doi:10.1016/0016-2361(91)90216-W.

Fuhr, B. J., L. L. Klein, and C. Reichert. 1986. Measurement of asphaltene flocculation in bitumen solutions. *J Can Pet Technol* 25 (5). doi:10.2118/86-05-02.

Goual, L., M. Sedghi, X. Wang, and Z. Zhu. 2014. Asphaltene aggregation and impact of alkylphenols. *Langmuir* 30 (19): 5394–5403. doi:10.1021/la500615k.

Hammami, A., C. H. Phelps, T. Monger-McClure, and T. M. Little. 2000. Asphaltene precipitation from live oils: An experimental investigation of onset conditions and reversibility. *Energy Fuels* 14 (1): 14–18. doi:10.1021/ef990104z.

Hirschberg, A., L. N. J. DeJong, B.A. Schipper, and J. G. Meijer. 1984. Influence of temperature and pressure on asphaltene flocculation. *SPE J* SPE-11202-PA, 24 (3): 283–293. doi:10.2118/11202-PA.

Ibrahim, H. H., and R. O. Idem. 2004. Correlations of characteristics of Saskatchewan crude oils/asphaltenes with their asphaltenes precipitation behavior and inhibition mechanisms: Differences between CO_2- and n-heptane-induced asphaltene precipitation. *Energy Fuels* 18 (5): 1354–1369. doi:10.1021/ef034044f.

Indo, K., J. Ratulowski, B. Dindoruk, J. Gao, J. Zuo, and O. C. Mullins. 2009. Asphaltene nanoaggregates measured in a live crude oil by centrifugation. *Energy Fuels* 23 (9): 4460–4469. doi:10.1021/ef900369r.

Jacobs, I. C. 1989. Chemical Systems for the control of asphaltene sludge during oilwell acidizing treatments. In Society of Petroleum Engineers. doi:10.2118/18475-MS.

Jamaluddin, A. K. M., J. Creek, C. S. Kabir, J. D. McFadden, D. D'Cruz, M. T. Joseph, N. Joshi, B. Ross. 2001. A comparison of various laboratory techniques to measure thermodynamic asphaltene instability. In. SPE-72154-MS. Kuala Lumpur, Malaysia: Society Petroleum Engineers. doi:10.2118/72154-MS.

Jamaluddin, A. K. M., J. Nighswander, N. Joshi, D. Calder, B. Ross. 2002. Asphaltenes characterization: A key to deepwater developments. In. SPE-77936-MS. Melbourne, Australia: Society Petroleum Engineers. doi:10.2118/77936-MS.

Kabir, C. S., and A. K. M. Jamaluddin. 2002. Asphaltene characterization and mitigation in south Kuwait's Marrat reservoir. *SPE Prod Facil* 17 (4): 251–258.

Karan, K., A. Hammami, M. Flannery, and B. A. Stankiewicz. 2003. Evaluation of asphaltene instability and a chemical control during production of live oils. *Pet Sci Technol* 21 (3–4): 629–645.

Kim, S. T., M.-E. Boudh-Hir, and G. A. Mansoori. 1990. The role of asphaltene in wettability reversal. In. Society of Petroleum Engineers. doi:10.2118/20700-MS.

Kokal, S. L., J. Najman, S. G. Sayegh, and A. E. George. 1992. Measurement and correlation of asphaltene precipitation from heavy oils by gas injection. *J Can Pet Technol* 31 (4). doi:10.2118/92-04-01.

Maqbool, T., A. T. Balgoa, and H. S. Fogler. 2009. Revisiting asphaltene precipitation from crude oils: A case of neglected kinetic effects. *Energy Fuels* 23 (7): 3681–3686. doi:10.1021/ef9002236.

Maqbool, T., P. Srikiratiwong, and H. S. Fogler. 2011. Effect of temperature on the precipitation kinetics of asphaltenes. *Energy Fuels* 25 (2): 694–700. doi:10.1021/ef101112r.

Mousavi-Dehghani, S. A., M. R. Riazi, M. Vafaie-Sefti, and G. A. Mansoori. 2004. An analysis of methods for determination of onsets of asphaltene phase separations. *J Pet Sci Eng* 42 (2–4): 145–156. doi:10.1016/j.petrol.2003.12.007.

Murgich, J., E. Rogel, O. León, and R. Isea. 2001. A molecular mechanics-density functional study of the adsorption of fragments of asphaltenes and resins on the (001) surface of Fe2o3. *Pet Sci Technol* 19 (3–4): 437–455. doi:10.1081/LFT-100000775.

Nassar, N. N., A. Hassan, L. Carbognani, F. Lopez-Linares, and P. Pereira-Almao. 2012. Iron oxide nanoparticles for rapid adsorption and enhanced catalytic oxidation of thermally cracked asphaltenes. *Fuel* 95: 257–262. doi:10.1016/j.fuel.2011.09.022.

Oh, K., and M. D. Deo. 2002. Effect of organic additives on the onset of asphaltene precipitation. *Energy Fuels* 16 (3): 694–699. doi:10.1021/ef010223q.

Oh, K., T. A. Ring, and M. D. Deo. 2004. Asphaltene aggregation in organic solvents. *J Colloid Interface Sci* 271 (1): 212–219. doi:10.1016/j.jcis.2003.09.054.

Ovalles, C. and C. E. Rechsteiner Jr. 2015. *Analytical Methods in Petroleum Upstream Applications*. Boca Raton, FL: CRC Press.

Peramanu, S., C. Singh, M. Agrawala, and H. W. Yarranton. 2001. Investigation on the reversibility of asphaltene precipitation. *Energy Fuels* 15 (4): 910–917. doi:10.1021/ef010002k.

Pfeiffer, J. Ph, and R. N. J. Saal. 1940. Asphaltic bitumen as colloid system. *J Phys Chem* 44 (2): 139–149. doi:10.1021/j150398a001.

Pina, A., P. Mougin, and E. Béhar. 2006. Characterisation of asphaltenes and modelling of flocculation–state of the art. *Oil & Gas Science and Technology - Revue de l'IFP* 61 (3): 319–343. doi:10.2516/ogst:2006037a.

Priyanto, S., G. Ali Mansoori, and A. Suwono. 2001. Measurement of property relationships of nano-structure micelles and coacervates of asphaltene in a pure solvent. *Chemical Engineering Science*, Festschrift in honour of Professor T.-M. Guo, 56 (24): 6933–6939. doi:10.1016/S0009-2509(01)00337-2.

Prunelet, A., M. Fleury, and J. P. Cohen Addad. 2004. Detection of asphaltene flocculation using NMR relaxometry. *Comptes Rendus Chim* 7 (3–4): 283–289. doi:10.1016/j.crci.2003.11.011.

Rassamdana, H., B. Dabir, M. Nematy, M. Farhani, and M. Sahimi. 1996. Asphalt flocculation and deposition: I. The onset of precipitation. *AIChE J* 42 (1): 10–22.

Schramm, L. L. 2000. *Surfactants: Fundamentals and Applications in the Petroleum Industry*. Cambridge: Cambridge University Press.

Serrano-Saldaña, E., A. Domínguez-Ortiz, H. Pérez-Aguilar, I. Kornhauser-Strauss, and F. Rojas-González. 2004. Wettability of solid/brine/n-dodecane systems: Experimental study of the effects of ionic strength and surfactant concentration. *Colloids Surf A Physicochem Eng Asp* 241 (1): 343–349. doi:10.1016/j.colsurfa.2004.04.025.

Sivaraman, A., A. K. M. Jamaluddin, Y. Hu, F. B. Thomas, and D. B. Bennion. 1998. Defining SLE and VLE conditions of hydrocarbon fluids containing wax and asphaltenes using acoustic resonance technology. In. New Orleans, LA: AIChE Spring National Meeting.

Spiecker, P. M., K. L. Gawrys, and P. K. Kilpatrick. 2003. Aggregation and solubility behavior of asphaltenes and their subfractions. *J Colloid Interface Sci* 267 (1): 178–193.

Sung, C. A., M. Tavakkoli, A. Chen, and F. M. Vargas. 2016. Prevention and control of corrosion-induced asphaltene deposition. In Offshore Technology Conference. doi:10.4043/27008-MS.

Tavakkoli, M., A. Chen, and F. M. Vargas. 2016a. Rethinking the modeling approach for asphaltene precipitation using the PC-SAFT equation of state. *Fluid Phase Equilibria* 416: 120–129.

Tavakkoli, M., A. Chen, C. A. Sung, K. M. Kidder, J. J. Lee, S. M. Alhassan, and F. M. Vargas. 2016b. Effect of emulsified water on asphaltene instability in crude oils. *Energy Fuels* 30 (5): 3676–3686. doi:10.1021/acs.energyfuels.5b02180.

Tavakkoli, M., M. R. Grimes, X. Liu, C. K. Garcia, S. C. Correa, Q. J. Cox, and F. M. Vargas. 2015. Indirect method: A novel technique for experimental determination of asphaltene precipitation. *Energy Fuels* 29 (5): 2890–2900. doi:10.1021/ef502188u.

Tavakkoli, M., S. R. Panuganti, V. Taghikhani, M. R. Pishvaie, and W. G. Chapman. 2014. Understanding the polydisperse behavior of asphaltenes during precipitation. *Fuel* 117: 206–217. doi:10.1016/j.fuel.2013.09.069.

Tharanivasan, A. K., W. Y. Svrcek, H. W. Yarranton, S. D. Taylor, D. Merino-Garcia, and P. M. Rahimi. 2009. Measurement and modeling of asphaltene precipitation from crude oil blends. *Energy Fuels* 23 (8): 3971–3980.

Trejo, F., G. Centeno, and J. Ancheyta. 2004. Precipitation, fractionation and characterization of asphaltenes from heavy and light crude oils. *Fuel* 83 (16): 2169–2175.

Vargas, F. M., M. Garcia-Bermudes, M. Boggara, S. Punnapala, M. I. L. Abutaqiya, N. T. Mathew, S. Prasad, A. Khaleel, M. H. Al Rashed, and H. Y. Al Asafen. 2014. On the development of an enhanced method to predict asphaltene precipitation. Offshore Technology Conference. doi:10.4043/25294-MS.

Wang, J., and J. S. Buckley. 2001. A two-component solubility model of the onset of asphaltene flocculation in crude oils. *Energy Fuels* 15 (5): 1004–1012. doi:10.1021/ef010012l.

Wang, J., and J. S. Buckley. 2003. Asphaltene stability in crude oil and aromatic solvents the influence of oil composition. *Energy Fuels* 17 (6): 1445–1451. doi:10.1021/ef030030y.

Wattana, P., D. J. Wojciechowski, G. Bolaños, and H. S. Fogler. 2003. Study of asphaltene precipitation using refractive index measurement. *Pet Sci Technol* 21 (3–4): 591–613. doi:10.1081/LFT-120018541.

Wiehe, I. A., H. W. Yarranton, K. Akbarzadeh, P. M. Rahimi, and A. Teclemariam. 2005. The paradox of asphaltene precipitation with normal paraffins. *Energy Fuels* 19 (4): 1261–1267. doi:10.1021/ef0496956.

Yarranton, H. W., H. Alboudwarej, and R. Jakher. 2000. Investigation of asphaltene association with vapor pressure osmometry and interfacial tension measurements. *Ind Eng Chem Res* 39 (8): 2916–2924. doi:10.1021/ie000073r.

Yudin, I. K., and M. A. Anisimov. 2007. Dynamic light scattering monitoring of asphaltene aggregation in crude oils and hydrocarbon solutions. *Asph Heavy Oils Pet*, O. C. Mullins, E. Y. Sheu, A. Hammami, and A. G. Marshall (Eds.), 439–68. New York: Springer.

4 Asphaltene Precipitation Modeling

C. Sisco, M. I. L. Abutaqiya, F. Wang,
J. Zhang, M. Tavakkoli, and F. M. Vargas

CONTENTS

Understanding the phase behavior of crude oils is important toward designing strategies that optimize production and minimize the risk of flow assurance problems that can lead to costly shutdowns and cleanups of producing wells. Fluid phase behavior is strongly dependent on pressure, temperature, and composition, and these properties change both during production (as the oil moves through the well from reservoir to surface) and throughout the life of a project (as a result of pressure depletion in the reservoir, gas injection, etc.). Because reservoir fluid sampling and high-pressure experiments are time-consuming, expensive, and sometimes inaccurate, using theoretical models to understand phase behavior is a valuable and inexpensive tool.

A phase diagram for an asphaltenic crude oil is shown in Figure 4.1a. The curves on the plot, called *phase boundaries*, denote conditions at which phase splitting or combining occurs. The uppermost asphaltene onset pressure (UAOP) and lowermost

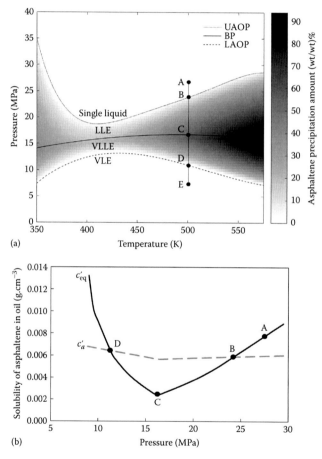

FIGURE 4.1 (a) Asphaltene phase envelope generated from a thermodynamic model. Shading denotes the asphaltene unstable region. (Adapted from Tavakkoli, M. et al., *Fluid Phase Equilibria*, 416, 120–129, 2016.) (b) Asphaltene solubility in oil as a function of pressure (at T = 500 K). Phase separation occurs when the asphaltene concentration (c'_a) exceeds the solubility limit of the oil (c'_{eq}). BP, bubble pressure; LAOP, lowermost asphaltene onset pressure; LLE, liquid-liquid equilibrium; UAOP, uppermost asphaltene onset pressure; VLE, vapor-liquid equilibrium; VLLE, vapor-liquid-liquid equilibrium.

asphaltene onset pressure (LAOP) curves bracket the asphaltene instability region. This region is important from an operational perspective because unstable (or precipitated) asphaltenes can deposit on pipe walls, reducing the efficiency of reservoir-fluid extraction and sometimes requiring shutdowns and solvent injection for cleaning deposits off of the pipe walls. The middle curve (BP) denotes the maximum pressure for vapor existence, and asphaltene precipitation is usually at its maximum along this curve.

Above the uppermost phase boundary (A), the crude oil exists as a single liquid phase with even volatile components like methane and nitrogen dissolved in the bulk oil. At the upper asphaltene onset pressure (B), the single-phase fluid becomes thermodynamically unstable and forms a second phase. The nature of that second phase—whether it should be regarded as liquid or solid—is somewhat contentious from a modeling perspective but inconsequential for the development of the thermo-dynamics framework. To preserve generality, the phase that separates from the bulk oil at point B is referred to as the *rich phase*—because it is richer in asphaltenes than the bulk phase—and the phase from which it separates is referred to as the *lean phase*. In Section 4.2 on thermodynamic models, the nature of the asphaltenes that precipitate from the bulk oil solvent will become relevant because asphaltene property calculations are dependent on the physical state of the phase.

An equivalent statement about thermodynamic instability of the single-phase fluid—and one that will be returned to frequently throughout this chapter—is that the upper asphaltene onset pressure is the highest pressure for which the asphaltene concentration in the oil (c_a') exceeds the equilibrium solubility of asphaltenes in the bulk oil (c_{eq}'). The maximum concentration of asphaltenes that a bulk oil solvent can solubilize is a function of pressure, temperature, and solvent composition, and at certain conditions, the concentration of asphaltenes exceeds the solubility limit of the oil. Asphaltenes that cannot be solubilized must then phase separate from the bulk oil, and the onset of asphaltene precipitation marks the beginning of the asphaltene unstable region. Figure 4.1b illustrates the asphaltene instability phenomenon from the solubility perspective, showing the effect of pressure on c_{eq}' at constant temperature. The pressures for which $c_a' = c_{eq}'$ are the UAOP and LAOP. The pressure for which asphaltenes are most insoluble in oil is the bubble pressure (BP).

Starting from the high-pressure region (A), the solubility of asphaltenes in the bulk oil phase decreases on depressurization until the asphaltene concentration exceeds the solubility limit of the oil (B). Asphaltene solubility continues to decrease until the rich and lean phases become thermodynamically unstable and a vapor phase rich in light components begins forming (C). For pressures between the BP and LAOP, three phases (i.e., vapor, lean phase, rich phase) exist at equilibrium. The lowest phase boundary (D) denotes the pressure at which the lean phase is capable of redissolving the rich phase. The redissolution process at the LAOP occurs because the components in the lean phase that contribute most to asphaltene instability are increasingly liberated to the vapor phase and the lean phase becomes rich in good asphaltene solvents. The high-pressure onset phase separation is driven primarily by volume expansion while the low-pressure onset phase recombination is driven by composition changes of the solvent.

Figure 4.1a and b are merely different graphical projections of the same thermodynamic phenomena; the saturation points lying on the line intersecting the PT-projection at 500 K are the same saturation points observed on the solubility plot.

The shaded area in Figure 4.1a denotes the asphaltene unstable region where the pressure and temperature are such that bulk oil is incapable of solubilizing all asphaltenes in the mixture. In other words, the asphaltene concentration is greater than the solubility of asphaltenes in oil ($c'_a > c'_{eq}(P, T, \{x_i\})$) everywhere in the asphaltene unstable region (between points B and D). The shading on the plot shows the relative amount of asphaltenes that precipitate as a function of the thermodynamic variables. The BP is usually the point at which asphaltene solubility in the bulk oil is lowest and the amount of asphaltenes precipitated is largest, as observed in Figure 4.1.

The amount of asphaltenes precipitated and not the precipitation onset is actually of primary interest because the deposition problem is driven mainly by the amount of asphaltenes that are available to deposit, not necessarily the conditions at which the phase separation occurs. The UAOP and LAOP only set the limits of the region in which asphaltenes might start depositing but do not inform on the potential severity of the problem. The shading in Figure 4.1 transmits much more relevant information about asphaltene solubility at various pressure and temperature conditions than the traditional phase diagrams containing only UAOP, BP. and LAOP curves. Experimental determination of the upper onset pressure is common whereas lower onsets are uncommon. Thus, when the more ambiguous *asphaltene onset pressure* (AOP) is used, it is usually referring to the onset above the BP, and the modifiers *upper* and *lower* are used only to avoid confusion when the distinction is not clear from the context.

A phase diagram accurately expressing the pressure-volume-temperature (PVT) relationship of a crude oil is indispensable in helping to avoid operating conditions that may lead to flow assurance problems or in predicting the location in a pipe where flow assurance problems might occur. Knowledge of the phase boundaries and phase amounts (especially asphaltene precipitation amount) as a function of pressure, temperature, and composition is important for designing strategies to efficiently extract oil from producing wells, and thermodynamic models are often used to predict the phase boundaries, phase amounts, and compositions for a crude oil mixture. Because the primary concern is to either prevent or mitigate the flow assurance problems associated with asphaltene precipitation, the results of thermodynamic modeling are regularly used in a flow simulator to model the aggregation and deposition tendencies of precipitated asphaltenes. The question then becomes how the phase behavior of asphaltenic crude oil mixtures can be modeled accurately.

In this chapter, the most commonly used models for describing crude oil phase behavior are reviewed, making clear both their strengths and weaknesses, and a step-by-step procedure for crude oil characterization and modeling with the perturbed chain form of the statistical associating fluid theory (PC-SAFT) is provided. First, though, it is necessary to review the concepts and equations of phase equilibrium so that the various models used to quantify phase behavior can be understood within the general thermodynamics framework.

4.1 THE MULTIPHASE EQUILIBRIUM PROBLEM

For a mixture of fixed overall composition, computing the number of phases at equilibrium and the distribution of components between those phases at fixed pressure and temperature is known as a PT-flash. Other flash specifications—such as fixed temperature and volume (TV), pressure and enthalpy (PH), entropy and volume (SV), and so on—can be formulated, but the specification of pressure and temperature is most common because these variables are easily manipulated experimentally, physical property models for thermodynamic calculations are often explicit in pressure and temperature, and the PT-flash problem can be formulated as an unconstrained minimization of the Gibbs energy function. This offers a level of assurance for solving the phase equilibrium problem that other specifications cannot readily provide. Every point on the phase diagram in Figure 4.1a represents the amount of asphaltenes in the rich phase predicted by a PT-flash. The PT-flash yields a rich phase everywhere between the LAOP and UAOP curves, and the intensity of the shading indicates the amount of asphaltenes present in the rich phase.

For a mixture of N_c components, distributed in N_p phases, the equilibrium state at fixed pressure, P, and temperature, T, is the state that minimizes the Gibbs energy, which is calculated by:

$$G(P, T, \boldsymbol{n}) = \sum_{i=1}^{N_c}\sum_{k=1}^{N_p} n_{ik}\mu_{ik} = RT\sum_{i=1}^{N_c}\sum_{k=1}^{N_p} n_{ik}\ln\hat{f}_{ik} \qquad (4.1)$$

where:
n_{ik} is the molar amount
μ_{ik} is the chemical potential
$\hat{f}_{ik}$ is the fugacity of component i in phase k

At fixed pressure and temperature, the chemical potential and fugacity can change only by changes in molar amounts, and the flash problem is solved by finding the molar amounts of each component in each phase that minimizes the Gibbs energy function, subject to satisfying all material balance constraints. At the minimum of the Gibbs energy function, all derivatives of the Gibbs energy with respect to the independent variables must be zero, yielding the following equilibrium conditions:

$$P_k = P_{k+1} = \ldots = P_{N_p} \qquad (4.2)$$

$$T_k = T_{k+1} = \ldots = T_{N_p} \qquad (4.3)$$

$$\hat{f}_{ik} = \hat{f}_{i(k+1)} = \ldots = \hat{f}_{iN_p} \qquad (4.4)$$

The equi-fugacity criteria are necessary conditions for phase equilibrium, whereas the minimum of the Gibbs energy satisfies both the necessary and sufficient condition for equilibrium. This means that among the many states that may satisfy the equilibrium relations, only the state that minimizes the Gibbs energy is the true equilibrium state. The general equation for calculating fugacity for component i in phase k is given by:

$$\hat{f}_{ik}(P, T, \boldsymbol{n}) = x_{ik}\gamma_{ik}\phi_i P = x_{ik}\hat{\phi}_{ik}P \qquad (4.5)$$

where:

   x_{ik} is the molar composition
   γ_{ik} is the activity coefficient
   $\hat{\phi}_{ik}$ is the fugacity coefficient of component i in phase k

The fugacity coefficient of pure component i (ϕ_i) is calculated at the temperature and pressure of the system and molar composition is related to molar amounts by:

$$x_{ik} = \frac{n_{ik}}{n_k} = \frac{n_{ik}}{\sum\limits_{i=1}^{N_c} n_{ik}} \tag{4.6}$$

The mixture fugacity coefficient ($\hat{\phi}_{ik}$) is the product of the activity coefficient (γ_{ik}) and the pure component fugacity coefficient (ϕ_i) and is defined by:

$$\ln \hat{\phi}_{ik}(P,T,\boldsymbol{n}) = \ln(\gamma_{ik}\phi_i) = \int\limits_0^P \frac{(\bar{Z}_{ik}-1)}{P}dP \tag{4.7}$$

where $\bar{Z}_{ik}$ is the partial molar compressibility factor in phase k, $\bar{Z}_{ik} = \partial(nZ_k)_{P,T,n_{j\neq i}}/\partial n_i$ and Z_k is the compressibility factor of phase k. The compressibility factor quantifies the deviation between the actual volume of a phase ($\hat{V}_k$) and its volume if it was an ideal gas ($\hat{V}_k^{ig}$).

$$Z_k(P,T,\boldsymbol{n}) = \frac{P\hat{V}_k}{RT} = \frac{\hat{V}_k}{\hat{V}_k^{ig}} \tag{4.8}$$

To derive the material balance for phase(s) in equilibrium, assume that a mixture of molar amount F and overall molar composition $\{z_i\}$ partitions into three phases: a vapor phase of molar amount V and composition $\{y_i\}$, a bulk oil phase of molar amount L_1 and composition $\{x_{i1}\}$, and an asphaltene-rich phase of molar amount L_2 and composition $\{x_{i2}\}$ (Figure 4.2).

The overall material balance and component material balances can be written as:

$$F = V + L_1 + L_2 \tag{4.9}$$

$$z_i F = y_i V + x_{i1}L_1 + x_{i2}L_2 \tag{4.10}$$

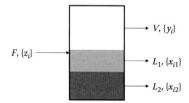

FIGURE 4.2   Material balance for three phases in equilibrium.

For the sake of simplification, we assume that no vapor phase exists at equilibrium (which is true for all pressures higher than the BP) and derive the material balance for liquid-liquid equilibrium. The overall and component material balances can be combined into a single expression, yielding:

$$z_i = x_{i1}(1-\beta) + x_{i2}\beta \qquad (4.11)$$

where β is the phase fraction of the asphaltene-rich phase, $\beta = L_2/F$. Substituting the equilibrium partition coefficients, $K_i = x_{i2}/x_{i1}$, and rearranging in terms of x_{i2} yields the Rachford-Rice equation, where the sum of mole fractions constraint allows N_c equations to be written in a single formula:

$$\sum_{i=1}^{N_c} x_{i2} = \sum_{i=1}^{N_c} \frac{z_i K_i}{1+\beta(K_i-1)} = 1 \qquad (4.12)$$

Partition coefficients are defined as a ratio of the molar composition of each component in two equilibrium phases. Rearranging the fugacity equation (Equation 4.5) in terms of the composition, the partition coefficients can be written in terms of the fugacity coefficients by making use of the equi-fugacity criteria:

$$K_i = \frac{x_{i2}}{x_{i1}} = \frac{\hat{\phi}_{i1} P/\hat{f}_{i1}}{\hat{\phi}_{i2} P/\hat{f}_{i2}} = \frac{\hat{\phi}_{i1}}{\hat{\phi}_{i2}} \qquad (4.13)$$

Compositions are then calculated from the material balances by:

$$x_{i1} = \frac{z_i}{1+\beta(K_i-1)} \qquad x_{i2} = K_i x_{i1} \qquad (4.14)$$

The solution to the Rachford-Rice equation for the PT-flash is found by estimating β, calculating compositions (Equation 4.14) and fugacity coefficients (from a physical property model like an equation of state [EOS]) and updating K_i (Equation 4.13) and β until the Rachford-Rice equation is satisfied. This approach for converging K_i is known as successive substitution or fixed-point iteration, and convergence is fairly reliable as long as the initial guesses for the partition coefficients and phase amounts are sufficiently close to the actual solution. When more than two equilibrium phases exist, a multiphase equilibrium algorithm must be used, such as that proposed by Michelsen (1994).

A special case of the liquid-liquid phase equilibrium problem is the UAOP calculation, where the phase fraction of the asphaltene-rich phase is set to zero and a search for the pressure that satisfies the two-phase Rachford-Rice equation is performed. For the determination of the UAOP, temperature and bulk phase composition are fixed ($\{x_{i1}\} = \{z_i\}$ when $\beta = L_2/F \cong 0$), so that only pressure and the composition of the rich phase can be varied. These variables are changed iteratively until the partition coefficients are found that satisfy the Rachford-Rice equation, and the corresponding pressure is designated the UAOP. The molar composition of the onset phase $\{x_{i2}\}$ at the UAOP is calculated from Equation 4.14 with the converged values of $\{K_i\}$, and the solubility of asphaltenes in the bulk oil phase is exactly equal to the concentration of asphaltenes in the overall mixture ($c'_{eq} = c'_a$) at the UAOP.

Assuming that appropriate care is taken in solving the phase equilibrium problem, how closely the results of these calculations match experimental and field observations for a given oil depends entirely on the accuracy of the physical property model in describing variations in the compressibility factor $Z(P,T,\boldsymbol{n})$ or volume $V(P,T,\boldsymbol{n})$ of a phase as a function of changes to the independent variables. The best physical property models for describing asphaltene phase behavior will be those that can predict well both liquid densities and partial molar compressibility factors for large nonspherical molecules. In the following sections, the most commonly used models for predicting thermodynamic properties of crude oils are reviewed and qualitatively evaluated, mainly in terms of their ability to accurately predict partition coefficients or solubility of asphaltenes in oil.

4.2 THERMODYNAMIC MODELS FOR CRUDE OILS

Mathematical models for thermodynamic calculations are usually formulated such that the PVT relationship of a pure component or mixture can be stated explicitly in terms of pressure, residual Helmholtz energy, or excess Gibbs energy:

$$
\begin{aligned}
&P = f(T,V,\boldsymbol{n}) \\
&A^{\text{res}} = f(T,V,\boldsymbol{n}) \rightarrow \text{Thermodynamic Properties} \\
&G^{\text{ex}} = f(P,T,\boldsymbol{n})
\end{aligned}
\tag{4.15}
$$

Regardless of the formulation, the goal of such models is to describe the functional dependence of fluid behavior on changes to easily measured independent variables. The PC-SAFT EOS, for example, is written in terms of the residual Helmholtz energy with temperature, volume. and mole numbers as the independent variables. These expressions for Helmholtz energy (to be discussed in Section 4.2.3.2) can then be differentiated with respect to the independent variables to yield any thermodynamic property of interest. A few commonly used properties are listed here:

$$
Z = -V\left(\frac{\partial(A^{\text{res}}/nRT)}{\partial V}\right)_{T,n} + 1
$$

$$
\ln\hat{\phi}_i = \left(\frac{\partial(A^{\text{res}}/RT)}{\partial n_i}\right)_{T,V} - \ln Z
$$

$$
\left(\frac{\partial P}{\partial n_i}\right)_{T,V} = -\left(\frac{\partial^2 A^{\text{res}}}{\partial V\,\partial n_i}\right)_T + \frac{RT}{V}
$$

$$
\left(\frac{\partial P}{\partial V}\right)_{T,n} = -\left(\frac{\partial^2 A^{\text{res}}}{\partial V^2}\right)_T - \frac{nRT}{V^2}
$$

$$
\bar{V}_i = \left(\frac{\partial V}{\partial n_i}\right)_{P,T} = -\left(\frac{\partial P}{\partial n_i}\right)_{T,V} \Big/ \left(\frac{\partial P}{\partial V}\right)_{T,n}
\tag{4.16}
$$

The fugacity coefficient calculation is of particular interest for its usefulness in calculating the Gibbs energy (Equation 4.1), fugacities (Equation 4.5), and partition coefficients (Equation 4.13), which are needed for determining the equilibrium state of a mixture.

Because crude oils contain thousands of components, ranging from highly volatile nonpolar components like nitrogen and methane to heavy polarizable components like resins and asphaltenes, it is difficult to produce models that reliably predict the phase behavior of systems containing diverse components across large ranges of pressure, temperature, and composition. These complications have prompted researchers to investigate many different methods for describing the phase behavior of asphaltenes in crude oil.

Asphaltene precipitation models are typically classified in two groups according to the mechanism by which asphaltenes are assumed to exist in solution and the factors that lead to their instability. The first group contains the micellar models, in which asphaltenes are assumed to be micelles stabilized by polar-polar interactions with resins. Resins are lighter than asphaltenes, but their molecular structures are similar. Under the micellar model assumptions, if the concentration of resins with respect to asphaltenes becomes too low, then asphaltenes are destabilized and precipitate from the bulk phase. The second approach for describing asphaltene phase behavior contains the solubility models. Solubility models assume that asphaltenes can be solubilized—either partially or fully—by the bulk oil and that asphaltene phase behavior is dominated by relatively weak van der Waals interactions instead of strong polar interactions. There is experimental evidence to suggest that polar interactions drive the formation of self-associated aggregates of asphaltene monomers in nonpolar solvents, but that the interactions between these aggregated monomers and other crude oil components are driven mainly by dispersion forces (Buenrostro-Gonzalez et al. 2002, 2004; Khaleel et al. 2015; Tavakkoli et al. 2016).

In the following discussion about asphaltene thermodynamics models, the simplifying assumptions are stated explicitly and evaluated qualitatively for their validity. For most asphaltene models, their primary goal is to predict the onset of asphaltene instability, which is often stated in terms of the solubility of asphaltenes in the bulk oil solvent. Even for the models that are too complicated for an explicit expression of asphaltene solubility, we discuss precisely how solubility can be calculated and infer the accuracy of the results from simple physical arguments.

4.2.1 Micellar Models

Colloidal models for asphaltene stability dominated mainstream asphaltene research from the 1940s until the early 2000s (Pfeiffer and Saal 1940; Mitchell and Speight 1972; Dickie and Yen 1967; Leontaritis and Mansoori 1988; Victorov and Firoozabadi 1996; Pan and Firoozabadi 1998; Cimino et al. 1993). The key assumptions in these models are that asphaltenes exist in the oil medium as suspended solid particles and are stabilized sterically by the presence of resins. The two major contributions responsible for keeping asphaltenes stable in solution are the short-range intermolecular repulsions between resin molecules adsorbed on neighboring asphaltene particles and the long range repulsions between asphaltene particles. The equi-fugacity criterion for resins (Equation 4.4) requires that the fugacities (or chemical potentials) be equal in the bulk oil solvent and on the asphaltene micelle at equilibrium.

$$\mu_{rs} = \mu_{rM} \tag{4.17}$$

μ_{rs} and μ_{rM} are the chemical potentials for resins in the oil solvent and resins adsorbed on the asphaltene micelles, respectively. From the Flory-Huggins theory (Flory 1942), and assuming resins are monodisperse, the volume fraction of resins dissolved in the oil solvent (φ_{rs}) is expressed as:

$$\ln \varphi_{rs} = \frac{\hat{V}_r}{\hat{V}_s} - 1 - \frac{\hat{V}_r}{RT}(\delta_s - \delta_r)^2 \tag{4.18}$$

where:
$\hat{V}_r$ and $\hat{V}_s$ are the molar volumes of resin and the oil solvent phase, respectively
δ_r and δ_s are the solubility parameters of resin and the oil solvent, respectively

The critical condition for asphaltene stability is when resins cover the whole surface of asphaltenes. The critical resin fraction, φ_{cr}, is the minimum volume fraction of resin necessary to keep the asphaltene particles covered at a given pressure and temperature. Asphaltene precipitation occurs when resin coverage is less than the critical resin fraction ($\varphi_{rs} < \varphi_{cr}$).

One unique characteristic of the colloidal models is that the chemical potential of a given resin at the critical condition is a constant at the same temperature and pressure above the BP, regardless of the solvent composition. This is because, at the asphaltene precipitation onset, the solid phase arrives at the same physical state where the colloidal asphaltene particles are covered with resins marginally below the minimum concentration that would keep asphaltenes stable in the colloidal suspension. This characteristic implies that once the critical chemical potential of a resin in one liquid composition is solved, the critical chemical potential of that resin in any liquid composition at the same temperature and pressure above the BP is also known.

Note that in the colloidal model described, which is often referred to as the steric-stabilization model, asphaltenes are stabilized sterically by resins adsorbed on their surface. The steric-stabilization mechanism in the colloidal models, however, deviates from the behavior of surfactants from which the colloidal models draw analogy upon. The surfactant micelles are thermodynamically stable in the fluid media in absence of any stabilizing agents. The steric-stabilization mechanism has not yet been supported by experimental evidence, and no observations have been made to verify the presence of resins between the asphaltene and oil interface (Cimino et al. 1993).

In contrast to the steric-stabilization models that view asphaltenes in oil as lyophobic colloids, an alternative conceptual model describes asphaltene in oil as lyophilic colloids. When asphaltenes in oil are viewed as lyophilic colloids, it is believed that thermodynamic equilibrium can be established for both the asphaltene monomers and micelles. Various experimental evidences have been presented to justify this model (Cimino et al. 1993). The lyophilic colloids-based models are also referred to as *solubilization models*. The relative solvation energies of the monomers and micelles, as well as the entropy and enthalpy changes resulting from the micellization process, determine the allocation of asphaltenes either as monomers precipitated from solution or as micelles stabilized by resins.

Victorov and Firoozabadi (1996) proposed a micellization model that describes asphaltene molecules in crude oil as micelles. They argued that in the absence of resins, asphaltenes would precipitate in their monomeric form because of their low solubility in the bulk oil. The micellar model uses the idea of aggregation equilibrium, which is valid for both monodisperse and polydisperse colloidal mixtures and applicable to both dilute and concentrated solutions. In their work, Victorov and Firoozabadi assumed that the amount of resin and asphaltene monomers were low such that the thermodynamics of dilute solutions can be applied. The micellar sizes were assumed to be fixed and only asphaltenes were allowed to precipitate.

Pan and Firoozabadi (1997, 1998) further refined the micellization model for asphaltene precipitation by modeling the precipitation process as a liquid-liquid phase split at reservoir conditions and liquid-solid phase split at ambient conditions. The precipitated heavy phase is assumed to consist of only resins and asphaltenes, and the main driving forces for precipitation come from two effects: the lyophobic effect that is governed by classical bulk phase thermodynamic equilibrium and the interfacial effect where colloid stability is improved by lowered interfacial tension because of adsorption of resins on the micellar core surface. Because the micellization models use equilibrium thermodynamics concepts, the implication is that the asphaltene precipitation process is reversible, meaning that precipitated asphaltenes can be redissolved when conditions are favorable for redissolution.

The key feature of Pan and Firoozabadi's micellization model is the introduction of a term for the Gibbs energy of micelle formation, ΔG_m^{00}, in the phase equilibrium calculation. ΔG_m^{00} accounts for the Gibbs energy change because of formation of a micelle in an infinitely dilute solution of crude oil without the asphaltene and resin components. This reversible process consists of the following steps:

1. Asphaltene molecules transfer from an infinitely dilute crude into a liquid state consisting of pure asphaltene, corresponding to an energy change of $(\Delta G_a^0)_{\mathrm{Tr}}$. The asphaltene molecules in the micellar core can deform, which corresponds to an energy change of $(\Delta G_a^0)_{\mathrm{Def}}$.
2. The micellar core forms an interface with the bulk phase, resulting in the generation of an interfacial energy $(\Delta G_m^0)_{\mathrm{Inf}}$.
3. Resin molecules then transfer from the infinitely dilute crude into the solvated shell, corresponding to the creation of a transfer energy $(\Delta G_r^0)_{\mathrm{Tr}}$.
4. The polar heads of resin molecules adsorb onto the micellar core surface, corresponding to the creation of the adsorption energy $(\Delta G_r^0)_{\mathrm{Adp}}$. The resin molecule deforms in the shell region, corresponding to the deformation energy $(\Delta G_r^0)_{\mathrm{Def}}$.

The summation of the energy changes as result of the processes mentioned gives the expression for ΔG_m^{00}:

$$\Delta G_m^{00} = \sum_{i=1}^{N_a} \left[(\Delta G_a^0)_{\mathrm{Tr}} + (\Delta G_a^0)_{\mathrm{Def}} \right] + (\Delta G_m^0)_{\mathrm{Inf}} + (\Delta G_r^0)_{\mathrm{Tr}} + (\Delta G_r^0)_{\mathrm{Adp}} + (\Delta G_r^0)_{\mathrm{Def}} \quad (4.19)$$

where N_a is the number of asphaltene subfractions and only one resin fraction has been assumed. The state of the system at equilibrium is determined by finding the global minimum of the Gibbs energy of the system, which includes the Gibbs energies of the liquid and solid. The optimization scheme takes considerable effort to find the global minimum of the Gibbs energy, whose value depends on the following independent variables, subject to material balance constraints:

1. Number of asphaltene molecules in the micellar core
2. Number of resin molecules in the micellar shell
3. Shell thickness of a micelle
4. Number of asphaltene monomers in the light liquid phase
5. Number of resin monomers in the light liquid phase
6. Number of micelles in the light liquid phase
7. Number of resin molecules in the precipitated phase

The amount of precipitate, the composition of the phases, and some features of the micellar structure can be determined from the calculations. The composition of the solvent liquid phase will be saturated with asphaltenes, meaning that for the given pressure and temperature conditions, the asphaltene concentration in the solvent phase is equal to the solubility of asphaltenes in that phase. Asphaltenes that exceed this solubility limit of the solvent are included among the precipitated phase. This approach, though rigorous in its description of thermodynamics, is prohibitive because of the number of independent variables that must be solved and the computationally expensive optimization procedure that has to be used to find the set of variables that satisfy the minimum Gibbs energy criterion for equilibrium.

4.2.2 SOLUBILITY MODELS: SOLUTION THEORIES

The main distinguishing feature of the solubility models compared to the micellar models discussed in Section 4.2.1 is their treatment of the dominant interactions that determine asphaltene stability. In the colloidal models, polar interactions between resins and asphaltenes are assumed to be the major contributor to asphaltene stability, whereas in the solubility models, asphaltene stability is a consequence of dispersion forces between it and the other components present in the oil. The solubility models to be discussed in this section differ from the EOS-based solubility models in Section 4.2.3, mainly in their treatment of nonasphaltene components. The models in this section consider the solvent as a *pseudo-pure* component with a characteristic solubility parameter and molar volume, which allows the asphaltene solubility equations to be written as if the crude oil is a binary mixture consisting of a bulk oil solvent and asphaltene solute.

4.2.2.1 Regular Solution Theory

Regular solution theory was originally developed by Scatchard (1931) and Hildebrand and Wood (1933) to describe the thermodynamics of solutions of nonpolar components. The simplifying assumption in regular solution theory is that excess entropy is zero, meaning that the entropy of mixing is equal to the ideal solution value. London's geometric mean for describing interactions between different molecules is assumed to be closely obeyed in the regular solution models. The validity of this assumption breaks down as the size difference between the molecules increases.

The activity of a single component in a binary mixture can be derived based on regular solution theory and the use of solubility parameters (δ), which then allows solubility calculations for asphaltenes in a bulk solvent. Griffith and Siegmund (1985) evaluated the compatibility of different residual fuel oils using the regular solution model to describe asphaltene solubility. They assumed the oil system to be a binary mixture of asphaltene and deasphalted oil solvent (maltene), and the activity of the asphaltene solute was assumed to be unity, yielding:

$$\ln x_a = -\frac{\hat{V}_a}{RT}\left[\varphi_s^2(\delta_s - \delta_a)^2\right] \tag{4.20}$$

where:
x_a is the mole fraction solubility of asphaltene in oil
$\hat{V}_a$ is the molar volume of asphaltene
δ_a and δ_s are the solubility parameters of asphaltene and deasphalted oil solvent, respectively
φ_s is the volume fraction of the deasphalted oil solvent

The main advantage of this approach is that asphaltene solubility can be calculated directly from pure component data. The thermodynamic definition of the solubility parameter is:

$$\delta = \sqrt{\frac{-U^{\text{res}}}{V}} \tag{4.21}$$

Solubility parameters are not easily measurable quantities, so they are often determined either from correlation to other properties for the deasphalted oil solvent (Koenhen and Smolders 1975; Buckley et al. 1998; Vargas and Chapman 2010) or from miscibility tests with heptane-toluene solutions of varying compositions for the asphaltene solute. Solubility parameters are especially difficult to estimate accurately at high pressures and temperatures, meaning that regular solution models are often incapable of reliable high-pressure AOP calculations. Another issue with the regular solution model for asphaltene solubility predictions is the inconsistency between the assumptions in regular solution theory and the nature of the oil system. The size and shape of asphaltene molecules is different from the rest of the molecules in the oil, and asphaltenes also have a tendency to self-associate. Thus, the assumptions implicit in the regular solution model are often inappropriate for describing asphaltene solubility in oil.

4.2.2.2 Hirschberg and de Boer Models

Hirschberg et al. (1984) developed a solubility model based on Flory-Huggins polymer solution theory to study asphaltene flocculation in two light crude oils under
temperature and pressure variations, precipitant mixing, and gas injection. The
Flory-Huggins model was developed for polymer solutions consisting of large polymer molecules and small solvent molecules (Huggins 1941; Flory 1942), which is a
suitable physical model to describe asphaltenes dissolved in oil. A simplified expression for the maximum volume fraction of asphaltene dissolved in the crude oil (φ_a^{eq})
is derived using the Hirschberg model, assuming that the asphaltene-rich phase is
pure asphaltene and the solvent-rich phase has a negligible amount of dissolved
asphaltene. The maximum volume fraction of asphaltene dissolved in oil (φ_a^{eq}) is
analogous to the maximum concentration of asphaltenes that can be solubilized in
the bulk oil phase (c_{eq}'), discussed in the introduction to this chapter.

$$\ln \varphi_a^{eq} = \frac{\hat{V}_a}{\hat{V}_s} - 1 - \frac{\hat{V}_a}{RT}(\delta_a - \delta_s)^2 \qquad (4.22)$$

where $\hat{V}_s$ is the molar volume of the deasphalted oil solvent, and other parameters
are defined as before. The first two terms on the right-hand side of Equation 4.22
account for the entropy change of mixing, and the last term is the Flory interaction
parameter, χ, which accounts for the enthalpy change of mixing from regular solution theory.

The oil is treated as a binary mixture of oil solvent and asphaltene solute with the
molar volumes and solubility parameters of both the oil solvent and asphaltene solute
as input parameters. Solvent properties can be calculated with liquid phase compositions obtained from a vapor-liquid equilibrium calculation, and asphaltene properties
are often fit to data from titration experiments. For high-temperature applications,
Hirschberg proposed an empirical correlation for the asphaltene solubility parameter
as a function of temperature. Recent work by Wang et al. (2016) offers another path
to calculate solubility parameters at elevated temperatures using a more fundamental
approach. The molar volume and solubility parameter for asphaltene were assumed to
be independent of pressure.

Hirschberg's model provides a decent match with experimental data of asphaltene
precipitation on both gas injection and pressure depletion. However, there are shortcomings as discussed in the original paper. The model predicts less precipitated asphaltene at high n-heptane dilution ratios for one of the studied crude oils because of the
separation of resins from asphaltenes. Also, the temperature and pressure dependence
of the input parameters are not captured well, especially for the asphaltene solubility
parameter and molar volume, yielding discrepancies between the model predictions
and experimental data at elevated temperatures and pressures. The model also poorly
predicts CO_2-induced asphaltene precipitation, likely as a consequence of applying the
geometric mean assumption to interactions between carbon dioxide molecules and
asphaltenes.

Various researchers have applied the Hirschberg solubility model to study asphaltene precipitation under different conditions of pressure, temperature, and solvent
composition (Burke et al. 1990; Kawanaka et al. 1991; Rassamdana et al. 1996;

Wang and Buckley 2001b; Kraiwattanawong et al. 2007). De Boer et al. (1995) developed a simple method to predict asphaltene precipitation tendency in different crude oils based on the Hirschberg model. They generated a screening map showing the asphaltene problems of a number of different oils with their properties correlated to various measurable quantities including crude oil density at reservoir conditions, the difference between reservoir pressure and BP, asphaltene supersaturation at the BP, and asphaltene solubility parameters. Light crude oils were predicted to have severe asphaltene precipitation problems because of their high asphaltene supersaturation at the BP, which is consistent with field observations.

4.2.2.3 Flory, Huggins, and Zuo

Hirschberg (1988) presented the first framework for understanding asphaltene gradients in oil reservoirs. In his model, the reservoir fluid was modeled as a binary mixture of asphaltene solute and maltene solvent in thermal and phase equilibrium within the reservoir column, using the Flory-Huggins theory. In addition, his model included a gravity term to account for the segregation of heavy molecules, such as asphaltenes.

Zuo and co-workers (2010; Freed et al. 2010) revisited Hirschberg's model and introduced variations in fluid properties with well-depth. The Flory-Huggins formalism with a gravitational term proposed by Hirschberg was retained and a cubic EOS was added to account for pressure and composition effects to the properties of the oil solvent. The Flory-Huggins-Zuo (FHZ) equation is derived describing the equilibrium amount of asphaltenes in solution (φ_a) at two depths (h_1 and h_2):

$$\ln \frac{[\varphi_a]_{h_2}}{[\varphi_a]_{h_1}} = \mathrm{Grav}(h_2,h_1) + \mathrm{Entr}(h_2,h_1) + \mathrm{Sol}(h_2,h_1) \qquad (4.23)$$

$$\mathrm{Grav}(h_2,h_1) = \frac{V_{\mathrm{pa}}}{k_b T} g(\rho_s - \rho_a)(h_2 - h_1)$$

$$\mathrm{Entr}(h_2,h_1) = \left[\left(\frac{\hat{V}_a}{\hat{V}_s} - 1\right)\right]_{h_2} - \left[\left(\frac{\hat{V}_a}{\hat{V}_s} - 1\right)\right]_{h_1}$$

$$\mathrm{Sol}(h_2,h_1) = \left[\frac{\hat{V}_a}{RT}(\delta_a - \delta_s)^2\right]_{h_1} - \left[\frac{\hat{V}_a}{RT}(\delta_a - \delta_s)^2\right]_{h_2}$$

where:
 V_{pa} is the particle size of asphaltenes in solution
 ρ is the mass density

The asphaltene gradient depends on three effects: gravity, entropy and solubility (enthalpy). The gravity and solubility effects tend to sharpen the asphaltene gradient, meaning that more asphaltenes are found deeper in the column when these terms dominate, whereas the entropy term tends to reduce the asphaltene gradient, which favors mixing between asphaltenes and other components present in the oil.

The FHZ model assumes that the properties of asphaltenes are independent of height and only a small amount of asphaltenes are present in the oil and have no effect on the oil properties. Thus, the maltene properties can be calculated from its composition using an EOS, and asphaltene properties—namely volume and solubility parameter—can be correlated to asphaltene molecular weight and density. The average molecular weight of asphaltene is tuned to downhole fluid analysis (DFA) results (Zuo et al. 2010), and the solubility parameter of asphaltene can be evaluated from a solubility profile analysis using an in-line filtration method (Rogel et al. 2015). The volume and solubility parameter of asphaltenes from different sources can be assumed invariant such that their effects on the asphaltene gradient calculation are negligible (Rogel et al. 2016). With this assumption, the FHZ equation can be applied to different oils using default asphaltene properties.

Based on a variety of case studies, the asphaltene gradient is driven by the gravity term for low gas-to-oil ratio (GOR) crudes and the solubility term for large GOR crudes (Zuo et al. 2010, 2013). The FHZ model successfully predicted two different tar mats formed by a late gas charge to an oil column (Zuo et al. 2013). The oil viscosity profile under biodegradation can also be predicted by combining the FHZ model with a simple diffusive model (Zuo et al. 2015). A dynamic FHZ model has been developed including time effects on the asphaltene gradient (Wang et al. 2015). The FHZ equation has the same limitations as the original Flory-Huggins model with additional error introduced because of the assumption that the size of asphaltene particles is constant throughout the entire column.

4.2.2.4 Asphaltene Instability Trend Method

Buckley et al. (1998) and Wang & Buckley (2003) found that the precipitation of asphaltenes from oil is dominated by London dispersion interactions that can be quantified by the difference of refractive index of asphaltene and the oil medium. Asphaltene precipitation occurs when the refractive index of the mixture drops below a characteristic value observed from titration experiments of different crude oils with n-alkanes. This characteristic refractive index is defined as the refractive index of the onset (P_{RI}), which varies with different crude oils and precipitants but is a weak function of the addition of aromatic solvents. A linear relationship was proposed for P_{RI} and the square root of the molar volume of the precipitant ($\hat{V}_p$), which allows an experimental approach to predict asphaltene precipitation of live oils at reservoir conditions (Wang and Buckley 2001a). The light components present in the live oil are considered to be asphaltene precipitants before the BP, and molar volumes of those light components resulting from compression can be calculated using PVT data or an EOS. P_{RI} of the live oil at reservoir conditions is estimated from its relationship with $\hat{V}_p$ and temperature and pressure effects on P_{RI} are neglected. The refractive index of the live oil at reservoir conditions can be evaluated using the refractive index of the stock tank oil (STO) and PVT data. The asphaltene precipitation onset is then predicted by comparing the refractive index of live oil and P_{RI} at different pressures. Figure 4.3 shows the refractive index and P_{RI} of a live oil sample as a function of pressure at the reservoir temperature (Gonzalez et al. 2012). When the refractive index of the oil drops below P_{RI}, the oil can no longer solubilize all asphaltenes, and they begin to precipitate. The AOP predicted by asphaltene instability trend (ASIST) depends on

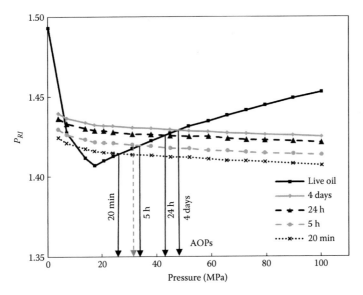

FIGURE 4.3 Asphaltene precipitation onset for an oil sample using ASIST at different aging times. Solid vertical lines indicate the AOPs for respective aging times. The dashed vertical line is the AOP measured by the SDS/HPM technique. (Adapted from Gonzalez, D.L. et al., *Energy & Fuels*, 26, 6218–6227, 2012.)

the aging time of the titration experiments at ambient condition. The onset pressures of various oil samples at the aging time of 5 hours matches with those measured using high-pressure microscopy and particle size analysis techniques.

The solubility parameter of these oil systems, which is required in most solubility models for asphaltene precipitation, is correlated with their refractive index directly (Wang and Buckley 2001b). Therefore, a similar method using solubility parameter to predict asphaltene instability at reservoir conditions has been developed (Creek et al. 2009). The injection of light gases, such as methane and ethane, is also studied by this approach. ASIST is an empirical method to predict asphaltene stability at reservoir conditions using ambient experimental results and PVT data. The predictions depend heavily on the measured P_{RI}, which is limited by the accuracy of the precipitation onset measurements.

4.2.3 SOLUBILITY MODELS: EQUATIONS OF STATE

The second group of solubility models contains the equations of state. Fewer simplifying assumptions are made for EOS models than for the models discussed in Section 4.2.2, and the crude oil is no longer treated as a binary mixture of a non-asphaltene solvent and asphaltene solute. Because the EOS treatment does not make many of the simplifying assumptions that are made for the models covered in Sections 4.2.1 and 4.2.2, asphaltene solubility in oil cannot be derived in a compact expression for these models. Instead, a return to the general thermodynamics concepts covered in the introduction to this chapter and in Section 4.1 is required to formulate a general solubility calculation methodology that can be applied to any EOS model.

When an EOS model is combined with a robust flash algorithm, such as the multiphase equilibrium formulation introduced by Michelsen (1994), the inputs are overall molar composition $\{z_i\}$ and some set of component-specific parameters that quantify the intermolecular interactions between the various components; and the outputs from the flash calculation are the equilibrium distributions of phase amounts and phase compositions. Additionally, any thermodynamic property of interest—such as those listed in Equation 4.16—can be calculated from differentiation of the model. The outputs of the PT-flash + EOS model can then be manipulated to produce any set of desired properties, including solubility of asphaltene in oil in terms of volume fraction (φ_a^{eq}) or mass per volume (c'_{eq}). Figure 4.1, for example, was generated using a PT-flash algorithm in conjunction with the PC-SAFT EOS.

4.2.3.1 Cubic Equation of State

The first reliable EOS for relating pressure, P, volume, V, and temperature, T, of dilute gases was the ideal gas law:

$$P = \frac{nRT}{V} = \frac{RT}{\hat{V}} \tag{4.24}$$

The assumptions implicit in the ideal gas law are that molecules have an infinitesimally small volume in relation to the total system volume, V, and that the molecules are sufficiently far apart that they do not act on each other. These assumptions are reasonable for dilute gases at low pressure but break down at moderate temperatures and pressures. Because understanding the phase behavior of complex liquid mixtures at high pressures (like reservoir fluids) is vital for modern industrial endeavors, subsequent equations of state had to be developed to describe condensation, liquid-liquid instability, critical points, and so on, behaviors which the ideal gas law cannot capture.

The first EOS capable of predicting such phenomena was proposed by Johannes van der Waals, stemming from the hypothesis that liquids could be regarded as compressed gases and that a single equation with multiple volume roots could be used to describe both vapor and liquid phases. The EOS of van der Waals (VDW) captures deviations from ideal gas behavior because of attractive and repulsive molecular interactions, relaxing both assumptions implicit in the ideal gas law. VDW in pressure-explicit form is given by the following:

$$P = \frac{RT}{\hat{V} - b} - \frac{a}{\hat{V}^2} \tag{4.25}$$

The parameter "a" accounts for molecular attraction, and "b" accounts for the volume of the molecules. Rearranging VDW in terms of molar volume yields:

$$\hat{V}^3 - \left(b + \frac{RT}{P}\right)\hat{V}^2 + \frac{a}{P}\hat{V} - \frac{ab}{P} = 0 \tag{4.26}$$

The third-order polynomial in volume is the defining characteristic of all cubic equations of state. Writing this instead in terms of the compressibility factor yields:

$$Z^3 - \left(\frac{bP}{RT} + 1\right)Z^2 + \frac{aP}{R^2T^2}Z - \frac{abP^2}{R^3T^3} = 0 \qquad (4.27)$$

The "a" and "b" parameters for VDW are calculated from the critical temperature and critical pressure of a pure component by the following expressions:

$$a = \frac{27}{64}\frac{R^2T_c^2}{P_c} \qquad b = \frac{1}{8}\frac{RT_c}{P_c} \qquad (4.28)$$

Subsequent equations of state based on the original VDW have since been developed with the only major modifications being the addition of temperature-dependence on the molecular attraction term and an additional parameter known as the *acentric factor* (ω) that accounts for molecular shape. The cubic EOS most often employed for predicting PVT properties of petroleum fluids are the Soave-Redlich-Kwong (SRK) and Peng-Robinson (PR) (Redlich and Kwong 1949; Soave 1972; Peng and Robinson 1976).

Instead of giving explicit expressions for SRK and PR, a generalized cubic EOS is presented that can be applied to any cubic EOS.

$$P = \frac{RT}{\hat{V} - b} - \frac{a(T)}{(\hat{V} + \delta_1 b)(\hat{V} + \delta_2 b)} \qquad (4.29)$$

With compressibility factor given by:

$$Z^3 + \sigma_2 Z^2 + \sigma_1 Z + \sigma_0 = 0 \qquad (4.30)$$

The coefficients of the compressibility factor equation are dependent on the EOS, and the parameters for VDW, SRK, and PR in the generalized cubic EOS are given in Table 4.1. The auxiliary variables needed for the coefficients of the compressibility factor calculation are given by:

$$A = \frac{a(T)P}{R^2T^2} \qquad B = \frac{bP}{RT} \qquad (4.31)$$

$$a(T) = \beta_1 \frac{\alpha(T)R^2T_c^2}{P_c} \qquad b = \beta_2 \frac{RT_c}{P_c} \qquad (4.32)$$

$$\alpha(T) = \left[1 + m\left(1 - \sqrt{T/T_c}\right)\right]^2 \qquad m_{\mathrm{VDW}} = 0$$

$$m_{\mathrm{SRK}} = 0.480 + 1.574\omega - 0.176\omega^2$$

$$m_{\mathrm{PR}} = 0.37464 + 1.54226\omega - 0.26992\omega^2 \qquad (4.33)$$

SRK and PR generally perform well in predicting the phase behavior of light non-polar substances but perform poorly in predicting liquid phase compressibility,

TABLE 4.1

Model Constants for Calculating Physical Properties with the Generalized Cubic EOS

EOS	δ_1	δ_2	σ_0	σ_1	σ_2	β_1	β_2
VDW	0	0	$-AB$	A	$-(1+B)$	27/64	1/8
SRK	1	0	$-AB$	$A-B-B^2$	-1	0.42748	0.08664
PR	$1+\sqrt{2}$	$1-\sqrt{2}$	$-AB+B^2+B^3$	$A-2B-3B^2$	$-(1-B)$	0.45724	0.07780

especially as the molecular size of the components increase. Because crude oils are mostly nonpolar, SRK and PR can regularly give acceptable predictions, but one additional parameter is often needed to correct for poor liquid volume predictions. This parameter is the Peneloux volume translation, c (Péneloux et al. 1982).

$$\hat{V}_{EOS+P} = \hat{V}_{EOS} - c \tag{4.34}$$

where $\hat{V}_{EOS}$ is the molar volume predicted by the cubic EOS without volume correction.

To extend the cubic EOS to mixtures, the following mixing rules are generally used:

$$\bar{a} = \sum_i \sum_j x_i x_j \sqrt{a_i a_j}\,(1-k_{ij}) \quad \bar{b} = \sum_i x_i b_i \quad \bar{c} = \sum_i x_i c_i \tag{4.35}$$

where:

x_i is the molar composition of component i in the mixture
k_{ij} is the binary interaction parameter
a_i, b_i, and c_i are calculated the same as for pure components

The input parameters for the cubic equations of state are molar composition (x_i), critical pressure (P_{c_i}), critical temperature (T_{c_i}), and acentric factor (ω_i) for each component present in a phase. The binary interaction parameters are optional input parameters; their values are typically tuned to match the BP predicted by the EOS model to experimental values of BP for a given binary mixture of components i and j. Conceptually, k_{ij} represents the deficiency in a model to accurately describe the intermolecular forces between component i and component j. Values near zero generally mean that the model captures well the intermolecular interactions. Negative values of k_{ij} suggest that the model underpredicts the intermolecular attractions, and positive values suggest the opposite.

Once the compressibility factor has been determined, fugacity coefficients can be calculated by:

$$\bar{I} = \frac{1}{\delta_1 - \delta_2} \ln\left(\frac{Z+\delta_1 B}{Z+\delta_2 B}\right)$$

$$\bar{a}_i = 2\sum_j \left[x_j \sqrt{a_i a_j}\,(1-k_{ij}) \right] - \bar{a}$$

$$\bar{q}_i = \frac{A}{B}\left(1 + \frac{\bar{a}_i}{\bar{a}} - \frac{b_i}{\bar{b}}\right)$$

$$\ln\hat{\phi}_i = \frac{b_i}{b}(Z-1) - \ln\left(|Z-B|\right) - \bar{q}_i\bar{I} \qquad (4.36)$$

Even when the Peneloux volume correction is included, the cubics do not perform well in predicting liquid densities of phases containing heavy components with complex molecular structures like resins and asphaltenes. Because of the shortcomings of the cubic equations of state in describing the phase behavior of complex molecules, more sophisticated EOS models have since been developed. However, the cubics do remain a popular model for industrial purposes because of their simplicity, their wide availability in modeling software, and low computational cost. For some applications, like reservoir simulations, the phase equilibrium problem has to be solved thousands of times, and the use of complicated equations of state can make these calculations much too costly for practical use. When computational speed is prohibitive, the cubic equations of state remain among the most popular options.

4.2.3.2 Perturbed-Chain Statistical Associating Fluid Theory Equation of State

The EOS that has perhaps gained most popularity for oil applications because of the shortcomings of SRK and PR is PC-SAFT. PC-SAFT is arguably the most widely used EOS model from the SAFT family, which have become the standard EOS models for describing the phase behavior of large, complex molecules like polymers and heavy constituents of crude oil like resins and asphaltenes. The SAFT EOS was originally developed by Chapman et al. (1988, 1990) and a variety of researchers have since proposed modifications to better describe certain classes of fluids. PC-SAFT is one such modification (Gross and Sadowski 2001), and it has shown particular promise in describing the phase behavior of crude oils. The SAFT EOS and its successors are derived from thermodynamic perturbation theory, which describes real fluids as perturbations from idealized and well-described reference systems. In a sense, the cubic equations of state are perturbations from the ideal gas law, in which the ideal gas reference fluid is a system of infinitesimally small noninteracting molecules and the "a" and "b" parameters quantify deviations, or perturbations, from ideal gas behavior. For SAFT, the reference fluid consists of nondeformable nonattractive spheres bonded covalently to form chains and the perturbation terms account for attractive and repulsive forces excluded from the reference terms (Figure 4.4).

The SAFT EOS family shares the same fundamental form of the EOS:

$$\frac{A^{res}(T,V,\boldsymbol{n})}{nRT} = \hat{a}^{res} = \hat{a}^{ref} + \hat{a}^{pert} \qquad (4.37)$$

where:
A^{res} is the residual Helmholtz energy
$\hat{a}^{res}$ is the reduced residual molar Helmholtz energy
$\hat{a}^{ref}$ and $\hat{a}^{pert}$ are the reference and perturbation contributions, respectively

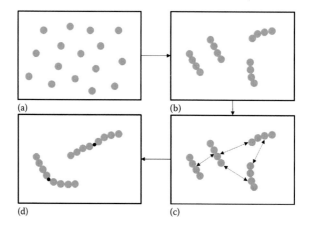

(a)   (b)

(d)   (c)

FIGURE 4.4 Pictorial representation of PC-SAFT: (a) Hard sphere molecules, (b) bonded covalently to form chains, (c) with weak, long range dispersion forces between chains, and (d) association forces to account for strong short-range directional forces. The PC-SAFT reference fluid consists of the hard sphere and chain terms, which together are called the *hard chain*. The dispersion term accounts for the attractive van der Waals forces that dominate phase behavior for nonpolar molecules. The association term accounts for the strong short-range forces in polar molecules and is neglected for nonpolar molecules.

For PC-SAFT, the reference and perturbation terms are given, respectively, as:

$$\hat{a}^{\mathrm{ref}} = \hat{a}^{\mathrm{hc}} = \hat{a}^{\mathrm{hs}} + \hat{a}^{\mathrm{chain}} \tag{4.38}$$

$$\hat{a}^{\mathrm{pert}} = \hat{a}^{\mathrm{disp}} + \hat{a}^{\mathrm{assoc}} \tag{4.39}$$

The hard sphere term was derived from molecular simulations performed on nondeformable spheres of varying sizes (Carnahan and Starling 1970; Mansoori et al. 1971), and the chain term was developed by Chapman to describe the interactions of spheres in the limit of infinite attraction. The hard sphere and chain terms are given by:

$$\hat{a}^{\mathrm{hs}} = \frac{6}{\pi\tilde{\rho}} \left[\frac{3\zeta_1\zeta_2}{1-\zeta_3} + \frac{\zeta_2{}^3}{(1-\zeta_3)^2\zeta_3} + \left(\frac{\zeta_2{}^3}{\zeta_3{}^2} - \zeta_0 \right) \ln(1-\zeta_3) \right] \tag{4.40}$$

$$\hat{a}^{\mathrm{chain}} = -\sum_{i=1}^{N_c} x_i (m_i - 1) \ln(g_{ii}^{\mathrm{hs}}) \tag{4.41}$$

The hard chain term is a summation of the hard sphere and chain terms. The parameter, m_i, is the number of monomer segments in the chain of component i, the radial distribution function g_{ij}^{hs} describes the relative spacing of the molecules in the mixture, the ζ_n variables are size parameters, and d_{ij} is an average segment diameter for components i and j:

$$g_{ij}^{hs} = \frac{1}{(1-\zeta_3)} + d_{ij}\left[\frac{3\zeta_2}{(1-\zeta_3)^2}\right] + d_{ij}^2\left[\frac{2\zeta_2^2}{(1-\zeta_3)^3}\right] \tag{4.42}$$

$$\zeta_{n\varepsilon\{0,1,2,3\}} = \frac{\pi}{6}\tilde{\rho}\sum_{i=1}^{N_c} x_i m_i d_i^n$$

$$d_{ij} = \frac{d_i d_j}{d_i + d_j}$$

$$d_i = \sigma_i\left[1 - 0.12\exp\left(-\frac{3\varepsilon_i}{k_B T}\right)\right]$$

where $\tilde{\rho}$ is the number density.

The perturbation term must then capture all physical forces not described by the reference fluid. In the case of nonpolar molecules, only a description of the weak dispersion forces is needed, whereas polar molecules must also include some additional term(s) to describe the strong intermolecular forces due to permanent dipoles. In PC-SAFT, the dispersion forces are assumed to follow a square-well interaction potential of fixed diameter and well-depth. The dispersion term is given by:

$$\hat{a}^{disp} = -2\pi\tilde{\rho}I_1\overline{m^2\varepsilon\sigma^3} - \pi\tilde{\rho}\overline{m}C_1 I_2\overline{m^2\varepsilon^2\sigma^3} \tag{4.43}$$

The I_1 and I_2 terms are power series' in reduced density, C_1 is a compressibility coefficient, and $\overline{m}$, $\overline{m^2\varepsilon\sigma^3}$, and $\overline{m^2\varepsilon^2\sigma^3}$ are defined to extend the PC-SAFT EOS to mixtures:

$$I_1 = \sum_{i=0}^{6} a_i\zeta_3^i \qquad I_2 = \sum_{i=0}^{6} b_i\zeta_3^i \tag{4.44}$$

$$C_1 = \left[1 + \overline{m}\frac{8\zeta_3 - 2\zeta_3^2}{(1-\zeta_3)^4} + (1-\overline{m})\frac{20\zeta_3 - 27\zeta_3^2 + 12\zeta_3^3 - 2\zeta_3^4}{[(1-\zeta_3)(2-\zeta_3)]^2}\right]^{-1}$$

$$\overline{m} = \sum_{i=1}^{N_c} x_i m_i$$

$$\overline{m^2\varepsilon^2\sigma^3} = \sum_{i=1}^{N_c}\sum_{j=1}^{N_c} x_i x_j m_i m_j \left(\frac{\varepsilon_{ij}}{k_B T}\right)\sigma_{ij}^3 \qquad \overline{m^2\varepsilon^2\sigma^3} = \sum_{i=1}^{N_c}\sum_{j=1}^{N_c} x_i x_j m_i m_j \left(\frac{\varepsilon_{ij}}{k_B T}\right)^2 \sigma_{ij}^3$$

The coefficients of the density polynomials (a_i and b_i) are provided in the paper of Gross and Sadowski (2001), and mixing rules are used to obtain σ_{ij} and ε_{ij}/k_B.

$$\sigma_{ij} = \frac{\sigma_i + \sigma_j}{2} \qquad \frac{\varepsilon_{ij}}{k_B} = \sqrt{\frac{\varepsilon_i}{k_B}\frac{\varepsilon_j}{k_B}}(1 - k_{ij}) \tag{4.45}$$

For nonpolar fluids, the PC-SAFT residual Helmholtz expression can be solved when molar composition (x_i), segment number (m_i), segment diameter (σ_i), and dispersion energy (ε_i/k_B), are specified. Gross and Sadowski (2001) listed the PC-SAFT parameters (m_i, σ_i, ε_i/k_B) for many common hydrocarbons and nonpolar light gases in their original paper and provide a complete description for the calculation of the residual Helmholtz energy.

For strongly polar fluids, dispersion forces alone are not sufficient to describe phase behavior, and an additional term is required in the perturbation term—association:

$$\hat{a}^{assoc} = \sum_i^{N_c} x_i \sum_{A_i} \left(\ln X^{A_i} - \frac{1}{2} X^{A_i} + \frac{1}{2} \right) \tag{4.46}$$

The unbonded monomer fraction of component i at site A (X^{A_i}) is calculated by:

$$X^{A_i} = \left[1 + \tilde{\rho} \sum_j^{N_c} x_j \sum_{B_j} X^{B_j} \Delta^{A_i B_j} \right]^{-1} \tag{4.47}$$

and the association strength is:

$$\Delta^{A_i B_j} = d_{ij}^3 g_{ij}^{hs} \kappa^{A_i B_j} \left[\exp\left(\frac{\varepsilon^{A_i B_j}}{k_B T} \right) - 1 \right] \tag{4.48}$$

The association term is designed to handle fluids with strong permanent dipoles, usually those that can hydrogen-bond like water and alcohols. Including the association term requires two additional tuning parameters: association volume κ^{AB} and association energy ε^{AB}/k_B.

4.2.3.3 Cubic-Plus-Association Equation of State

The Cubic-Plus-Association (CPA) EOS is, as the name suggests, a combination of the classical cubic EOS and Chapman's association term originally developed for SAFT. In terms of the reduced residual Helmholtz, CPA is given as a physical contribution (cubic EOS) and chemical contribution (association). Some researchers prefer to use SRK for the physical contribution (Kontogeorgis et al. 1996), whereas others prefer PR (Li and Firoozabadi 2010b). Here, the physical contribution is given by SRK:

$$\hat{a}^{res} = \hat{a}^{SRK} + \hat{a}^{assoc} \tag{4.49}$$

$$\hat{a}^{SRK} = \ln\left[\frac{\hat{V}}{\hat{V} - b} \right] + \frac{a(T)}{RTb} \ln\left[\frac{\hat{V}}{\hat{V} + b} \right] \tag{4.50}$$

The reduced Helmholtz for association retains the same expression as shown for PC-SAFT (Equation 4.46), but the radial distribution function and energy parameter are modified:

$$g_{ij}^{CPA} = \frac{1}{1-1.9\eta} \qquad \eta = \left(\frac{1}{4\hat{V}}\right)b \tag{4.51}$$

$$a(T) = a_0\left[1 + c_1\left(1 - \sqrt{T/T_c}\right)\right]^2 \tag{4.52}$$

The tuning parameters for the SRK term are b, a_0, c_1.

Unfortunately, the addition of the association term means that CPA is no longer cubic in volume like the classical cubics, so the numerical efficiency that makes the cubics attractive is lost. Additionally, there are five tuning parameters for CPA, which are fit to liquid density and vapor pressure data (like PC-SAFT), instead of the three parameters for cubics fit to the critical point. Thus, CPA improves dramatically on the liquid-volume predictions produced by the cubics, partly because of a retuning of the model parameters and partly because of the additional association term and more fitting parameters. Many academic works have demonstrated the improved performance of CPA over the classical cubics in predicting the phase behavior of asphaltenic crude oils and polar compounds like water and alcohols (Kontogeorgis et al. 2007; Yan et al. 2009; Tsivintzelis et al. 2014; Liang et al. 2014; Arya et al. 2015, 2016, Li and Firoozabadi 2010a, 2010b).

4.2.4 Intermolecular Forces in Crude Oil Systems

There are two major factors to consider when applying mathematical models to physical systems: speed and accuracy. These considerations are often contradictory because what is gained in accuracy is usually paid for by an increase in model complexity and subsequent increase in computational time. For applications requiring thousands of calls to a physical property model routine, speed becomes the dominant consideration. For applications requiring only a few calls to these routines, accuracy is more important. General phase behavior calculations, such as those shown in Figure 4.1, can be completed without too much computational strain, so any EOS discussed in this chapter is suitably fast to handle these applications. Because this type of calculation is the main objective of this chapter, speed is not a limiting factor, and the choice of EOS is down to which model best represents the physical characteristics of the systems under consideration.

Fluid phase behavior is driven by the strength and directionality of the intermolecular forces between the various components present in a given system, and the primary goal of EOS models is to describe these intermolecular forces as a function of easily measured variables like pressure, temperature, volume, and composition. Intermolecular forces arise because of the chemical structures of the components present in a mixture; nonpolar components interact only by weak dispersion forces due to electron polarizability, whereas polar components have permanent dipoles with stronger interactions that require more energy to overcome the bonds that hold their molecules in the liquid state. The vast difference in normal boiling points of methane (111.7 K) and water (373.1 K)—nonpolar and polar molecules, respectively, with similar molecular weights—illustrates the importance of distinguishing the types of intermolecular forces present in a system and their relative strengths. This example also suggests that an EOS capable of accurately modeling methane is unlikely to accurately model water.

The intermolecular forces that characterize the phase behavior of asphaltenes are more nuanced than that of methane or water because asphaltenes are a solubility class—meaning they consist of many different molecular structures with different sizes and intermolecular forces—and they are more polarizable than alkanes and aromatics but with no permanent dipoles like exist for alcohols and carboxylic acids. Asphaltenes tend to self-associate even in good solvents and rarely, if ever, exist in monomeric form. Their aromatic cores interact by pi-pi bonding and stack to form nano-aggregates that are much heavier than the asphaltene monomer. This stacking phenomenon of asphaltene monomers is because of strong short-range forces that are characteristic of association, but the aggregated monomers—once formed—tend to interact with other components by dispersion forces instead of association (Buckley et al. 1998; Buenrostro-Gonzalez et al. 2002, 2004). This distinction offers two reasonable modeling approaches, one in which the preaggregated asphaltene monomer is modeled as an associating component and one in which the aggregated asphaltene is modeled as a nonassociating component. Various researchers have shown that PC-SAFT is capable of modeling asphaltene precipitation for extreme ranges of pressure and temperature without the association term (Ting et al. 2003; Gonzalez et al. 2007; Panuganti et al. 2012, 2013; Punnapala and Vargas 2013; Tavakkoli et al. 2016), whereas the cubic equations of state are generally incapable of accurately modeling the liquid-liquid phase split of asphaltene precipitation from bulk oil without the association term. This suggests that asphaltene precipitation modeling with CPA yields results consistent with experimental data because of the additional tuning parameters and not necessarily because CPA is properly capturing the physics of asphaltene phase behavior. Given sufficient parameters tuned to enough experimental data, it is unsurprising that CPA often performs as well as PC-SAFT. However, the benefit of an EOS model is in its predictive capability. PC-SAFT has shown to predict asphaltene precipitation phenomena at different gas-injection concentrations accurately with parameters tuned to only two or three experimental AOP points. Thus, for the application of asphaltene precipitation, PC-SAFT without association is our preferred model. Cubics can often perform satisfactorily for vapor-liquid equilibrium (VLE) applications, but their shortcomings in describing liquid phase compressibility becomes especially pronounced in liquid-liquid equilibrium (LLE) applications.

4.3 CRUDE OIL CHARACTERIZATION FOR EQUATION OF STATE MODELS

Crude oils contain thousands of components with vastly different chemical structures. Obtaining compositions for all these components is experimentally infeasible, and even if it could be done, applying an EOS to such a system would be computationally unmanageable. To efficiently apply an EOS model to predict the phase behavior of a crude oil, the crude oil must be characterized into a reasonable number of pseudo-components. The process of defining a set of representative pseudo-components along with their composition and EOS parameters is called *characterization*. The EOS parameters for classical cubics are T_C, P_C, ω, for non-associating PC-SAFT are m, σ, and ε/k_B and for CPA are a_0, b, c_1, κ^{AB}, and ε^{AB}/k_B.

The compositional analyses performed by service laboratories usually list the compositions with approximately 30–50 components. The compositions of nonhydrocarbon gases like H_2S, CO_2 and N_2, and light hydrocarbon components (lighter than C_6) are reported in detail. Therefore, direct application of an EOS can be easily implemented for these components. Components including C_6 and heavier (C_{6+} fractions) are usually reported as boiling cuts corresponding to single carbon numbers (i.e., C_7, C_8,...etc). Katz and Firoozabadi (1978) recommended the definition of a boiling cut for a specific carbon number n such that the lower boiling limit is 0.5 K higher than the boiling point of the previous n-alkane ($n-1$), whereas the higher boiling limit is 0.5 K higher than the n-alkane with carbon number n. For example, the C_9 pseudo-fraction represents components with boiling range between 399.25 K and 424.45 K, where 399.25 K corresponds to a boiling point that is 0.5 K higher than that of n-C_8 and 424.45 K corresponds to a normal boiling point that is 0.5 K higher than that of n-C_9. Katz and Firoozabadi (1978) reported boiling ranges, average density, and average molecular weight for each carbon fraction up to C_{35}. Service laboratories usually perform the compositional analysis up to C_{35} because the average properties for these pseudo-components can be determined from the work of Katz and Firoozabadi (1978). The composition of the remaining portion of the crude oil (pseudo-fraction C_{36+}) along with its density and molecular weight are determined mathematically because these properties for the whole sample are measured.

Composition measurements are performed after flashing the reservoir fluid sample at standard pressure and temperature (0.1 MPa and 288.15 K), and the relative amounts of gas and liquid are measured and reported as GOR in units of standard cubic feet of gas per stock tank barrel of oil (scf/stb). The compositions are reported for flashed gas and flashed liquid separately. Then, the reservoir fluid composition is determined by mathematically combining the measured gas and liquid compositions using the measured GOR. A detailed sample PVT report is provided in Table 4.2 for reservoir fluid U8.

The task from an EOS modeling perspective is to characterize the C_{7+} fractions in the crude oil and assign EOS parameters to the resultant pseudo-components. Various crude oil characterization procedures using different equations of state have been proposed over the past few decades. Pedersen's Single Carbon Number characterization procedure is among the most popular of these procedures because of its simplicity; it was originally developed for the cubic EOS (Pedersen et al. 1992) and recently extended to handle PC-SAFT parameter estimation (Pedersen et al. 2012). Other characterization methods make use of probability distribution functions such as the gamma function to represent the composition of the C_{7+} fraction (Whitson 1983; Quiñones-Cisneros et al. 2005). The use of distribution functions have been implemented primarily with cubic EOS models. Other characterization methods based on the saturate, aromatic, resin, and asphaltene (SARA) analysis of the STO have also been proposed for use with PC-SAFT (Ting 2003;

TABLE 4.2
Compositional Analysis for Reservoir Fluid U8

Component	M	Flashed Liquid[a]		Flashed Gas		Reservoir Fluid[b]	
	g.mol^{-1}	mol%	wt%	mol%	wt%	mol%	wt%
N_2	28.04	0.000	0.000	0.990	0.767	0.400	0.086
CO_2	44.01	0.000	0.000	2.680	3.262	1.090	0.368
H_2S	34.08	0.000	0.000	1.700	1.602	0.680	0.178
C_1	16.04	0.012	0.001	46.980	20.839	19.201	2.361
C_2	30.07	0.091	0.014	10.562	8.783	4.369	1.007
C_3	44.10	0.559	0.126	11.689	14.256	5.106	1.726
iC_4	58.12	0.417	0.124	3.325	5.345	1.605	0.715
nC_4	58.12	1.507	0.448	7.655	12.303	4.019	1.791
iC_5	72.15	1.604	0.592	3.419	6.822	2.345	1.298
nC_5	72.15	2.463	0.909	3.868	7.719	3.038	1.681
C_6	86.11	5.529	2.435	3.688	8.784	4.778	3.155
Mcyclo-C_5	84.16	0.985	0.424	0.000	0.000	0.000	0.000
Benzene	78.11	0.308	0.123	0.000	0.000	0.000	0.000
Cyclo-C_6	84.16	0.885	0.381	0.000	0.000	0.000	0.000
C_7	100.20	5.833	2.989	3.480	9.519	53.359	85.634
Mcyclo-C_6	98.19	1.533	0.770	0.000	0.000	0.000	0.000
Toluene	92.14	1.114	0.525	0.000	0.000	0.000	0.000
C_8	114.23	6.674	3.899	0.000	0.000	0.000	0.000
C_2-benzene	106.17	0.613	0.333	0.000	0.000	0.000	0.000
m, p xylene	106.17	1.394	0.757	0.000	0.000	0.000	0.000
o-xylene	106.17	0.599	0.325	0.000	0.000	0.000	0.000
C_9	128.26	5.976	3.920	0.000	0.000	0.000	0.000
1,2,4 TMB	120.19	0.748	0.460	0.000	0.000	0.000	0.000
C_{10}	142.29	6.884	5.009	0.000	0.000	0.000	0.000
C_{11}	147	6.692	5.031	0.000	0.000	0.000	0.000
C_{12}	161	5.554	4.573	0.000	0.000	0.000	0.000
C_{13}	175	4.972	4.450	0.000	0.000	0.000	0.000
C_{14}	190	4.289	4.167	0.000	0.000	0.000	0.000
C_{15}	206	3.877	4.084	0.000	0.000	0.000	0.000
C_{16}	222	3.239	3.677	0.000	0.000	0.000	0.000
C_{17}	237	2.706	3.280	0.000	0.000	0.000	0.000
C_{18}	251	2.434	3.124	0.000	0.000	0.000	0.000
C_{19}	263	2.355	3.167	0.000	0.000	0.000	0.000
C_{20}	275	2.003	2.817	0.000	0.000	0.000	0.000
C_{21}	291	1.772	2.637	0.000	0.000	0.000	0.000
C_{22}	305	1.553	2.422	0.000	0.000	0.000	0.000
C_{23}	318	1.341	2.181	0.000	0.000	0.000	0.000
C_{24}	331	1.198	2.028	0.000	0.000	0.000	0.000
C_{25}	345	1.087	1.918	0.000	0.000	0.000	0.000
C_{26}	359	0.981	1.801	0.000	0.000	0.000	0.000
C_{27}	374	0.909	1.739	0.000	0.000	0.000	0.000

(*Continued*)

TABLE 4.2 (*Continued*)
Compositional Analysis for Reservoir Fluid U8

Component	M	Flashed Liquid[a]		Flashed Gas		Reservoir Fluid[b]	
	g.mol⁻¹	mol%	wt%	mol%	wt%	mol%	wt%
C_{28}	388	0.826	1.639	0.000	0.000	0.000	0.000
C_{29}	402	0.745	1.532	0.000	0.000	0.000	0.000
C_{30}	416	0.685	1.457	0.000	0.000	0.000	0.000
C_{31}	430	0.612	1.346	0.000	0.000	0.000	0.000
C_{32}	444	0.547	1.242	0.000	0.000	0.000	0.000
C_{33}	458	0.479	1.122	0.000	0.000	0.000	0.000
C_{34}	472	0.439	1.060	0.000	0.000	0.000	0.000
C_{35}	486	0.392	0.974	0.000	0.000	0.000	0.000
C_{36+}	895	2.616	11.973	0.000	0.000	0.000	0.000

[a] Also known as stock tank oil (STO) or dead oil.
[b] Recombined based on "zero-flash" gas-to-oil ratio (GOR) = 68.1 Sm³/m³ (382 scf/stb).

Gonzalez 2008; Panuganti et al. 2012; Punnapala and Vargas 2013; Tavakkoli et al. 2016). In Section 4.3.1, the SARA-based characterization method using PC-SAFT EOS is discussed in detail.

As mentioned previously, the aim of crude oil characterization is to lump the components listed in the original compositional analysis into a reasonable number of pseudo-components that are capable of describing a complex crude oil mixture well enough to produce accurate modeling results. Choosing a number of *reasonable* pseudo-components to represent the behavior of crude oil depends mainly on the objective of the modeling study and the type of crude oil under investigation. Generally, heavy crude oils require more pseudo-components than light crude oils to accurately capture phase behavior. Heavy crude oils are defined as those having an API gravity of 30 or lower (Pedersen et al. 2015). If the objective is modeling vapor-liquid equilibrium for a reservoir fluid (e.g., BP), it is not necessary to characterize the heavy end in detail because the BP is dominated by the light ends. Díaz et al. (2011) showed that the modeling results of the BP of Athabasca bitumen with solvent mixtures using 5 or 40 heavy-end pseudo-components yield similar accuracy using the Advanced PR EOS. Furthermore, Abutaqiya et al. (2017) investigated light crude oils from the Middle East and showed that by characterizing the STO as a single liquid fraction using PC-SAFT, accurate modeling results can be obtained for BPs of oil blends with different injection gases, as well as other volumetric properties for reservoir fluids (e.g., density, oil formation factor, etc.).

If the objective is modeling dew points, liquid drop out in gas condensates, or asphaltene precipitation, more detailed characterization of the heavy ends is required to accurately capture these phenomena. Pedersen et al. (2015) showed that the choice of 6 or 22 pseudo-components to characterize the heavy ends significantly affects the predictions of liquid drop out from gas condensates using the SRK EOS. For the purpose of modeling asphaltene precipitation, different authors have used different numbers of pseudo-components to represent the crude oil. If the purpose of the asphaltene precipitation study is to model the AOP, usually 3–8 pseudo-fractions are used to represent the maltenes, whereas a single pseudo-fraction is used to represent the asphaltenes (Gonzalez 2008; Panuganti et al. 2012; Punnapala and Vargas 2013; Hustad et al. 2013). For the purpose of modeling the amount of asphaltene precipitation because of titration with n-alkane precipitants, it is necessary to characterize the asphaltene pseudo-fraction as several cuts to account for asphaltene polydispersity. Tavakkoli et al. (2016) used 4 asphaltene pseudo-fractions in modeling asphaltene precipitation from light crude oils, while Díaz et al. (2011) used 6 asphaltene pseudo-fractions in modeling asphaltene precipitation from bitumen.

4.3.1 PERTURBED-CHAIN STATISTICAL ASSOCIATING FLUID THEORY CHARACTERIZATION WITH SATURATE, AROMATIC, RESIN, AND ASPHALTENE ANALYSIS

Modeling asphaltene precipitation using PC-SAFT and the SARA analysis of crude oil has been in continuous development over the years (Ting 2003; Gonzalez 2008; Panuganti et al. 2012; Punnapala and Vargas 2013; Tavakkoli et al. 2016). In general, these developments followed the same lumping procedure but differed in the methodology for parameter estimation. Because SARA analysis is used to characterize the STO, this characterization procedure is referred to as the *SARA-based method*. The description of the SARA-based method in this section follows the latest development as implemented by Punnapala and Vargas (2013) for monodisperse asphaltenes and Tavakkoli et al. (2016) for polydisperse asphaltenes. The two approaches differ only in the characterization of the asphaltenes pseudo-fraction.

A schematic of the SARA-based characterization is shown in Figure 4.5. In this method, the flashed gas and flashed liquid from the PVT report (e.g., Table 4.2) are characterized separately and then mathematically combined using the experimental GOR to form the live reservoir fluid.

4.3.1.1 Characterization of Flashed Gas

The flashed gas is characterized as seven components including: N_2, CO_2, H_2S, methane, ethane, propane, and heavy gas (C_{4+}). The gas-phase components are largely responsible for accuracy in the BP calculation and considering each flashed gas component individually enhances the predictive capabilities of the model (Panuganti et al. 2012).

4.3.1.2 Characterization of Flashed Liquid

The liquid fraction is characterized based on the flashed liquid composition and the SARA analysis. The flashed liquid (also called *STO* or *dead oil*) is characterized as three pseudo-components: Saturates, Aromatics + Resins (A + R), and Asphaltenes. The plus fraction in the flashed liquid is defined depending on the level of detail provided in the PVT report. For example, in Table 4.2, detailed compositional analysis is provided up to C_9 pseudo-fraction. Therefore, the plus fraction is defined as C_{10+}. The components lighter than C_{10} in the flashed liquid are distributed based on their chemical structure into Saturates, A+R, and Asphaltenes. The normal alkanes, branched alkanes, and cycloalkanes are all lumped into the saturates pseudo-fraction. The Aromatics are lumped into the A+R pseudo-fraction. Boiling

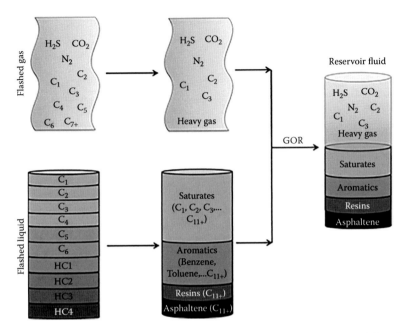

FIGURE 4.5   Schematic for a SARA-based crude oil characterization procedure. (Adapted from Khaleel, A. et al., In Society of Petroleum Engineers, doi:10.2118/177941-MS, 2015.)

cuts lower than C_{10} are assumed paraffinic and are lumped into the Saturates. At this point, all components lighter than C_{10} are already distributed into their respective pseudo-fraction. The C_{10+} pseudo-fraction is distributed into the three groups such that the weight fraction of each group matches the reported SARA analysis. The molecular weight value for C_{10+} in Asphaltenes is assumed to be 1700 g/mol as an initial guess only for the purpose of calculating molecular weights and molar compositions of Saturates and A+R. This value can be tuned later to match experimental data (usually AOP experiments). The molecular weight values for C_{10+} in Saturates and A+R are set as an initial guess such that the calculated STO molecular weight matches the reported value. The molecular weight of C_{10+} in Saturates and A+R are not modified further because it was observed that simulation results are only slightly affected by variations in these values. To standardize the characterization procedure, a constant ratio between the molecular weight of C_{10+} in Saturates and the molecular weight of C_{10+} in A+R is set to 0.9. In Section 4.4, a detailed step-by-step example is demonstrated on how to perform the aforementioned lumping.

After characterizing the flashed gas and flashed liquid, they are mathematically combined using the experimental GOR to determine the live oil composition. Once this task is completed, the composition and molecular weight of each component in the live oil is known. The next step is to determine the PC-SAFT parameters for each of these components.

4.3.1.3 Simulation Parameters Estimation

When only nonpolar components are present in a mixture, the association term of PC-SAFT is neglected. Because the phase behavior of asphaltenes is largely determined by the polarizability of the molecules—not the polarity—association can often be safely neglected (Buckley 1999; Buckley et al. 1998). With this assumption, three PC-SAFT parameters are required for each component: the number of segments per molecule (m_i), the temperature-independent diameter of each molecular segment (σ_i), and the dispersion energy (ε_i/k_B).

The PC-SAFT parameters for pure nonpolar hydrocarbons and light gases are taken from the work of Gross and Sadowski (2001). The parameters for the Heavy Gas, Saturates, A+R and Asphaltene pseudo-components are calculated using published correlations by Gonzalez et al. (2007). They correlated the three PC-SAFT parameters for different hydrocarbon series (i.e., paraffins, polynuclear aromatics [PNA], benzene-derivatives, etc.) as a function of molecular weight. Figure 4.6 shows the variation of PC-SAFT parameters as a function of molecular weight for n-alkanes and PNA. The Heavy Gas and Saturates pseudo-fractions are assumed to consist purely of n-alkanes. Therefore, their PC-SAFT parameters are calculated from the correlation of Gonzalez et al. (2007) for n-alkanes. For A+R and Asphaltenes, Punnapala and Vargas (2013) defined an aromaticity parameter (γ) that

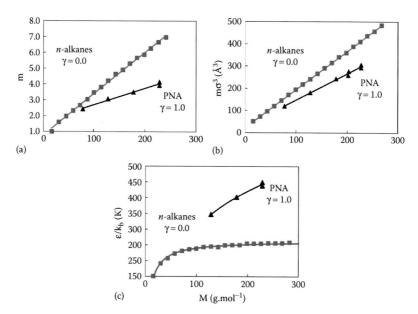

FIGURE 4.6 Variation of PC-SAFT parameters as a function of molecular weight for n-alkanes and PNA: (a) m, (b) mσ^3, and (c) ε/k_b. (Adapted from Gonzalez, D.L., Modeling of asphaltene precipitation and deposition tendency using the PC-SAFT equation of state, Rice University, http://hdl.handle.net/1911/22140, 2008.)

TABLE 4.3

Correlations for PC-SAFT Parameters of Heavy Gas, Saturates, Aromatics + Resins, and Asphaltenes

Parameter	Heavy Gas and Saturates	A+R and Asphaltenes
m	$0.0257M_i + 0.8444$	$(1-\gamma_i)(0.0257M_i + 0.8444) + \gamma_i(0.0101M_i + 1.7296)$
$\sigma(\text{Å})$	$4.047 - \dfrac{4.8013\,ln(M_i)}{M_i}$	$(1-\gamma_i)\left(4.047 - \dfrac{4.8013\,ln(M_i)}{M_i}\right) + \gamma_i\left(4.6169 - \dfrac{93.98}{M_i}\right)$
$\varepsilon/k_B\,(K)$	$\exp\left(5.5769 - \dfrac{9.523}{M_i}\right)$	$(1-\gamma_i)\exp\left(5.5769 - \dfrac{9.523}{M_i}\right) + \gamma_i\left(508 - \dfrac{234100}{M_i^{1.5}}\right)$

varies from 0 to 1, where 0 corresponds to n-alkanes and 1 to PNA. The aromaticity of A+R ($\gamma_{\text{A+R}}$) is tuned to match the bubble pressure and density of the crude oil. The aromaticity of Asphaltenes (γ_{Asph}) as well as its molecular weight are tuned to match experimental values for onset of precipitation (AOP). The mathematical correlations for PC-SAFT parameter estimation are shown in Table 4.3. With the overall molar composition and EOS parameters specified, the phase behavior modeling can be performed and potential flow assurance problems assessed.

4.3.2 ASPHALTENE POLYDISPERSITY

Whereas light hydrocarbon components like methane, ethane, and propane are defined precisely by their chemical structures, asphaltenes are a solubility class containing all constituents of the STO that are insoluble in n-pentane and soluble in toluene. The physicochemical properties of the asphaltene fraction are a function of the relative amounts of the various types of asphaltenes contained therein and their respective structures. Because asphaltenes are a broad solubility class, the various molecules included in the asphaltene fraction can have different structures and properties, and the relative amounts of the various molecules is dependent on the crude oil source from which the asphaltenes came.

In the characterization procedure presented in Section 4.3.1, the asphaltene component was assumed to be monodisperse and the EOS parameters were fit only to AOP data. In reality, asphaltenes are polydisperse, and the AOP is only representative of the most thermodynamically unstable asphaltenes that are the first to precipitate from the bulk oil phase. When the Asphaltene component is assumed monodisperse, all asphaltenes are bound to follow the behavior of the most unstable constituents of the asphaltene fraction because of the chosen fitting criteria. If predicting AOP is the only asphaltene property of interest, then the monodisperse Asphaltene component often provides a sufficient description of the physics. However, if asphaltene amounts in the asphaltene instability region (between UAOP and LAOP) are of interest, then it is advantageous to consider a polydisperse Asphaltene component that better captures the physicochemical differences of the various asphaltene subfractions.

Tavakkoli et al. (2016) proposed a characterization scheme that mirrors the one introduced in Section 4.3.1 but treats the Asphaltene component as a polydisperse mixture using a molecular weight distribution. Additionally, the PC-SAFT parameter fitting for Asphaltenes is performed with onset data generated by the indirect method instead of the customary high-pressure NIR spectroscopy method. In this scheme, Asphaltenes are extracted from the STO and characterized based on the precipitating solvent. Asphaltenes that are soluble in n-C_6 but insoluble in n-C_5 are categorized as C_{5-6} asphaltenes; C_{6-7} asphaltenes include those that are soluble in n-C_7 but insoluble in n-C_6; and C_{7-8} asphaltenes include those that are soluble in n-C_8 but insoluble in n-C_7. C_{8+} asphaltenes include all asphaltenes soluble in toluene (as per the definition of an asphaltene) but insoluble in n-C_8 or any lighter normal alkane.

Considering the additional complexity of a polydisperse Asphaltenes component means that more parameters are required as each subfraction requires its own model parameter values. More tuning parameters usually means that more experimental data for tuning those parameters will be required, and when experimental data are prohibitively expensive, this can be an undesirable consequence. More tuning parameters also can hint at a deficiency in the model where the tuning parameters act more as empirical corrections than descriptions of actual physical phenomena. Because Tavakkoli et al. (2016) use a relatively inexpensive experimental method for generating asphaltene onset data, cost is not an issue. Additionally, the PC-SAFT parameters for all Asphaltenes subfractions are still calculated by the correlations in Table 4.3, and the aromaticity values for all subfractions are kept constant. Thus, the PC-SAFT parameters for the Asphaltenes subfractions are a function of their respective molecular weight only.

Tavakkoli et al. (2016) used the gamma distribution function to describe the molecular weight distribution of the asphaltenes subfractions.

$$f(r) = \frac{1}{M_m \Gamma(\alpha)} \left[\frac{\alpha}{(\bar{r}-1)} \right]^{\alpha} (r-1)^{\alpha-1} \exp \left[\frac{\alpha(1-r)}{(\bar{r}-1)} \right] \tag{4.53}$$

where:

r and $\bar{r}$ are given by M/M_m and $\bar{M}/M_m$, respectively

M is the molecular weight of the asphaltene subfraction

M_m is the molecular weight of the preaggregated asphaltene monomer

The distribution is discretized into infinitesimal increments of constant Δr and the mass fraction w_i and $\bar{r}_i$ parameter are calculated by Equations 4.54 and 4.55. Once $\bar{r}_i$ is known, the molecular weight of each asphaltene subfraction is calculated by Equation 4.56:

$$w_i = \frac{\int_{r_i}^{r_{i+1}} f(r)dr}{\int_{r_1}^{r_n} f(r)dr} \tag{4.54}$$

$$\bar{r}_i = \frac{\displaystyle\int_{r_i}^{r_{i+1}} rf(r)dr}{\displaystyle\int_{r_i}^{r_{i+1}} f(r)dr} \qquad (4.55)$$

$$M_i = \bar{r}_i M_m \qquad (4.56)$$

The benefit of this mass distribution scheme is that it adds only one fitting parameter, α, to the asphaltene characterization procedure regardless of the number of asphaltene subfractions chosen.

4.4 ASPHALTENE PRECIPITATION MODELING EXAMPLES

The modeling of asphaltene precipitation using the SARA-based approach can be categorized according to the way the asphaltenes pseudo-fraction is treated: monodisperse in molecular weight or polydisperse. The monodisperse approach treats the asphaltenes pseudo-fraction as a single component with a single molecular weight. The polydisperse asphaltene approach treats the asphaltenes as multiple subfractions with different molecular weights. This section demonstrates example calculations on how the characterization is performed using each approach.

4.4.1 MONODISPERSE ASPHALTENE

The monodisperse asphaltene approach has been used by several authors for modeling asphaltene onset in reservoir fluids (Ting 2003; Gonzalez et al. 2007, 2008; Gonzalez 2008; Punnapala and Vargas 2013; Panuganti et al. 2013; AlHammadi et al. 2015). The example calculation shown here is based on the latest development as reported by Punnapala and Vargas (2013).

The properties of the crude oil used in this sample calculation are shown in Table 4.4. The live oil data in Table 4.4 represents the whole crude oil sample (flashed gas + flashed liquid), whereas the STO represents only the liquid that remains after flashing the reservoir sample at zero-flash conditions (0.1 MPa and 288.15 K). The GOR is usually measured at zero-flash conditions. The SARA-based characterization method treats the flashed liquid and flashed gas separately then mathematically combines them based on the zero-flash GOR to form the reservoir fluid.

The steps for characterizing the crude oil using the SARA-based approach are summarized as follows:

1. Characterize the flashed gas by lumping all C_{4+} fractions into a single pseudo-component (Heavy Gas).
2. Characterize the flashed liquid as Saturates, Aromatics + Resins, and Asphaltene pseudo-components using the SARA analysis.
3. Combine the characterized gas and liquid phases using the zero-flash GOR to obtain the reservoir fluid composition.

TABLE 4.4
Crude Oil U8 Properties

Reservoir Fluid

GOR (Sm^3/m^3)	68.1
T^{res} (K)	400
BP (MPa) at T^{res}	7.61
$M_{res. fl.}$(g/mol)	130.4

Stock Tank Oil (STO)

Saturates wt%	80.64
Aromatics wt%	17.44
Resins wt%	1.47
Asphaltenes wt%	0.45
M_{STO}(g/mol)	195.5
ρ_{STO}(kg/m^3)	823.9

4. Obtain PC-SAFT simulation parameters for all non-Asphaltene pseudo-components.
5. Obtain PC-SAFT simulation parameters for the Asphaltene pseudo-component.

Step 1: Characterize the flashed gas by lumping all C$_{4+}$ fractions into a single pseudo-component (Heavy Gas).

Table 4.5 shows the composition of the flashed gas from the PVT report before and after characterization. The average molecular weight of Heavy Gas is calculated from:

TABLE 4.5
Composition of the Flashed Gas and Characterized Flashed Gas

Component	Flashed Gas		Component	Characterized Gas	
	g.mol^{-1}	mol%		g.mol^{-1}	mol%
N$_2$	28.01	0.99	N$_2$	28.01	0.99
CO$_2$	44.01	2.68	CO$_2$	44.01	2.68
H$_2$S	34.08	1.70	H$_2$S	34.08	1.70
C$_1$	16.04	46.98	C$_1$	16.04	46.98
C$_2$	30.07	10.56	C$_2$	30.07	10.56
C$_3$	44.10	11.69	C$_3$	44.10	11.69
iC$_4$	58.12	3.33	Heavy Gas	71.50	25.44
nC$_4$	58.12	7.65			
iC$_5$	72.15	3.42			
nC$_5$	72.15	3.87			
C$_6$	86.18	3.69			
C$_{7+}$	99.00	3.48			

$$M_{HG} = \frac{\sum_{i>C_3} x_i M_i}{\sum_{i>C_3} x_i} \qquad (4.57)$$

where x_i is the molar composition of components heavier than C_3 in the flashed gas.

Step 2: Characterize the flashed liquid as Saturates, Aromatics + Resins, and Asphaltene pseudo-components using the SARA analysis.

To characterize the flashed liquid, first the plus fraction is identified based on available data. For crude oil U8, the plus fraction is defined as C_{10+}. Then, the components are distributed according to their chemical structure into Saturates, A+R, and Asphaltenes. Table 4.6 lists the flashed liquid composition and the distribution of each petroleum fraction into Saturates, A+R, and Asphaltenes assuming a basis of 100 g of STO.

As observed from Table 4.6, it is assumed that the boiling cuts up to C_9 consist of saturates only. The plus fraction C_{10+} is treated as a mixture of Saturates, A+R, and Asphaltenes. The plus fraction is distributed such that the reported SARA analysis in Table 4.4 is satisfied. To calculate amount of C_{10+} in the A+R pseudo-fraction, the following calculation is performed:

$$C_{10+}(A+R) = 18.91 - 0.12 - 0.52 - 0.33 - 0.76 - 0.33 - 0.46 = 16.39$$

where the number 18.91 in this calculation is the A+R mass composition from the SARA analysis and all other numbers are the mass amounts of pure aromatics present. Because the plus fraction (C_{10+}) is distributed into Saturates, A+R, and Asphaltenes, the molar mass of C_{10+} in each of these fractions is expected to be different. The molar mass of C_{10+} in the Asphaltene pseudo-fraction is a tuning parameter to be optimized during Step 5. An initial value of 1700 g/mol is used for the purpose of calculating molar compositions in Step 2 and tuning parameters in Step 4. The molar mass of C_{10+} for the Saturates and A+R pseudo-components is set such that the overall molar mass of C_{10+} in the STO is equal to the reported value (257.1 g/mol). To standardize the calculation of the molar mass of C_{10+} for the saturates and A+R pseudo-components, a fixed ratio of 0.9 is used:

$$M_{C_{10+}}^{Sat} = 0.9\left(M_{C_{10+}}^{A+R}\right)$$

The choice of this ratio is not expected to significantly affect the simulation results and can be used for all modeling studies. With this ratio, the molar mass of C_{10+} in Saturates (Sat) and Aromatics + Resins (A+R) can be calculated from:

$$M_{C_{10+}}^{A+R} = \left\{\frac{w_{C_{10+}}^{Sat}}{w_{C_{10+}}^{STO}} + 0.9\left(\frac{w_{C_{10+}}^{A+R}}{w_{C_{10+}}^{STO}}\right)\right\} / \left\{\frac{0.9}{M_{C_{10+}}^{STO}} - \frac{0.9}{M_{C_{10+}}^{Asph}}\left(\frac{w_{C_{10+}}^{Asph}}{w_{C_{10+}}^{STO}}\right)\right\} \qquad (4.58)$$

TABLE 4.6

Composition of Flashed Liquid and Distribution of Each Petroleum Cut into Saturates (Sat), Aromatics + Resins (A+R), and Asphaltenes (Asph) Assuming 100 g of STO

Component	Flashed Liquid		Characterized Liquid (wt%)		
	g.mol^{-1}	wt%	$w_{i,Sat}$	$w_{i,A+R}$	$w_{i,Asph}$
N_2	28.01	0.00	—	—	—
CO_2	44.01	0.00	—	—	—
H_2S	34.08	0.00	—	—	—
C_1	16.04	0.00	—	—	—
C_2	30.07	0.01	0.01	—	—
C_3	44.10	0.13	0.13	—	—
i-C_4	58.12	0.12	0.12	—	—
n-C_4	58.12	0.45	0.45	—	—
i-C_5	72.15	0.59	0.59	—	—
n-C_5	86.18	0.91	0.91	—	—
C_6	86.12	2.43	2.43	—	—
M-cyclo-C_5	84.16	0.42	0.42	—	—
Benzene	78.11	0.12	—	0.12	—
Cyclo-C_6	84.16	0.38	0.38	—	—
C_7	100.20	2.99	2.99	—	—
M-cyclo-C_6	98.19	0.77	0.77	—	—
Toluene	92.14	0.52	—	0.52	—
C_8	114.23	3.90	3.90	—	—
C_2-Benzene	106.17	0.33	—	0.33	—
m&p-Xylene	106.17	0.76	—	0.76	—
o-Xylene	106.17	0.33	—	0.33	—
C_9	128.26	3.92	3.92	—	—
1,2,4 TMB	120.19	0.46	—	0.46	—
C_{10+}	257.10	80.45	63.61	16.39	0.45
Total	**195.5**	**100**	**80.64**	**18.91**	**0.45**
$M\ C_{10+}$/g.mol^{-1}			250.6	278.5	1700

which in this case becomes:

$$M_{C_{10+}}^{A+R} = \frac{\left\{ \dfrac{63.61}{80.45} + 0.9\left(\dfrac{16.39}{80.45} \right) \right\}}{\left\{ \dfrac{0.9}{257.1} - \dfrac{0.9}{1700}\left(\dfrac{0.45}{80.45} \right) \right\}} = 278.49 \ \textbf{g.mol}^{-1}$$

$$M_{C_{10+}}^{Sat} = 0.9\left(M_{C_{10+}}^{A+R} \right) = 0.9(278.5) = 250.64 \ \textbf{g.mol}^{-1}$$

TABLE 4.7

Composition and Molecular Weight of the Pseudo-Components in the Characterized Liquid

Component	Characterized Flashed Liquid		
	wt%	g.mol^{-1}	mol%
Saturates	80.64	188.41	83.67
A+R	18.91	227.10	16.28
Asphaltenes	0.45	1700	0.05175

After distributing the pure components and the plus fraction into each pseudo-component, the molecular weight and molar compositions for Saturates, A+R, and Asphaltenes can be calculated. These results are shown in Table 4.7.

Step 3: Combine the characterized gas and liquid phases using the zero-flash GOR to obtain the reservoir fluid composition.

To combine the two phases, GOR is converted from Sm^3/m^3 to mol gas/mol liquid using the experimental properties reported as follows:

$$\frac{\text{mol gas}}{\text{m}^3 \text{of STO}} = \text{GOR}\left(\frac{\text{Sm}^3}{\text{m}^3}\right)*42.302\frac{\text{mol gas}}{\text{Sm}^3} = 68.1\left(\frac{\text{Sm}^3}{\text{m}^3}\right)*42.302\frac{\text{mol gas}}{\text{Sm}^3}$$

$$= 2880.7\frac{\text{mol gas}}{\text{m}^3 \text{of STO}}$$

$$\frac{\text{mol liquid}}{\text{m}^3 \text{ of STO}} = \frac{\rho_{STO}\left(\frac{\text{kg}}{\text{m}^3}\right)}{M_{STO}\left(\frac{\text{kg}}{\text{kmol}}\right)}*1000\left(\frac{\text{mol}}{\text{kmol}}\right) = \frac{823.9\left(\frac{\text{kg}}{\text{m}^3}\right)}{195.5\left(\frac{\text{kg}}{\text{kmol}}\right)}*1000\left(\frac{\text{mol}}{\text{kmol}}\right)$$

$$= 4214.3\frac{\text{mol liquid}}{\text{m}^3 \text{of STO}}$$

$$\frac{\text{mol gas}}{\text{mol liquid}} = \frac{2880.7}{4214.3} = 0.6835$$

$$\frac{\text{mol gas}}{\text{mol total}} = \frac{2880.7}{4214.3+2880.7} = 0.4060$$

Thus, 40.6% by moles of the reservoir fluid is gas and 59.4% is liquid at 0.1 MPa and 288.15 K, and the compositions of the gas and liquid phases are given in Tables 4.5 and 4.6, respectively. Knowing the molar amounts of the gas and liquid phases as well as the composition for each phase, the live oil composition can

TABLE 4.8
**Molecular Weight and Molar Composition
of the Characterized Reservoir Fluid**

Component	M	z_i
	g.mol^{-1}	mol%
N_2	28.01	0.40
CO_2	44.01	1.09
H_2S	34.08	0.69
C_1	16.04	19.06
C_2	30.07	4.29
C_3	44.10	4.74
Heavy Gas	71.78	10.32
Saturates	188.41	49.72
A+R	227.10	9.67
Asphaltenes	1700*	0.03075

*Initial estimate.

be calculated. The molar composition of the live oil is shown in Table 4.8, where the molecular weight of Asphaltenes is estimated and will be optimized in Step 5. Because the molar compositions are a function of molecular weight and the molecular weight of Asphaltene is a tuning parameter, the molar compositions should be readjusted during the optimization of the Asphaltene molecular weight in Step 5.

Step 4: Obtain PC-SAFT simulation parameters for all non-Asphaltene pseudo-components.

PC-SAFT parameters for pure components are available from the literature (Gross and Sadowski 2001), whereas those for Heavy Gas, Saturates, A+R and Asphaltene pseudo-components are calculated from correlations in Table 4.3. Heavy Gas and Saturates are assumed to be composed of pure alkanes. PC-SAFT parameters for the A+R pseudo-component can be obtained using its calculated molecular weight and fitting its aromaticity parameter (γ_{A+R}), which is usually performed over BP, STO density, and saturation density. However, saturation density is not reported for this case, so the fitting is performed over BP and STO density. The binary interaction parameters used in this example are taken from Punnapala and Vargas (2013) and are reported in Table 4.9.

Figure 4.7 demonstrates the effect of γ_{A+R} on the predictions of BP and STO density. The error is reported as an equally weighted average of the average absolute percent deviation (AAPD) of BP and STO density:

$$AAPD = \left[\frac{|BP_{Exp} - BP_{Model}|}{BP_{Exp}} + \frac{|\rho_{STO\,Exp} - \rho_{STO\,Model}|}{\rho_{STO\,Exp}} \right] \times 100 / 2 \qquad (4.59)$$

The γ_{A+R} that minimizes the error in the prediction of BP and ρ_{STO} is 0.498. Once the optimum γ_{A+R} is found, the PC-SAFT parameters for all pseudo-components

TABLE 4.9

k_{ij} Values Used for the Crude Characterization

Component	N_2	CO_2	H_2S	C_1	C_2	C_3	Heavy Gas	Sat	A+R	Asph.
N_2	—	0	0.09	0.03	0.04	0.06	0.075	0.14	0.158	0.16
CO_2		—	0.0678	0.05	0.097	0.1	0.12	0.13	0.05	0.10
H_2S			—	0.062	0.058	0.05	0.07	0.09	0.015	0.015
C_1				—	0	0	0.03	0.03	0.029	0.07
C_2					—	0	0.02	0.012	0.025	0.06
C_3						—	0.015	0.01	0.01	0.01
Heavy Gas							—	0.005	0.012	0.01
Saturates								—	0.007	−0.004
A+R									—	0
Asph.										—

Source: Punnapala, S. and Vargas, F.M., *Fuel*, 108, 417–429, 2013.

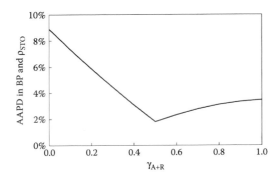

FIGURE 4.7 Effect of aromaticity of the A+R fraction on the predictions of bubble pressure (BP) and STO density (ρ_{STO}).

(except Asphaltene) are known and are shown in Table 4.10. If the density of the liquid at the BP and reservoir temperature is available, the optimization is performed with higher preference for matching saturation density in the overall objective function. It is generally recommended that the fitting is performed such that the error in saturation density does not exceed 0.5%, whereas the error in the BP does not exceed 10%. These limits are chosen for two reasons: (1) the relatively higher uncertainty in measuring BP as compared to saturation density and (2) to ensure good predictions of high-temperature and high-pressure properties for the crude oil.

TABLE 4.10

PC-SAFT Parameters for All Non-Asphaltene Components

Component	M	z_i	m	σ	ε/k_B	γ
	g.mol^{-1}	mol%		Å	K	
N_2	28.01	0.40	1.205	3.313	90.96	—
CO_2	44.01	1.09	2.073	2.785	169.21	—
H_2S	34.08	0.69	1.652	3.074	227.34	—
C_1	16.04	19.06	1.000	3.704	150.03	—
C_2	30.07	4.29	1.607	3.521	191.42	—
C_3	44.10	4.74	2.002	3.618	208.11	—
Heavy Gas	71.78	10.32	2.689	3.761	231.42	0.000
Saturates	188.41	49.72	5.686	3.914	251.23	0.000
A+R	227.10	9.67	5.356	4.067	346.10	0.498

Step 5: Obtain PC-SAFT simulation parameters for the Asphaltene pseudo-component.

Because there are two fitting parameters for the Asphaltene pseudo-component, it is imperative that a minimum of two AOP data points be used to obtain unique simulation parameters. The available data for AOP experiments on the reservoir fluid U8 can be found in Table 4.11. The optimization is performed by scanning a range of γ_{Asph} between 0.35 and 0.65 and M_{Asph} between 2,000 and 11,000 g.mol^{-1}. The maximum recommended step size of the scanning is 0.005 for γ_{Asph} and 25 g.mol^{-1} for M_{Asph}. The choice of the scanning range for both parameters may be different for different crudes. Ideally, a larger range of γ_{Asph} (e.g., 0–1) and M_{Asph} (e.g. 500–20,000 g.mol^{-1}) should be used, but the computational time required for such scanning makes this procedure prohibitive. An initial scanning range can be used to optimize the parameters and updated if necessary. Updating the range should be performed if the error plots indicate that a global minimum may exist outside the chosen range.

Figure 4.8 shows the effect of the two Asphaltene parameters on the predictions of the AOP data points. The error is expressed as the average absolute percent deviation

TABLE 4.11

Experimental Data for Asphaltene Onset and Bubble Pressures for the Reservoir Fluid U8

T (K)	AOP (MPa)	BP (MPa)
399.8	—	7.61
355.4	—	6.75
338.7	15.33	6.25
322.0	29.12	5.78

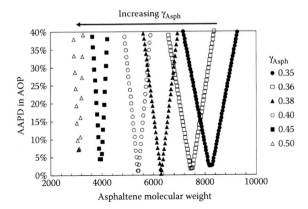

FIGURE 4.8 Error plot for optimizing the asphaltene parameters for crude oil U8. Optimized parameters are $M_{Asph} = 6300$, $\gamma_{Asph} = 0.38$. (AAPD is calculated based on the Equation 4.59.)

(AAPD) between the model and the experimental data points of UAOP at 338.7 and 322.0 K. It can be seen from Figure 4.8 that the choice of the scanning range shows an absolute minimum at an intermediate aromaticity value, that is, 0.38. If the minimum occurred at an aromaticity near the lower or upper limit of the scanning range (in this case 0.35 or 0.65), then the minimum found might not be the global minimum and the range should be expanded. The optimized asphaltene parameters from Figure 4.8 are $M_{Asph} = 6300$ and $\gamma_{Asph} = 0.38$. The final set of simulation parameters for the live oil U8 are shown in Table 4.12. The asphaltene phase envelope and BP curve produced with these parameters are shown in Figure 4.9.

TABLE 4.12
Final Set of Simulation Parameters for the Characterized Crude Oil

Component	M	z_i	m	σ	ε/k_B	γ
	g.mol^{-1}	mol%		Å	K	
N_2	28.01	0.40	1.205	3.313	90.96	—
CO_2	44.01	1.09	2.073	2.785	169.21	—
H_2S	34.08	0.69	1.652	3.074	227.34	—
C_1	16.04	19.06	1.000	3.704	150.03	—
C_2	30.07	4.29	1.607	3.521	191.42	—
C_3	44.10	4.74	2.002	3.618	208.11	—
Heavy Gas	71.78	10.32	2.689	3.761	231.42	0.000
Saturates	188.41	49.72	5.686	3.914	251.23	0.000
A+R	227.10	9.67	5.356	4.067	346.10	0.498
Asphaltenes	6300	0.008298	125.74	4.254	356.45	0.380

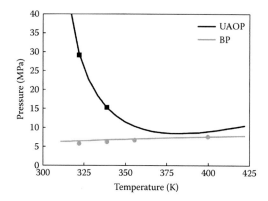

FIGURE 4.9   AOP and BP curves for U8. Saturation curves were generated with PC-SAFT. Points are experimental saturation data. BP, bubble pressure; UAOP, uppermost asphaltene onset pressure.

4.5   FINAL REMARKS

The goal of this chapter was to familiarize the reader with some of the numerous methods for modeling asphaltene precipitation in crude oil mixtures (Section 4.2) by first laying the phase equilibrium thermodynamics groundwork on which the methods rely (Section 4.1). Our favored approach for modeling asphaltene instability in crude oil (and phase behavior in general) at both ambient conditions and under high pressures and high temperatures are the EOS models. EOS models do not make many assumptions that simpler models make and are often flexible enough to handle the vastly different components that can be found in the reservoir. The intermolecular forces that determine fluid phase behavior in asphaltenic crude oils were also described (Section 4.2.4), which then motivate the use of certain modeling approaches that are consistent with experimental evidence.

The largest improvements in asphaltene phase behavior modeling are now likely to come from the characterization methodology, both in how asphaltene properties are measured in the laboratory and how they are described in the modeling approach (Sections 4.3 and 4.4). Because asphaltenes are by definition a solubility class, researchers have known that consideration of polydispersity in modeling asphaltenes would undoubtedly improve predictions. The question, however, was how to obtain the experimental information to tune the simulation parameters. The high-pressure asphaltene onset experiments are expensive to perform and the results can be unreliable. These experiments also reveal the phase behavior of only the most unstable asphaltenes, so an approach was needed to describe the behavior of all asphaltene fractions that would not add a tremendous financial burden to the already expensive experiments being performed. The indirect method offers one such method, providing a description of the thermodynamics of different fractions of the asphaltenes for a significantly reduced cost to the traditional high-pressure experiments (Section 4.3.2). However, indirect method experiments are currently performed only at ambient pressures and PC-SAFT simulation parameters fit to ambient data are then used to predict phase behavior for reservoir pressures and temperatures, an extrapolation which likely carries non-negligible errors.

REFERENCES

Abutaqiya, M. I. L., S. R. Panuganti, and F. M. Vargas. 2017. Efficient algorithm for the prediction of pressure-volume-temperature properties of crude oils using the perturbed-chain statistical associating fluid theory equation of state. *Industrial & Engineering Chemistry Research* 56: 6088–6102. doi:10.1021/acs.iecr.7b00368.

AlHammadi, A. A., F. M. Vargas, and W. G. Chapman. 2015. Comparison of cubic-plus-association and perturbed-chain statistical associating fluid theory methods for modeling asphaltene phase behavior and pressure–volume–temperature properties. *Energy & Fuels* 29 (5): 2864–2875.

Arya, A., N. von Solms, and G. M. Kontogeorgis. 2015. Determination of asphaltene onset conditions using the cubic plus association equation of state. *Fluid Phase Equilibria* 400: 8–19. doi:10.1016/j.fluid.2015.04.032.

Arya, A., N. von Solms, and G. M. Kontogeorgis. 2016. Investigation of the gas injection effect on asphaltene onset precipitation using the cubic-plus-association equation of state. *Energy & Fuels* 30 (5): 3560–3574. doi:10.1021/acs.energyfuels.5b01874.

Buckley, J. S. 1999. Predicting the onset of asphaltene precipitation from refractive index measurements. *Energy & Fuels* 13 (2): 328–332. doi:10.1021/ef980201c.

Buckley, J. S., G. J. Hirasaki, Y. Liu, S. Von Drasek, J-X. Wang, and B. S. Gill. 1998. Asphaltene precipitation and solvent properties of crude oils. *Petroleum Science and Technology* 16 (3–4): 251–285. doi:10.1080/10916469808949783.

Buenrostro-Gonzalez, E., S. I. Andersen, J. A. Garcia-Martinez, and C. Lira-Galeana. 2002. Solubility/molecular structure relationships of asphaltenes in polar and nonpolar media. *Energy & Fuels* 16 (3): 732–741. doi:10.1021/ef0102317.

Buenrostro-Gonzalez, E., C. Lira-Galeana, A. Gil-Villegas, and J. Wu. 2004. Asphaltene precipitation in crude oils: Theory and experiments. *AIChE Journal* 50 (10): 2552–2570. doi:10.1002/aic.10243.

Burke, N. E., R. E. Hobbs, and S. F. Kashou. 1990. Measurement and modeling of asphaltene precipitation (Includes Associated Paper 23831). *Journal of Petroleum Technology* 42 (11): 1–440.

Carnahan, N. F., and K. E. Starling. 1970. Thermodynamic properties of a rigid-sphere fluid. *The Journal of Chemical Physics* 53 (2): 600–603. doi:10.1063/1.1674033.

Chapman, W. G., K. E. Gubbins, G. Jackson, and M. Radosz. 1990. New reference equation of state for associating liquids. *Industrial & Engineering Chemistry Research* 29 (8): 1709–1721. doi:10.1021/ie00104a021.

Chapman, W. G., G. Jackson, and K. E. Gubbins. 1988. Phase equilibria of associating fluids. *Molecular Physics* 65 (5): 1057–1079. doi:10.1080/00268978800101601.

Cimino, R., S. Correra, A. Del Bianco, and T. P. Lockhart. 1993. Asphaltenes: Fundamentals and applications. In *Asphaltenes. Fundamentals and Applications*, 97–126. New York: Plenum Press.

Creek, J., J. Wang, and J. S. Buckley. 2009. Verification of asphaltene-instability-trend (ASIST) predictions for low-molecular-weight alkanes. *SPE Production & Operations* 24 (2). doi:10.2118/125203-PA.

De Boer, R. B., K. Leerlooyer, M. R. P. Eigner, and A. R. D. Van Bergen. 1995. Screening of crude oils for asphalt precipitation: Theory practice and the selection of inhibitors. *SPE Production & Facilities* 10 (01): 55–61.

Díaz, O. C., J. Modaresghazani, M. A. Satyro, and H. W. Yarranton. 2011. Modeling the phase behavior of heavy oil and solvent mixtures. *Fluid Phase Equilibria* 304 (1): 74–85. doi:10.1016/j.fluid.2011.02.011.

Dickie, J. P., and T. Fu. Yen. 1967. Macrostructures of the asphaltic fractions by various instrumental methods. *Analytical Chemistry* 39 (14): 1847–1852. doi:10.1021/ac50157a057.

Flory, P. J. 1942. Thermodynamics of high polymer solutions. *The Journal of Chemical Physics* 10 (1): 51. doi:10.1063/1.1723621.

Freed, D. E., O. C. Mullins, and J. Y. Zuo. 2010. Theoretical treatment of asphaltene gradients in the presence of GOR gradients. *Energy & Fuels* 24 (7): 3942–3949. doi:10.1021/ef1001056.

Gonzalez, D. L. 2008. Modeling of asphaltene precipitation and deposition tendency using the PC-SAFT equation of state. Rice University. http://hdl.handle.net/1911/22140.

Gonzalez, D. L., G. J. Hirasaki, J. Creek, and W. G. Chapman. 2007. Modeling of asphaltene precipitation due to changes in composition using the perturbed chain statistical associating fluid theory equation of state. *Energy & Fuels* 21 (3): 1231–1242. doi:10.1021/ef060453a.

Gonzalez, D. L., F. M. Vargas, G. J. Hirasaki, and W. G. Chapman. 2008. Modeling study of CO$_2$-induced asphaltene precipitation. *Energy & Fuels* 22 (2): 757–762. doi:10.1021/ef700369u.

Gonzalez, D. L., F. M. Vargas, E. Mahmoodaghdam, F. Lim, and N. Joshi. 2012. Asphaltene stability prediction based on dead oil properties: Experimental evaluation. *Energy & Fuels* 26 (10): 6218–6227. doi:10.1021/ef300837y.

Griffith, M. G., and C. W. Siegmund. 1985. Controlling compatibility of residual fuel oils. January. doi:10.1520/STP35281S.

Gross, J., and G. Sadowski. 2001. Perturbed-chain SAFT: An equation of state based on a perturbation theory for chain molecules. *Industrial & Engineering Chemistry Research* 40 (4): 1244–1260. doi:10.1021/ie0003887.

Hildebrand, J. H., and S. E. Wood. 1933. The derivation of equations for regular solutions. *The Journal of Chemical Physics* 1 (12): 817. doi:10.1063/1.1749250.

Hirschberg, A., L. N. J. DeJong, B. A. Schipper, and J. G. Meijer. 1984. Influence of temperature and pressure on asphaltene flocculation. *SPE Journal*, SPE-11202-PA, 24 (03): 283–293. doi:10.2118/11202-PA.

Hirschberg, A. 1988. Role of asphaltenes in compositional grading of a reservoir's fluid column. *Journal of Petroleum Technology* 40 (01): 89–94. doi:10.2118/13171-PA.

Huggins, M. L. 1941. Solutions of long chain compounds. *The Journal of Chemical Physics* 9 (5): 440. doi:10.1063/1.1750930.

Hustad, O. S., N. Jia, K. S. Pedersen, A. I. Memon, and S. Leekumjorn. 2013. High pressure data and modeling results for phase behavior and asphaltene onsets of GoM oil mixed with nitrogen. In Society of Petroleum Engineers. doi:10.2118/166097-MS.

Katz, D. L., and A. Firoozabadi. 1978. Predicting phase behavior of condensate/crude oil systems using methane interaction coefficients. *Journal of Petroleum Technology* 30 (11): 1649–1655. doi:10.2118/6721-PA.

Kawanaka, S., S. J. Park, G. Ali Mansoori. 1991. Organic deposition from reservoir fluids: A thermodynamic predictive technique. *SPE Reservoir Engineering* 6 (02): 185–192.

Khaleel, A., M. Abutaqiya, M. Tavakkoli, A. A. Melendez-Alvarez, and F. M. Vargas. 2015. On the prediction, prevention and remediation of asphaltene deposition. In Society of Petroleum Engineers. doi:10.2118/177941-MS.

Koenhen, D. M., and C. A. Smolders. 1975. The determination of solubility parameters of solvents and polymers by means of correlations with other physical quantities. *Journal of Applied Polymer Science* 19 (4): 1163–1179.

Kontogeorgis, G. M., G. K. Folas, N. Muro-Suñé, N. von Solms, M. L. Michelsen, and E. H. Stenby. 2007. Modelling of associating mixtures for applications in the oil & gas and chemical industries. *Fluid Phase Equilibria* 261 (1–2): 205–211. doi:10.1016/j.fluid.2007.05.022.

Kontogeorgis, G. M., E. C. Voutsas, I. V. Yakoumis, and D. P. Tassios. 1996. An equation of state for associating fluids. *Industrial & Engineering Chemistry Research* 35 (11): 4310–4318. doi:10.1021/ie9600203.

Kraiwattanawong, K., H. S. Fogler, S. G. Gharfeh, P. Singh, W. H. Thomason, and S. Chavadej. 2007. Thermodynamic solubility models to predict asphaltene instability in live crude oils. *Energy & Fuels* 21 (3): 1248–1255. doi:10.1021/ef060386k.

Leontaritis, K. J., and G. Ali Mansoori. 1988. Asphaltene deposition: A survey of field experiences and research approaches. *Journal of Petroleum Science and Engineering* 1 (3): 229–239. doi:10.1016/0920-4105(88)90013-7.

Li, Z., and A. Firoozabadi. 2010a. Modeling asphaltene precipitation by n-alkanes from heavy oils and bitumens using cubic-plus-association equation of state. *Energy & Fuels* 24 (2): 1106–1113. doi:10.1021/ef9009857.

Li, Z., and A. Firoozabadi. 2010b. Cubic-plus-association equation of state for asphaltene precipitation in live oils. *Energy & Fuels* 24 (5): 2956–2963. doi:10.1021/ef9014263.

Liang, X., I. Tsivintzelis, and G. M. Kontogeorgis. 2014. Modeling water containing systems with the simplified PC-SAFT and CPA equations of state. *Industrial & Engineering Chemistry Research* 53 (37): 14493–14507. doi:10.1021/ie501993y.

Mansoori, G. A., N. F. Carnahan, K. E. Starling, and T. W. Leland Jr. 1971. Equilibrium thermodynamic properties of the mixture of hard spheres. *The Journal of Chemical Physics* 54 (4): 1523–1525. doi:10.1063/1.1675048.

Michelsen, M. L. 1994. Calculation of multiphase equilibrium. *Computers & Chemical Engineering* 18 (7): 545–550.

Mitchell, D. L., and J. G. Speight. 1972. The solubility of asphaltenes in hydrocarbon solvents. *Energy & Fuels* 52 (2): 149–152.

Pan, H., and A. Firoozabadi. 1997. Thermodynamic micellization model for asphaltene precipitation from reservoir crudes at high pressures and temperatures. In Society of Petroleum Engineers. doi:10.2118/38857-MS.

Pan, H., and A. Firoozabadi. 1998. Thermodynamic micellization model for asphaltene aggregation and precipitation in petroleum fluids. *SPE Production & Facilities* 13 (2): 118–127.

Panuganti, S. R., M. Tavakkoli, F. M. Vargas, D. L. Gonzalez, and W. G. Chapman. 2013. SAFT model for upstream asphaltene applications. *Fluid Phase Equilibria* 359: 2–16. doi:10.1016/j.fluid.2013.05.010.

Panuganti, S. R., F. M. Vargas, D. L. Gonzalez, A. S. Kurup, and W. G. Chapman. 2012. PC-SAFT characterization of crude oils and modeling of asphaltene phase behavior. *Fuel* 93: 658–669. doi:10.1016/j.fuel.2011.09.028.

Pedersen, K. S., A. L. Blilie, and K. K. Meisingset. 1992. PVT calculations on petroleum reservoir fluids using measured and estimated compositional data for the plus fraction. *Industrial & Engineering Chemistry Research* 31 (5): 1378–1384. doi:10.1021/ie00005a019.

Pedersen, K. S., P. L. Christensen, and J. A. Shaikh. 2015. *Phase Behavior of Petroleum Reservoir Fluids*, 2nd ed. Boca Raton, FL: CRC Press.

Pedersen, K. S., S. Leekumjorn, K. Krejbjerg, and J. Azeem. 2012. Modeling of EOR PVT data using PC-SAFT equation. In Abu Dhabi, UAE: Society of Petroleum Engineers.

Péneloux, A., E. Rauzy, and R. Fréze. 1982. A consistent correction for Redlich-Kwong-soave volumes. *Fluid Phase Equilibria* 8 (1): 7–23. doi:10.1016/0378-3812(82)80002-2.

Peng, D. Y., and D. B. Robinson. 1976. A new two-constant equation of state. *Industrial & Engineering Chemistry Fundamentals* 15 (1): 59–64.

Pfeiffer, J. P., and R. N. J. Saal. 1940. Asphaltic bitumen as colloid system. *Journal of Physical Chemistry* 44 (2): 139–149. doi:10.1021/j150398a001.

Punnapala, S., and F. M. Vargas. 2013. Revisiting the PC-SAFT characterization procedure for an improved asphaltene precipitation prediction. *Fuel* 108: 417–429. doi:10.1016/j.fuel.2012.12.058.

Quiñones-Cisneros, S. E., S. I. Andersen, and J. Creek. 2005. Density and viscosity modeling and characterization of heavy oils. *Energy & Fuels* 19 (4): 1314–1318. doi:10.1021/ef0497715.

Rassamdana, H., B. Dabir, M. Nematy, M. Farhani, and M. Sahimi. 1996. Asphalt flocculation and deposition: I. The onset of precipitation. *AIChE Journal* 42 (1): 10–22.

Redlich, O., and J. N. S. Kwong. 1949. On the thermodynamics of solutions. V. An equation of state. Fugacities of gaseous solutions. *Chemical Reviews* 44 (1): 233–244. doi:10.1021/cr60137a013.

Rogel, E., C. Ovalles, K. D. Bake, J. Y. Zuo, H. Dumont, A. E. Pomerantz, and O. C. Mullins. 2016. Asphaltene densities and solubility parameter distributions: Impact on asphaltene gradients. *Energy & Fuels* 30 (11): 9132–9140. doi:10.1021/acs.energyfuels.6b01794.

Rogel, E., M. R., J. Vien, and T. Miao. 2015. Characterization of asphaltene fractions: Distribution, chemical characteristics, and solubility behavior. *Energy & Fuels* 29 (4): 2143–2152. doi:10.1021/ef5026455.

Scatchard, G. 1931. Equilibria in non-electrolyte solutions in relation to the vapor pressures and densities of the components. *Chemical Reviews* 8 (2): 321–333.

Soave, G. 1972. Equilibrium constants from a modified redlich-kwong equation of state. *Chemical Engineering Science* 27 (6): 1197. doi:10.1016/0009-2509(72)80096-4.

Tavakkoli, M., A. Chen, and F. M. Vargas. 2016. Rethinking the modeling approach for asphaltene precipitation using the PC-SAFT equation of state. *Fluid Phase Equilibria* 416: 120–129. doi:10.1016/j.fluid.2015.11.003.

Ting, D. 2003. Thermodynamic stability and phase behavior of asphaltenes in oil and of other highly asymmetric mixtures. PhD, Houston, TX: Rice University.

Ting, D. P., G. J. Hirasaki, and W. G. Chapman. 2003. Modeling of asphaltene phase behavior with the SAFT equation of state. *Petroleum Science and Technology* 21 (3–4): 647–661. doi:10.1081/LFT-120018544.

Tsivintzelis, I., S. Ali, and G. M. Kontogeorgis. 2014. Modeling phase equilibria for acid gas mixtures using the cubic-plus-association equation of state. 3. Applications relevant to liquid or supercritical CO2 transport. *Journal of Chemical & Engineering Data*, April. doi:10.1021/je500090q.

Vargas, F. M., and W. G. Chapman. 2010. Application of the one-third rule in hydrocarbon and crude oil systems. *Fluid Phase Equilibria* 290 (1–2): 103–1008. doi:10.1016/j.fluid.2009.12.004.

Victorov, A. I., and A. Firoozabadi. 1996. Thermodynamic micellization model of asphaltene precipitation from petroleum fluids. *AIChE Journal* 42 (6): 1753–1764. doi:10.1002/aic.690420626.

Wang, F., T. J. Threatt, and F. M. Vargas. 2016. Determination of solubility parameters from density measurements for non-polar hydrocarbons at temperatures from (298–433) K and pressures up to 137 MPa. *Fluid Phase Equilibria* 430: 19–32. doi:10.1016/j.fluid.2016.09.021.

Wang, J., and J. S. Buckley. 2003. Asphaltene stability in crude oil and aromatic solvents the influence of oil composition. *Energy & Fuels* 17 (6): 1445–1451. doi:10.1021/ef030030y.

Wang, J., and J. S. Buckley. 2001a. An experimental approach to prediction of asphaltene flocculation. In Soc. Pet. Eng. doi:10.2118/64994-MS.

Wang, J., and J. S. Buckley. 2001b. A two-component solubility model of the onset of asphaltene flocculation in crude oils. *Energy & Fuels* 15 (5): 1004–1012. doi:10.1021/ef010012l.

Wang, K., J. Y. Zuo, Y. Chen, and O. C. Mullins. 2015. The dynamic Flory-Huggins-Zuo equation of state. *Energy* 91: 430–440. doi:10.1016/j.energy.2015.08.063.

Whitson, C. H. 1983. Characterizing hydrocarbon plus fractions. *SPE Journal* 23 (04): 683–694.

Yan, W., G. M. Kontogeorgis, and E. H. Stenby. 2009. Application of the CPA equation of state to reservoir fluids in presence of water and polar chemicals. *Fluid Phase Equilibria* 276 (1): 75–85. doi:10.1016/j.fluid.2008.10.007.

Zuo, J. Y., D. Freed, O. C. Mullins, D. Zhang, and A. Gisolf. 2010. Interpretation of DFA color gradients in oil columns using the Flory-Huggins solubility model. In Society of Petroleum Engineers. doi:10.2118/130305-MS.

Zuo, J. Y., R. Jackson, A. Agarwal, B. Herold, S. Kumar, I. De Santo, H. Dumont, C. Ayan, M. Beardsell, and O. C. Mullins. 2015. Diffusion model coupled with the Flory–Huggins–Zuo equation of state and Yen–Mullins model accounts for large viscosity and asphaltene variations in a reservoir undergoing active biodegradation. *Energy & Fuels* 29 (3): 1447–1460. doi:10.1021/ef502586q.

Zuo, J. Y., O. C. Mullins, D. Freed, H. Elshahawi, C. Dong, and D. J. Seifert. 2013. Advances in the Flory–Huggins–Zuo equation of state for asphaltene gradients and formation evaluation. *Energy & Fuels* 27 (4): 1722–1735. doi:10.1021/ef301239h.

5 Experimental Determination of Asphaltene Deposition

J. Kuang, N. Rajan Babu, J. Hu, A. Chen, M. Tavakkoli, and F. M. Vargas

CONTENTS

In Chapters 3 and 4, various experimental techniques and modeling strategies to evaluate the asphaltene precipitation tendency were reviewed. The destabilization of asphaltenes is a thermodynamic process that is driven by changes in pressure, temperature, or composition of the reservoir fluids. Asphaltene deposition, however, is a more complex process that is controlled not only by the thermodynamic conditions, but also by the transport phenomena, surface properties, particle sizes, and so on. (Akbarzadeh et al. 2007, 2012; Juyal et al. 2013; Ghahfarokhi et al. 2017). It is widely acknowledged that asphaltene precipitation is a necessary but not sufficient condition for the asphaltene deposition process. Although asphaltene precipitation has been investigated extensively, the mechanism of asphaltene deposition has not yet been fully understood (Juyal et al. 2013; Vilas Boas Favero et al. 2016).

Asphaltene deposition is a flow assurance problem that can potentially deteriorate because of the current tendency to produce from deep-water environment and the implementation of enhanced oil recovery (EOR) based on gas injection (Wang et al. 2004; Vargas 2009; Khaleel et al. 2015). Thus, it is crucial to understand the mechanism of asphaltene deposition during oil production as well as the driving forces that lead to this costly problem for the oil and gas industry. A reliable experimental technique for studying asphaltene deposition in the lab can offer valuable insights into the development of advanced simulation tools to predict asphaltene deposition under various oil production schemes (Buckley 2012; Hoepfner et al. 2013). It can also be used to test the performance of asphaltene deposition inhibitors and solvents to prevent and remediate asphaltene deposition problems, respectively.

This chapter will review the current experimental techniques used to investigate asphaltene deposition on metal surfaces and in porous media. The experimental assessment of the effect of water, electrolytes, corrosion, and coating on asphaltene deposition will also be discussed.

5.1 ASPHALTENE DEPOSITION EXPERIMENTS

5.1.1 DETERMINATION OF ASPHALTENE DEPOSITION IN WELLBORES AND PIPELINES

5.1.1.1 Capillary Deposition Flow Loop

There have been several investigations done over the years, dedicated to study asphaltene deposition and its underlying mechanism using capillary tube systems. Asphaltene deposition is examined by performing pressure drop measurements across a pressure- and temperature-controlled capillary tube in which fluid is injected at a low, constant flow rate. Pressure drop across the capillary tube is observed as a result of the formation of a deposit layer on the tube walls reducing the cross-sectional area of the tube, and thereby, restricting the flow of the fluid. Asphaltene deposition is induced in the capillary tube by the injection of an n-alkane as the precipitant along with the oil sample by means of two different pumps. The average hydrodynamic thickness of the deposited layer under laminar conditions can be inferred from the increase of pressure drop with time by using the Hagen-Poiseuille relationship, which is valid for laminar flow regimes and where the deposition layer thickness is small, compared to the radius of the capillary tube (Wang et al. 2004). The surface of the capillary plays an important role in building up of the first adsorbing layer (surface–asphaltene interactions). For deposition to occur, the surface has to be adsorbent toward asphaltenes. Multilayer adsorption involves interactions between asphaltenes molecules. The capillary tube deposition test has been one of the most reliable quantitative experimental techniques to study asphaltene stability and deposition mechanisms in wellbores and pipelines.

Broseta et al. (2000) presented the capillary flow method for the first time to determine asphaltene deposition. It was found that the capillary deposition method allows the detection of deposits in fluids with a low asphaltene content (~0.04 wt %). Deposition rates as low as 0.1 μm/h could be measured with a capillary of internal diameter of 500 μm and length of 15 m.

Wang et al. (2004) performed capillary deposition experiments in stainless-steel capillary tubes. The tests were used to study the influences of factors including temperature, degree of asphaltene instability, and precipitant molar volume on asphaltene deposition from mixtures of stock tank oils (STOs) and n-alkanes. Asphaltene deposition was induced by co-injection of oil and n-alkane precipitants. Pressure drop across the capillary tube was used to estimate the amount and distribution of deposit formation. It was found that the rate of deposition was unaffected by flow rate and capillary tube length over a temperature range of 293 K and 333 K for the studied oil samples. It was also observed that asphaltenes aggregated by addition of higher molar volume n-alkanes deposited more materials than those aggregated by lower molar volume n-alkanes. Hence, the rate of deposition was greater for higher molar volume precipitants than for lower ones for the same crude oil.

Nabzar and Aguiléra (2008) also performed capillary tube deposition experiments and determined that there are critical shear conditions under which asphaltenes would not deposit, highlighting the importance of hydrodynamics. The effect of shear on asphaltene deposited was analyzed to understand the deposition mechanism. Nabzar and Aguiléra (2008) stated that at low shear rates, deposition follows the colloidal deposition scaling of diffusion-limited deposition. It was also found that as the shear rate increases, asphaltenes pass through a shear-limited deposition process, and at high enough shear rates, there is no detectable deposition.

Hoepfner et al. (2013) investigated the deposition of asphaltenes in capillary tubes by the addition of an n-alkane precipitant to crude oil samples. Electron microscopy was adopted to analyze the deposited asphaltenes. Arterial asphaltene deposition was directly observed in the laboratory by Hoepfner et al. (2013). The scanning electron microscope (SEM) images showed that the deposited layer is significantly thicker at the capillary inlet when compared to the outlet, which clearly demonstrated the non-uniformity in the deposition profile and that the deposition was occurring preferentially near the capillary inlet. Micrographs of the capillary effluent showed that asphaltene aggregates that grow to 0.5 μm or greater do not participate in deposition process. This observation revealed that the asphaltene deposition process is most likely dominated by the submicrometer asphaltene aggregates. The larger asphaltene aggregates do not tend to deposit.

5.1.1.1.1 Experimental Setup

A schematic diagram of the capillary deposition test apparatus, used by Wang et al. (2004) to study asphaltene deposition, is shown in Figure 5.1. The deposition of asphaltene occurs in a long capillary tube (~16–30 m) of 0.05 to 0.1 cm inner diameter. Two high-pressure syringe pumps are used to inject oil sample and n-alkane precipitant at constant flow rate and under laminar flow conditions. The oil stream from pump 2 (Figure 5.1) is mixed with the precipitant stream from pump 1 by flowing through a mixing node within an ultrasonic bath to ensure complete mixing.

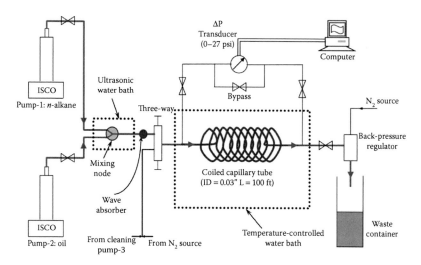

FIGURE 5.1 Capillary deposition experimental setup used to measure the deposition profile of asphaltene. ID, inner diameter; L, length; N_2, nitrogen. (Reprinted from Wang, J. et al., *J. Disper. Sci. Technol.* 25, 287–298, 2004.)

A pneumatic back-pressure regulator is connected to the outlet of the capillary tube to control the down-stream pressure. A pressure transducer is used to continuously measure the pressure drop across the capillary tube and its analog outputs are simultaneously recorded by a computer. The capillary tube is immersed in a water bath to maintain isothermal conditions (temperature variations within ± 0.5 K).

At the end of each experimental run, the remaining liquid inside the capillary (i.e. the mixture of oil and precipitant) is gradually displaced from the tube by pumping in nitrogen gas at constant pressure and weighed by an electronic balance. The recorded weight of effluent as a function of time was then used to calculate in-situ deposition thickness. The deposits are then recovered by pumping toluene and tetrahydrofuran (THF) alternately through the tube until the effluent is completely colorless. Toluene and tetrahydrofuran are removed from the mixture by evaporation. Finally, the deposits are heated to 393 K to ensure that no traces of solvent remained before the final weighing (Wang et al. 2004).

Alternatively, an oil-immiscible, viscous fluid such as glycerin could be injected from one end of the capillary under constant flow rate after the remaining oil-precipitant mixture is flushed out by nitrogen. Pressure buildup at the injection port is continuously recorded with a pressure transducer connected to a computer, while the capillary outlet is open to the atmosphere. Under constant flow rate, the variation of pressure drop is determined by the local effective diameter of the capillary, which is used to retrieve in-situ deposition thickness (Wang et al. 2004; Kurup et al. 2011).

Under constant flow rate, the deposition profile could also be obtained indirectly by monitoring the variation of pressure drop as asphaltenes deposit in the tube (Wang et al. 2004; Kurup et al. 2011).

5.1.1.1.2   Capabilities and Limitations

One of the advantages of capillary deposition experiments is that the scaling of fluid flow in a capillary tube to that of fluid flow in wellbores and transportation pipelines is relatively simple. The laminar boundary layer near the wall surface of the wellbores or pipelines plays a crucial role during the deposition process. This boundary layer is similar to the flow of fluid in a capillary tube and hence, the capillary deposition test may provide valuable insight into the asphaltene deposition in production tubings (Kurup et al. 2011).

Asphaltene precipitation and deposition in capillary tubes are initiated by the addition of a precipitant, which does not mimic the realistic conditions in the oil field. During production from an oil reservoir, asphaltenes start precipitating out of the oil and subsequently deposit on the pipe wall surfaces as a result of changes in pressure. Using a depressurization technique to induce asphaltene precipitation and deposition could help create a scenario similar to that of a field case. Also, a review of the studies on measuring asphaltene deposition reveals that live oil samples have not been tested using the capillary deposition method.

Asphaltene deposition rates are considerably low while performing asphaltene deposition tests on oil samples with low asphaltene content. Hence, to obtain significant amounts of deposits, a large volume of sample is required. In such a case, employing capillary deposition tests becomes unfeasible because it consumes as much as 1 liter of oil per experiment or even more. Thus, this has created the necessity for developing new laboratory experiments to study asphaltene deposition that would consume a less amount of the expensive oil sample, comparatively.

Also, the driving force for precipitation used in capillary deposition tests is small (closer to the onset of asphaltene precipitation) to prevent sudden and complete plugging of the capillary tubes. This precipitates mostly the most insoluble asphaltenes, which might not be a representative of the entire asphaltene distribution (Sections 2.2.2.2 and 4.3.2).

Asphaltene deposition in capillary tubes is measured indirectly from pressure drop measurements or displacement tests, in which a constant pressure nitrogen source is used to displace the remaining liquid inside the capillary through the outlet (Wang et al. 2004). Recently, SEM imaging was used to study asphaltene deposition in a capillary tube (Hoepfner et al. 2013). Each of these techniques has its own advantages and disadvantages. The pressure drop measurement may not be an accurate method for quantifying the mass of asphaltene deposits and determination of the deposition profile. The surface characteristic properties of the deposited asphaltenes might be lost when the displacement test is performed. Although the SEM images at the inlet and outlet of the capillary tube helped in confirming the non-uniform deposition profile along the length of the capillary tube, the in-situ deposition profile cannot be calibrated. These concerns create the need for investigating asphaltene deposition in other experimental techniques based on other geometries as well.

5.1.1.2 Quartz Crystal Microbalance

In many stages during oil production, various phenomena occurring at the solid–liquid interface can cause severe issues, such as fouling, coking, and wettability alteration (Abudu and Goual 2009). One of these phenomena is asphaltene deposition and fouling in oil reservoirs and production tubings. As previously discussed, the exact mechanism of asphaltene deposition and the fundamentals of asphaltene–surface interactions are not well understood. One of the experimental devices that has been previously used to study asphaltene–surface interaction is the quartz crystal microbalance (QCM) with dissipation.

5.1.1.2.1 Experimental Setup

The dissipative QCM is a device that measures the adsorption of materials to a surface, and it can be used for a variety of applications that range from the characterization of biointerfaces—proteins, cells, surfactants—to the crude oil chemistry (Dudášová et al. 2008). The QCM uses a piezoelectric material, in this case quartz, as an ultrasensitive weighing instrument. The piezoelectric quartz crystal is sandwiched between two electrodes as shown in Figure 5.2.

The QCM can simultaneously measure the changes in the resonance frequency, f, the dissipation, D, the viscoelastic and frictional losses in the system (Ekholm et al. 2002). The frequency and dissipation are measured at different points of operation, where typically f is measured before the driving oscillator is disconnected, and D is determined after disconnecting the driving field and measuring the dampening of the oscillation as the vibrational amplitude decays (Ekholm et al. 2002).

Pierre and Jacques Curie first discovered the piezoelectric effect in 1880 (Manbachi and Cobbold 2011). Mechanical stress on a piezoelectric material will cause an electric polarization. The reverse process is also true; that is, applying an electrical field will cause a mechanical deformation in the material. The QCM uses this effect and causes the quartz crystal to oscillate by applying an alternating-current voltage to the electrodes. The resonance frequency is obtained when the thickness of the plate is an odd integer, n, of half-wavelengths of the induced wave (Dudášová et al. 2008). The resonance frequency is dependent on the total oscillating mass, including the crystal, the solvent, and any adsorbed material. When a film attaches to the sensor crystal, the frequency decreases. This decrease is proportional to the mass of the attached film if the film is thin and rigid (Dudášová et al. 2008).

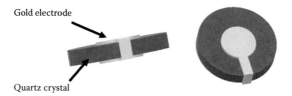

Gold electrode

Quartz crystal

FIGURE 5.2 Schematic of dissipative quartz crystal microbalance. Gold electrodes sandwich the piezoelectric material (quartz). (Reprinted from Ekholm, P. et al., *J. Colloid Interface Sci.*, 247, 342–350, 2002.)

The linear relationship between the change in mass and the change in frequency was demonstrated by Sauerbrey (1959). The change in mass is given by:

$$\Delta m = -\frac{\rho_q t_q \Delta f}{f_0 n} = -\frac{\rho_q v_q \Delta f}{2 f_0^2 n} = -\frac{C \Delta f}{n} \tag{5.1}$$

where:
   ρ_q and v_q are the specific density and shear wave velocity in the quartz
   t_q is the thickness of the quartz crystal
   f_0 is the fundamental resonance frequency ($n = 1$)
   C is a constant for the crystal

The Sauerbrey equation is valid under the following conditions (Dudášová et al. 2008):

a. The mass adsorbed is distributed evenly over the crystal
b. Δm is much smaller than the crystal
c. The mass adsorbed is rigidly attached

In the case that the last condition is not fulfilled, the dissipation factor is proportional to the power dissipation in the oscillatory system and will give information about the rigidity of the film. Dissipation is given by:

$$D = \frac{E_{\text{dissipated}}}{2 \pi E_{\text{stored}}} \tag{5.2}$$

where:
   $E_{\text{dissipated}}$ is the energy dissipated during a single oscillation
   E_{stored} is the energy stored in the oscillating system
   D is the dissipation resulting from the changes in contributions from frictional and viscous losses (Dudášová et al. 2008)

In a liquid environment, the losses are primarily from friction, and the dissipation factor has been previously derived by Stockbridge (Stockbridge and Warner 1962):

$$\Delta D = \frac{1}{\rho_q t_q} \sqrt{\frac{\rho_f \mu}{2 \pi f}} \tag{5.3}$$

where μ and ρ_f are the viscosity and density of the fluid. If the adsorbed film is viscous, then energy is dissipated because of the oscillatory motion induced in the film (Dudášová et al. 2008). Furthermore, frictional energy is also dissipated when the adsorbed film slips on the electrode. Thus, a rigidly attached layer will not change dissipation, and the Sauerbrey equation applies. Otherwise, a loose layer will cause dissipation to increase. The theory behind QCM is more thoroughly discussed by Rodahl (1995).

One of the first studies for the investigation of asphaltene adsorption using the QCM device was performed by Ekholm et al. (2002). Their study was primarily concerned with the adsorption of asphaltenes and resins onto hydrophilic surfaces. The asphaltenes and resins were first extracted from the crude oil by fractionation with various solvents (as discussed in Section 2.1.2). Then, various solutions with concentrations ranging from 25 to 1,000 ppm of asphaltenes or resins in n-heptane and toluene were prepared. The solutions were then injected into the measurement cell via syringe. Then, the frequency and dissipation factor shifts were measured (Ekholm et al. 2002).

By using the QCM experiments, Ekholm et al. (2002) were able to determine that 2.3 mg/m^2 of resins were quickly adsorbed onto the gold surface after a few minutes in a solution of n-heptane containing 5,000 ppm of redissolved resins. Moreover, the negligible change in the dissipation factor demonstrated that the adsorbed resins formed a rigidly attached film consisting of individual molecules on the surface, rather than aggregates. At higher concentrations, there was an increase in dissipation that indicated the formation of another layer, but this layer was easily desorbed whereas the binding energy of the first layer was strong. The amount of resins adsorbed on the surface decreases as more toluene was added (Ekholm et al. 2002). At 50:50 n-heptane to toluene ratio, only 0.6 mg/m^2 was adsorbed on the surface, and hardly any adsorption occurs in pure toluene. These results suggest that solubility of the resin is a key factor in determining the extent of adsorption (Ekholm et al. 2002).

With regard to asphaltenes, adsorption occurred to a larger extent, ranging from 4 to 5 mg/m^2 (Abudu and Goual 2009). The adsorbed amount was larger than non-associating polymers (1–3 mg/m^2), indicating the adsorption of asphaltene aggregates. Furthermore, a subsequent injection of resins after adsorption of asphaltenes showed no change in the mass, regardless of concentration. This demonstrates that after the asphaltenes have been adsorbed, there is insignificant interaction energy between the resins and asphaltenes, and thus, desorption of asphaltenes is unlikely. However, when the asphaltenes and resins fractions were mixed before adsorption, the final adsorbed amount was greater than the individual pure fractions (approximately 6.8 mg/m^2). Ekholm et al. (2002) concluded that it was likely that mixed aggregates were formed in the bulk solution and were then adsorbed to the surface.

In a study by Dudášová et al. (2008), QCM was used to monitor the adsorption or deposition of asphaltenes on to surfaces with different types of coatings, such as silica, alumina, and titanium. The authors expected that the hydrophilicity of the crystal surface would have a significant effect on the amount of adsorbed asphaltenes. However, no clear trend was determined based on the hydrophilicity. On the other hand, when the adsorbed amounts of asphaltenes were plotted against the asphaltene size, a rough correlation was found. This indicates that the particle size can also be a factor influencing the adsorption of asphaltenes.

Building on studies similar to the previous ones by Ekholm et al. (2002) and Dudášová et al. (2008), Abudu and Goual (2009) used a liquid loading correction factor when using the QCM device. Three main contributors to the frequency and dissipation are considered in their study:

a. Mass loading
b. Liquid loading
c. Liquid trapping

Abudu and Goual (2009) then used the QCM to study the adsorption kinetics of asphaltenes. They concluded that the primary adsorbed species had diameters of 0.5–1.6 nm at low concentrations (129–278 ppm of asphaltenes). At a higher concentration of 835 ppm, the primary adsorbed species were nano-aggregates with diameters of 2.6–5.6 nm. Finally, the authors showed that the aging process seemed to have opposite effects in toluene and n-alkanes because the film became more rigid in toluene and more viscoelastic in n-alkanes.

Tavakkoli et al. (2014a, 2014b) also performed a series of experiments using model oil and crude oil systems. In the model oil system, the temperature, asphaltene polydispersity, solvent, depositing surface, and flow rate were varied to understand the depositional tendency of asphaltenes using the QCM (Tavakkoli et al. 2014b). A couple of the concluding remarks from the study include: (1) Viscosity of the adsorbed layer decreases with an increase in temperature. (2) Polydispersity is an integral determinant in the deposition onto a gold surface. C_{7+} asphaltene mass reaches equilibrium more quickly than C_{5+} asphaltenes. (3) A rusted steel surface has a high absorption of asphaltenes during the initial time but eventually shows a decrease in the adsorbed asphaltenes. (4) Asphaltene adsorption curves reach equilibrium sooner at higher flow rates. At very long times and high flow rates, the rate of adsorption is independent of flow rate.

In the crude oil system, the primary purpose was to investigate crude oil asphaltene–surface interactions on various surfaces using the QCM device. The following conclusions are drawn from the study performed by Tavakkoli et al. (2014a): (1) Modeling of the experimental data shows that adsorption kinetics controls the adsorption process initially. After a period of time, the process is then governed by diffusion and convective transfer. (2) After asphaltene precipitation, the transport of asphaltenes over the crystal in the flow module likely follows a multistep process, including precipitation, aggregation, diffusion, advection, and deposition; a model is proposed to predict the amount of deposited mass beyond the onset of asphaltene precipitation in a QCM experiment with preliminary success in model and crude oil systems. (3) In the presence of rust (iron oxide), the deposited mass decreases on a carbon steel surface. However, compared to a rusted surface, carbon steel and gold surfaces both have more deposited mass.

5.1.1.2.2 Capabilities and Limitations

There are a few other techniques for studying the adsorption process, such as UV-depletion techniques and ellipsometry. When compared to UV-depletion techniques, the QCM is superior because (1) It is sensitive to the effect of different solvents; (2) It can give insight on adsorption kinetics and direct surface studies; (3) Information is displayed in real time; and (4) Information about the surface can also be elucidated (Dudášová et al. 2008). However, QCM fails in accounting for porosity, layers, and shape effects that typically occur in particles. Moreover, flow rates in QCM devices are generally small and will not result in enough shears to accurately represent deposition in the turbulent conditions (Dudášová et al. 2008). In general, the quantitative measurements of the actual amount adsorbed are fairly robust in both QCM and UV-depletion techniques (Dudášová et al. 2008). Compared to the results obtained from the ellipsometry, the QCM data are less representative for the extended layers of adsorption (Ekholm et al. 2002). However, ellipsometry cannot be performed on non-transparent systems, which is not an issue for QCM devices (Ekholm et al. 2002).

5.1.1.3 RealView Deposition Cell

Zougari et al. (2005) introduced the method of measuring asphaltene deposition using a novel laboratory-scale flow assurance tool called the organic solid deposition and control (OSDC) device. It is a high-pressure deposition cell for generating organic solid deposits under a wide range of operating conditions. This equipment in its batch mode was used to measure the deposition rate of waxes and asphaltenes from live fluids under turbulent flow conditions (Zougari et al. 2005, 2006). This technique uses a closed cell based on the Taylor-Couette flow principles and takes into account important parameters including shear, pressure, temperature, and flow time. It was designed to simulate the hydrodynamic and thermal characteristics encountered in typical oil production lines. The rapid rotational movement of a spindle at the center of the device produces a flow regime similar to that of a flow field within a production tubing and deposits waxes or asphaltene on the cylinder surface under changes in pressure, temperature, and composition. The OSDC has two concentric cylinders where the inner cylinder, which rotates at a very high speed, provides the necessary driving force for the flow of fluid in the apparatus, similar to that of a pump in the flow of fluid in a pipe and the outer cylinder, which is stationary, acts as the pipe wall surface.

However, the experimental measurements were inconclusive for asphaltene deposition from crudes with low asphaltene content (Akbarzadeh et al. 2009). For these fluids, the amount of deposit obtained from a batch experiment, with 150 cm^3 fluid in the cell, was often very small (less than 15 mg). This results in a large relative uncertainty in the measured data, and therefore interpretation is difficult. The other problem in batch deposition is depletion. A decrease in the amount of depositing asphaltenes in the cell over a short period of time (typically 2 hours) will yield averaged deposition rates that are not representative of those in the field. To overcome these limitations, Akbarzadeh et al. (2009) modified the high-pressure batch deposition cell to a flow-through system, called the RealView deposition cell. In a flow-through test, approximately 900 cm^3 of fluid is passed through the deposition cell. More fluid is used compared to a batch run and also as there is a constant influx of a fresh fluid. Hence, no or minimal depletion occurs in the flow-through system, more deposit is generated, and therefore, it provides more representative data for interpretation. RealView cell can be operated in both batch and flow-through modes (Akbarzadeh et al. 2012). A similarity between the flow of fluid in the RealView cell and pipeline is achieved by setting the speed of rotation of the inner cylinder spindle in such a way that the wall shear stress in the RealView cell is equivalent to that experienced in the pipeline (Eskin et al. 2011).

Zougari et al. (2006) devised OSDC, a novel laboratory-scale flow assurance tool to assess the potential and severity of organic solids deposition problems from hydrocarbon fluids at realistic production and transportation conditions. Tests were performed on live oil samples, at high-pressure, high-temperature (HPHT) conditions for testing wax and asphaltene deposition. Wax deposition rates from

OSDC were comparable to the results from pilot-scale flow loops by using a sample volume, which is at least an order of magnitude lower. Reproducibility of the experiments was also established under consistent test conditions.

Akbarzadeh et al. (2009) used a flow-through high-pressure RealView cell to measure the deposition rate of asphaltenes from an oil field in the Gulf of Mexico at reservoir pressure and temperature conditions. The effects of shear, residence time of the fluid in the cell, pressure, and chemical injection on the asphaltene deposition rate were investigated. While analyzing the effect of shear, corresponding to the production rate in the oil field, on the deposition rate, a strong shear dependency was found such that more deposit was formed at lower shear, that is, lower production rates in the real oil field and vice-versa. Their experiments helped in providing a link between laboratory-scale testing and field production.

Eskin et al. (2011) developed a detailed analysis for the applicability of RealView to study and measure asphaltene deposition in wellbore and pipelines. Based on experimental results from both batch and flow-through deposition tests in a RealView cell, predictions for asphaltene deposition in production pipelines were made that included both asphaltene particle size distribution evolution and asphaltene transport to the wall. Asphaltene particle size was introduced as an important variable in the asphaltene deposition mechanism, as it was proposed by Vargas et al. (2010). An interplay between shear and adhesion forces on depositing asphaltenes determines whether asphaltene adheres to the surface or not. Accordingly, critical particle size was included as a parameter, and asphaltene particles or aggregates larger than this size were considered not to deposit and instead were carried on along with the fluid flow. These modeling efforts allow a valuable interpretation of the experimental results.

The ability of RealView cell to assess the impacts of shear, run time, residence time, pressure, chemical inhibitor, and surface roughness on the deposition of asphaltenes from five fluids from the Gulf of Mexico was thoroughly investigated by Akbarzadeh et al. (2012). In a batch deposition test, with progress in time, asphaltene aggregates, formed in the fluid as a result of pressure reduction, grew to their critical sizes and did not deposit thereafter. Although the amount of deposit increased over time, the growth rate of the deposits was slowed down for longer run periods. Unlike batch deposition tests, in flow-through deposition tests, the amount of deposit grew continuously with respect to time as fresh oil was being passed through the deposition cell. Surface roughness also impacted the rate of asphaltene deposition in the RealView cell. More deposit was generated using a rough surface compared to a smooth surface. Using the results from the laboratory-scale experiments, the deposition profile was qualitatively predicted for realistic oil production wells and compared with field observations reported by Haskett and Tartera (1965).

5.1.1.3.1 Experimental Setup

A typical RealView deposition cell is shown in Figure 5.3.

To initiate a batch deposition test, the cell is initially filled with the sample of stock tank oil (STO). The STO is then displaced with live fluid, initially placed in a 1-L storage bottle, at reservoir temperature and at a pressure above its asphaltene

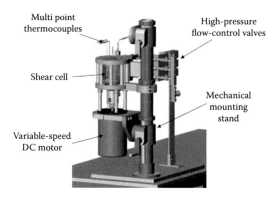

FIGURE 5.3 RealView Deposition cell. (Reprinted from Akbarzadeh, K. et al., *Energy Fuels*, 26, 495–510, 2012.)

onset pressure. The displacement continues until all STO (~150 cm³) and some of the live oil that has made contact with it are removed. The initial high pressure (above the asphaltene onset pressure) in the device is then reduced either abruptly or gradually to the desired test pressure, which is typically 0.7 MPa above the bubble point of fluid where the maximum amount of precipitated asphaltenes exists in the fluid (Akbarzadeh et al. 2012). Then, the spindle is switched on, and its rotational speed is set in such a way that the desired wall shear stress is achieved. Then, the batch test is run for a set time.

When the run is completed, the oil inside the device is drained at the test pressure, and the generated deposit that is formed on the cell wall is recovered using dichloromethane. The resulting solution is then introduced into a rotary evaporator where the solvent is evaporated, and the deposit is collected and weighed. The mass of the residue represents the total amount of deposit (deposited asphaltenes + occluded oil). The asphaltene content of the residue is then measured by using a modified IP-143 with *n*-heptane to determine the separate amounts of the deposited asphaltenes and the occluded oil (Eskin et al. 2011). Finally, the mass of deposited asphaltenes is determined by subtracting the calculated amount of asphaltenes related to the occluded oil from the asphaltene content of the residue.

Similar to the batch system, the first step in a flow-through test is to fill the cell with a live oil sample. The oil sample is then set at a specific flow rate into the top of the cell, forcing the live fluid to exit the cell through the outlet located at the bottom cap. The pressure, temperature, rotational speed, and flow rate are controlled throughout the test. A back-pressure regulator and two pumps are used to control the flow in and out of the cell, thereby regulating the pressure. After completion of each test, the remaining fluid in the cell is drained out by helium injection while the spindle is still rotating. This minimizes the settlement of asphaltenes on the bottom cap (Akbarzadeh et al. 2012). The deposited materials are then recovered with dichloromethane and analyzed as explained previously.

5.1.1.3.2 Capabilities and Limitations

RealView allows the testing of fluids at pressures up to 105 MPa. It can also function in a wide temperature range, from 269 K to 423 K. It can be operated such that the contained fluid is tested under either laminar or turbulent conditions. It allows the flexibility to test various surface materials and roughness, including actual pipeline materials and new surface materials and coatings. It allows controlled shear at the wall to enable scalability of deposition results (Zougari et al. 2006).

A major advantage of using this setup is that live oil samples could be tested under different flow regimes, run time, and shear. Studies using this technique revealed that, under flow or production conditions, asphaltene precipitation does not necessarily lead to asphaltene deposition, whereas when asphaltene precipitation is induced by solvent (at STO conditions), it usually leads to asphaltene deposition regardless of the type or content of asphaltenes. Furthermore, the deposition of asphaltene from solvent-induced precipitation led to much higher deposit in all studied cases unlike some live oil cases.

In a flow loop experiment, predominantly used to measure wax deposition, a huge volume of oil sample is required and to create a turbulent flow regime, higher pressure drop and higher flow rate need to be created, and hence, it requires a high-capacity pump. Asphaltene aggregates passing through such a pump will be fragmented, which will significantly affect the particle size distribution, and in turn, impact deposition. To avoid those problems, the RealView cell can be employed, and also, it requires relatively small quantities of expensive live oil samples.

Although, the flow-through setup eliminates the limitation of the batch system for generating enough deposit from low asphaltene content samples, a 4-hour flow-through test with an oil flow rate of 3 cm³/min (50 minute average residence time for fluid in the cell) will require approximately 900 cm³ of live fluid, compared to the approximate 150 cm³ of oil needed for a batch test (Akbarzadeh et al. 2012).

Another limitation of the RealView experiments is that water has to be removed; thus, it is not possible to investigate the effect of brine on asphaltene deposition tendency or its effect on the performance of asphaltene deposition inhibitors.

5.1.1.4 Packed Bed Column

5.1.1.4.1 Experimental Setup

5.1.1.4.1.1 Ambient Conditions Vilas Boas Favero et al. (2016) developed a packed bed apparatus to investigate the mechanism of asphaltene deposition on the metallic surface. Their glass packed bed column has an inner diameter of 10 mm and a height of 132 mm. It is packed with stainless steel spheres to form square horizontal lattices constructed by four spheres of 4 mm (inner diameter [ID]) and one sphere of 3 mm (ID) in between the lattices (Vilas Boas Favero et al. 2016). During a deposition test, oil and precipitant (*n*-heptane) are premixed to keep the size of the destabilized asphaltenes below 500 nm. Then, a peristaltic pump is used to maintain the upward flow of the mixture throughout the packed bed column. After a specified amount of run time, the column is drained at a flow rate of 0.05 g/min. Then, the retained materials are collected by injecting chloroform through the bed. The mass of the retained materials are quantified after evaporating the chloroform. The mass of the deposits is determined by subtracting the mass of occluded oil from the mass of the retained materials (Vilas Boas Favero et al. 2016). With this packed bed deposition apparatus, Vilas Boas Favero et al. (2016) determined that a higher concentration of unstable asphaltenes leads to a higher rate of asphaltene deposition. Additionally, their experimental results show that asphaltene deposition in the viscous flow regime is governed by the diffusion-limited deposition model.

Recently, a multi-section packed bed system to study asphaltene deposition on metallic surfaces was developed (Kuang et al. 2017). This new asphaltene deposition apparatus consists of a polytetrafluoroethylene (PTFE) multi-section column packed with carbon steel spheres to increase the surface area onto which asphaltenes can deposit. Unlike the system used by Vilas Boas Favero et al. (2016), the multi-section packed bed system has the capability of assembling several discrete column sections that can be independently analyzed. Figure 5.4 shows the details of the multi-section design. Every column section is connected using straight connectors (nylon tube fitting, 1.27 cm outer diameter [OD]), ferrules (ethylene tetrafluoroethylene [ETFE], 0.32 cm ID), external O-rings (ETFE, 1.27 cm OD), and black O-rings (nitrile rubber, 1.27 cm OD) to provide sealing protection against leakages. To separate the spheres packed in each section, a molded circular mesh (PTFE, 0.114 × 0.064 cm opening size) is used as a support between straight connectors of each section. At the end of each experiment, the column is drained by gravity and disassembled beginning from the top section to the bottom section; this allowed for easy procurement of spheres for quantification of deposition.

Figure 5.5 shows the schematic plot of the packed bed column to investigate asphaltene deposition under the dynamic conditions. The oil and precipitant are directed through the perfluoroalkoxy (PFA) tubing (0.159 cm OD, 0.051 cm ID) at constant volume flow rates by two syringe pumps (Harvard Apparatus Pump

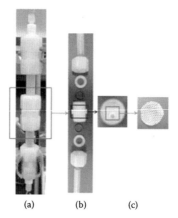

(a)   (b)   (c)

FIGURE 5.4 Column detail: (a) PTFE column, (b) detailed Nylon straight connector, and (c) PTFE mesh location. PTFE, polytetrafluoroethylene. (Reprinted from Kuang, J. et al., Novel way to assess the performance of asphaltene dispersants on the prevention of asphaltene deposition, In Preparation, 2017.)

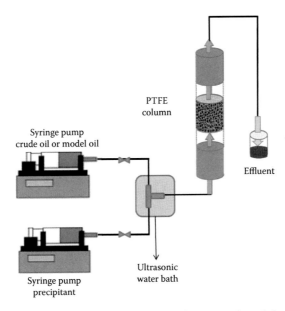

FIGURE 5.5 Schematic of the PTFE packed bed column to study asphaltene deposition under dynamic conditions. PTFE, polytetrafluoroethylene. (Reprinted from Kuang, J. et al., Novel way to assess the performance of asphaltene dispersants on the prevention of asphaltene deposition, In Preparation, 2017.)

11 Elite). The two fluids are combined in an ETFE three-way junction (0.159 cm OD, 0.051 cm ID), which is kept in an ultrasonic water bath operating at 40 kHz and kept at a constant temperature to promote good mixing. The mixture is then injected from the bottom section of the column to the top section, and the effluent is collected thereafter. Every section is packed pre-weighed with carbon steel spheres. The mass of the deposits is determined by the increased mass of the spheres after each run.

5.1.1.4.1.2 High Pressure, High Temperature Conditions To investigate the asphaltene deposition phenomenon under HPHT conditions, a stainless steel packed bed column setup was developed. The 316 stainless steel column has an OD of 1.27 cm and a height of 12.95 cm. It is packed with 975 to 985 carbon steel spheres, which have a diameter of 0.238 cm, and the resulting total surface area of the spheres is 174 cm². The pore volume and permeability of this stainless steel packed column is 5.80 cm³ and 362.36 D.

The configuration of the HPHT deposition test apparatus is shown in Figure 5.6. The operating temperature (293–573 K) can be controlled inside an explosion-proof oven, and the downstream pressure (0–20.68 MPa) can be manipulated by a back-pressure regulator connected to the outlet of the column. First, the column packed with carbon steel spheres is connected to the 0.318 cm ID stainless steel tubing in the oven. Two high-pressure liquid chromatography (HPLC) pumps are used to co-inject crude oil and the

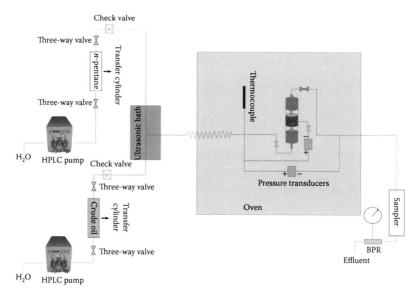

FIGURE 5.6 Schematic of the stainless steel packed bed column to study asphaltene deposition at HPHT conditions. BPR, back-pressure regulator; HPLC, high-pressure liquid chromatography; HPHT, high-pressure, high-temperature.

precipitant at constant flow rates. The pumps can inject fluids at flow rates over the range of 0–10 cm³/min, with a maximum working pressure of 41.37 MPa. Different driving forces (oil to precipitant ratios) for deposition can be achieved by adjusting the flow rates of the crude oil and the precipitant. The fluids are mixed through a T-junction within an ultrasonic bath to facilitate complete mixing. The mixture is then injected into the packed bed column, which is maintained at the desired operating pressure and temperature. Pressure transducers are used to monitor the pressure drop across the entire column and between each section. All pressure transducers are connected to a computer, where all the pressure data are continuously collected. After flowing through the packed bed from the bottom to the top, the effluent is then passed through a sampler and a back-pressure regulator. The co-injection is stopped after reaching the desired number of pore volumes. In case of serious plugging, the co-injection is discontinued automatically when the inlet pressure of the pumps reaches a maximum predefined pressure as a safety precaution.

5.1.1.4.2 Deposit Characterization

As introduced in Section 5.1.1.1, asphaltene deposition in the capillary deposition flow loop is measured indirectly from the pressure drop across the system (Broseta et al. 2000; Wang et al. 2004). The displacement test was then developed to estimate the deposition profile in capillaries by measuring the evolution of pressure after a viscous fluid is injected into the system (Wang and Buckley 2006). Hoepfner et al. (2013) and Chaisoontornyotin et al. (2016) discovered that the axial deposition profile was highly non-uniform based on the electron microscopy images. They also showed that asphaltenes were more likely to deposit instantaneously when the oil was destabilized by the precipitant (Hoepfner et al. 2013). Therefore, the quantification of the thickness and mass of the deposited materials may not be accurately determined based on the pressure drop data. In some cases, high pressure drops resulting from the localized plugging cannot be necessarily attributed to the formation of a large amount of asphaltene deposition over the entire tube.

The new packed bed deposition apparatus enables the direct quantification of both the deposited materials and asphaltenes by the change in mass of the spheres before and after the deposition process (Kuang et al. 2017). First, the mass of the clean spheres packed in the column is pre-weighed. After running the oil and precipitant through the packed bed column for a specified experimental duration, the packed bed is drained by gravity. The mixed liquid remaining inside the column is slowly displaced with nitrogen at a rate of 27.58 kPa. Then, the spheres with deposits are collected and heated to 393 K to ensure that the deposited materials are completely dried. The mass of the deposited materials with any occluded oil is calculated by subtracting the mass of the pre-weighed spheres from the mass of the spheres with dried deposits. To quantify the actual amount of asphaltene deposition on the spheres, the deposited materials with occluded oil are pre-rinsed with a neutral solvent to remove the occluded oil trapped in the deposits. A neutral solvent can neither redissolve the deposited asphaltenes nor precipitate asphaltenes from the occluded oil. The occluded oil in the neutral solvent is removed by filtration, and the deposits on the filter paper are then washed in a Soxhlet extractor (Figure 2.19) to remove any non-asphaltenic materials remaining in the deposits. The deposited asphaltenes on the spheres are collected by a second wash in a Soxhlet extractor using toluene. Finally, the mass of actual asphaltene deposition is quantified after toluene is evaporated at 393 K.

5.1.1.4.3 Capabilities and Limitations

Compared to the conventional setups to investigate asphaltene deposition, the packed bed column deposition apparatus offers the following distinct characteristics: (1) It requires a small amount of sample to investigate the deposition phenomena; (2) The multi-section design allows the determination of the axial deposition profile; (3) The amount of deposition can be accurately quantified by direct mass balance; (4) It enables the capability of characterizing the properties of the deposited asphaltenes using techniques such as SEM, X-ray powder diffraction (XRD), Fourier transform infrared spectroscopy (FTIR), elemental analysis, and so on; (5) Unlike the RealView deposition cell, the packed bed column can be used to investigate the effect of water on asphaltene deposition and on the performance of asphaltene inhibitors; (6) The packed bed design also allows the feasibility of investigating a variety of variables affecting the deposition process, such as material composition, surface area, surface roughness, corrosion, and so on.

On the other hand, the current design of the packed bed column to study asphaltene deposition still needs future improvements to address the following limitations: the setup has not been tested to simulate the deposition process under turbulent flow regimes, and the current deposition is driven by the addition of precipitants, whereas asphaltene deposition in the production tubing is induced by pressure depletion.

TABLE 5.1

Summary of the Major Techniques Used for the Determination of Asphaltene Deposition in the Laboratory

Geometry	Material	Pore Volume	Asphaltene Destabilization Method	Asphaltene Quantification Method	Capabilities	Limitations	References
Capillary	Stainless steel	0.74 cm³[a] 6.3 cm³[b] 0.015 cm³[c]	Precipitant	Pressure drop, mass balance	Visualize the onset of deposition by pressure drop; determines the axial deposition profile.	Requires large amount of sample; unable to study live oil samples; limited to laminar-flow regime	Broseta et al. (2000)[a] Wang et al. (2004)[b] Hoepfner et al. (2013)[c]
Pipe	Stainless steel	71 cm³[d] 129 cm³[e]	Precipitant	Pressure drop	Feasible to investigate both laminar- and turbulent flow conditions; determines the rate of deposition from the thermal properties	Unable to study live oil samples; requires large amount of sample	Jamialahmadi et al. (2009)[d] Ghahfarokhi et al. (2017)[e]
RealView deposition cell	Stainless steel	150 cm³[f,g]	Depressurization	Mass balance	Feasible to study live oil samples; models similar thermal and hydrodynamic conditions in the production tubing; determines the amount of deposition by direct mass balance	Requires large amount of sample; small amount of deposition in the batch setup; relatively complicated and expensive	Zougari et al. (2006)[f] Akbarzadeh et al. (2009)[g]
Packed bed column	Stainless steel, carbon steel	4.9 cm³[h] 6.3 cm³[i]	Precipitant	Mass balance	Requires very small amount of sample; determines the amount of deposition by direct mass balance; determines the axial deposition profile; obtains and characterizes the deposits; feasible to study the effects of surface properties	Unable to study live oil samples; limited to laminar-flow regime	Vilas Boas Favero et al. (2016)[h] Kuang et al. (2017)[i]

Note: The pore volume of each setup is taken from References a–i.

5.1.1.5   Comparisons of Different Techniques

The current experimental approaches to investigate asphaltene deposition in well-bores and pipelines can be categorized into four main geometries: capillary-circular, pipe-circular, Taylor-Couette cell, and packed bed column. Table 5.1 summarizes the main features for each geometry as well as the comparisons of the capabilities and limitations on the experimental determination of asphaltene deposition based on various references.

5.1.2   Determination of Asphaltene Deposition in Porous Media

5.1.2.1   Core-Flooding Test

Asphaltene precipitation, flocculation, and deposition may occur at several different steps during oil production, such as from production or transportation of crude oils. In porous reservoirs, asphaltene deposition may cause a significant reduction in permeability and change in rock wettability, leading to formation damage and severe production loss (Bagheri et al. 2011). Some processes, such as enhanced oil recovery, use either carbon dioxide or natural gas injection to produce more oil. However, studies indicate that these recovery processes may exacerbate asphaltene deposition issues in reservoirs. Injection of carbon dioxide could lead to asphaltene precipitation and deposition. It would ultimately alter the wettability and plug the reservoir (Zanganeh et al. 2012). These changes could potentially result in an adverse effect on enhanced oil recovery processes. Thus, it is imperative to investigate effects of various parameters on asphaltene deposition in the context of porous media.

Several core-flooding techniques have been previously developed to determine the effects of asphaltene deposition on formation damage and permeability reduction in porous media. Ali and Islam (1998) determined the effect of asphaltene precipitation and adsorption on low permeability carbonate rocks, where they identified three different plugging regimes during asphaltene deposition and propagation in a porous medium: monotonous steady state, quasi–steady state, and continuous plugging. Srivastava et al. (1999) examined the impact of carbon dioxide injection on asphaltene deposition in a core-flooding experiment, demonstrating that the deposition was dependent on the pore topography of the core matrix tested. In a series of core-flooding experiments focusing on particle size, Sim et al. (2005) showed that if the size of the particulates was smaller than 1/7 or 1/10 of the pore-throat diameter, then the particulates would not cause permeability reduction. Kord et al. (2012) designed dynamic and static asphaltene deposition tests performed on core samples using live crude oils to investigate the effects of fractures and fluid velocity on the severity of damage.

Despite the substantial amount of work done to study asphaltene deposition in porous media, the mechanisms of impairment caused by asphaltenes in oil reservoirs are still in active investigations. Moreover, the flow rates employed in the studies were usually low, around 0.3 m/d. However, the main driving force for asphaltene precipitation—pressure depletion—typically occurs in the near-well-bore region. Thus, the formation damage resulting from asphaltene deposition should be dominant near the wellbore region, where the fluid velocity is much greater than 0.3 m/d. Although previous studies have shed some insight, further research and modification to core-flooding experiments are necessary to establish a fundamental understanding of the effect of asphaltene deposition in porous media.

5.1.2.1.1 Experimental Setup

In the core-flooding experiment, a mixture of crude oil and asphaltene precipitant, that is, *n*-pentane or *n*-heptane, is injected into the core plug. Asphaltenes precipitate and then deposit in porous media. An increase in pressure drop value, which is continously monitored in different segments of the core plug, is an indication of the formation damage caused by asphaltene deposition.

The different parts of the core-flooding setup depicted by Figure 5.7 are explained in greater detail:

1. *High-Pressure Liquid Chromatography Pumps:* These pumps are used to inject water into the transfer vessels to move the pistons inside the vessels. The maximum working pressure of the pumps used by Tavakkoli et al. (2017) is 41.37 MPa. The pumps are connected to a computer system, and all flow rate and pressure data are recorded by the computer.
2. *Transfer Cylinders:* There is a piston inside the transfer cylinder that is compatible with the charged liquid (crude oil sample or *n*-pentane/*n*-heptane). With this piston, contact between the charged liquid and the working fluid inside the pump (water) is avoided. In this setup, the charged liquids are the

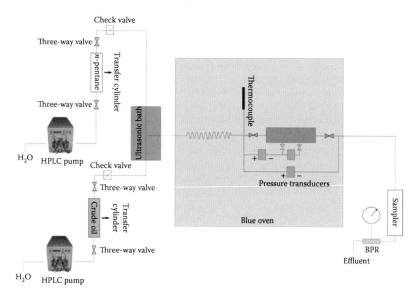

FIGURE 5.7 Schematic of the core-flooding apparatus designed. BPR, back-pressure regulator; HPLC, high-pressure liquid chromatography. (Reprinted from Tavakkoli, M. et al., Probing asphaltene deposition tendency in porous media using core flooding experiments, In Preparation, 2017.)

crude oil sample in one transfer cylinder and the asphaltene precipitant, that is, *n*-pentane or *n*-heptane, in the other cylinder.

3. *Check Valves*: They prevent a backflow going inside the transfer cylinder.
4. *Ultrasonic Bath*: The crude oil sample and the precipitant are mixed in a T-junction, which is kept in an ultrasonic bath. An ultrasonic bath is used to have a better mixing of oil and precipitant.
5. *Blue M. Oven*: An explosion-proof Blue M. Oven is used to reach the temperature of 353 K for the core-flood experiments.
6. *Thermocouple*: The thermocouple probe is in contact with the mixture of oil and precipitant inside the oven and before the inlet of the core plug to accurately monitor the temperature during the experiment.
7. *Core Holder*: Core holder is the heart of the core-flooding setup. It contains the core plug, which represents the porous media, and can withstand high pressures and temperatures. Figure 5.8 shows a core holder with three distinct sections. Core holder preparation will be discussed in detail in the next subsection titled Core Holder Preparation.
8. *Pressure Transducers*: Three different segments of the core holder are shown in Figure 5.8. Pressure transducers are used to monitor the pressure drop in these different segments: between the inlet and the first internal pressure tap (first segment pressure drop), between the two internal pressure taps (second segment pressure drop), and between the inlet and the outlet of the core (overall pressure drop). The pressure drop in the third segment of the core (between the second internal pressure tap and the outlet) can be calculated using the previously measured pressure drop values. All pressure transducers are connected to a computer, where all the pressure data is collected.
9. *Sampler*: The sampler is first filled with water. During the experiment, the mixture of oil and precipitant is going into the sampler through the top port. Water then comes out from the port at the bottom of the sampler and is produced through the back-pressure regulator.
10. *Back-Pressure Regulator*: The asphaltene precipitants used in this study are *n*-pentane and *n*-heptane. At 353 K, the minimum pressure for keeping *n*-pentane in the liquid phase and miscible with the crude oil used is around 0.4 MPa. To be on the safe side, the mixture of oil and precipitant are kept at the minimum pressure of 0.7 MPa using the back-pressure regulator.

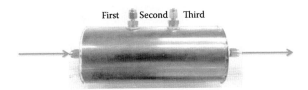

First   Second   Third

FIGURE 5.8   Three different segments of the core holder. (Reprinted from Tavakkoli, M. et al., Probing asphaltene deposition tendency in porous media using core flooding experiments, In Preparation, 2017.)

5.1.2.1.1.1 Core Holder Preparation A new technology has been developed for the preparation of the core holder because of the limitations of conventional metallic core holders (Tavakkoli et al. 2017). Metallic core holders are typically expensive to make, especially holders that have internal pressure taps. These internal pressure taps are required to study the effects of asphaltene deposition in porous media because monitoring pressure changes in distinct sections of the core plug is used to indirectly determine where asphaltene deposition occurs in the core. Another limitation of the conventional metallic holders is that because of the high cost, the number of experiments that can be performed for a given time is limited. Moreover, the metallic core holder can only be used for a single size and shape of the core plug and cannot be easily modified for other geometries and sizes. Additionally, a pump is required to apply an overburden pressure for the conventional holder. These disadvantages are all addressed with the design proposed by Tavakkoli et al. (2017).

The new core holder is casted using a special epoxy resin. This special resin is able to withstand HPHT conditions. Each core holder costs about two orders of magnitude less expensive than the corresponding metallic core holders. Furthermore, the new core holder can be used without overburden pressure and can be customized to different sizes and geometries. The explicit details of the design are presented by Tavakkoli et al. (2017).

5.1.2.1.1.2 Measuring Basic Properties of Porous Media The following steps describe the procedure to measure the relevant porous media properties such as pore volume (PV), porosity (ϕ), permeability (K), and water and oil saturations (S_{wi} and S_0).

> *Step 1*: The core is evacuated by connecting one end to the vacuum pump and leaving the other end closed. There is a perfect vacuum in the system without any leakage.
>
> *Step 2*: The inlet of the vacuumed core is then connected to the graduate burette filled with brine (e.g., 35,000 ppm NaCl in deionized [DI] water). The brine solution in the burette has already been vacuumed to remove any air bubbles in the solution. Then, the inlet valve is opened and the core is filled with the brine solution by atmospheric pressure. After filling the core with brine, it is evacuated again to remove any air that may have been trapped during the filling process. The pore volume and porosity values can then be calculated based on the volume of the brine left in the burette. All dead volumes should be taken into consideration during the calculations of the core pore volume. Figure 5.9 shows the schematic of the setup used for measuring the pore volume and porosity of the core plug.
>
> *Step 3*: The brine solution is then injected into the core and the permeability of the system is calculated by measuring the pressure drop value against the flow rate using Darcy's law (Hubbert 1956).
>
> *Step 4*: The crude oil sample is injected into the core at a very low flow rate (0.3 m/d) to displace the brine in the system. The initial water saturation (S_{wi}), the original oil in place (OOIP), and the oil saturation (S_0) are obtained in this step.

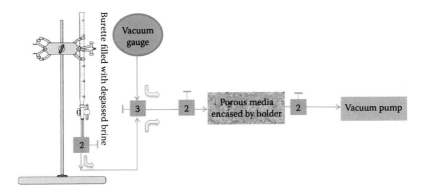

FIGURE 5.9 Diagram of the experimental setup used for determining pore volume and porosity of porous media. (Reprinted from Tavakkoli, M. et al., Probing asphaltene deposition tendency in porous media using core flooding experiments, In Preparation, 2017.)

> *Step 5*: The crude oil sample is then injected into the core, and the pressure-drop value versus the flow rate is measured. The oil permeability in the presence of brine is then calculated using Darcy's law (Hubbert 1956).

5.1.2.1.1.3 Aging of the Core Plugs   After measuring the basic properties of the porous medium and filling the cores with the oil sample, the core plugs can then be aged at an elevated temperature for a specific aging time to restore the wettability. Because with the new technology multiple core holders can be fabricated thanks to its low cost, the amount of time required to age the core plugs is significantly reduced in comparison to a single metallic core holder (2 months for aging four core plugs compared with 2 weeks for aging four core plugs at the same time).

5.1.2.1.1.4 Preprocessing of Crude Oil and Determination of Basic Properties   Before starting the core-flooding experiment, the crude oil is first centrifuged for 1 hour at 10,000 rpm (revolutions per minute), which corresponds to 10,947 relative centrifuge force (RCF) in the Sorvall ST16 centrifuge. This is done to remove any sediment, sand particles, or water present in the system. The water content is then measured using a Metrohm Karl Fischer titration apparatus model 870 KF Tirtrino plus. Furthermore, the molecular weight is measured using a Cryoscope Cryette WR™, which determines the molecular weight based on the change in freezing point of a solvent, such as xylene, as the crude oil sample is added. Finally, the saturates, aromatics, resins, and asphaltenes (SARA) fractionation of the sample, using the standard ASTM method D-2007–98, is used to determine the amount of each fraction (ASTM 1998).

5.1.2.1.1.5 Core-Flooding Experiment Operation   The experiment proceeds as follows. The crude oil and precipitant are injected into a T-junction using a pump and thoroughly mixed. The mixture is then sent through a filter and injected into the core holder containing the porous media. After leaving the core plug, it finally passes through the sampler and is collected as an effluent.

It is important to mention that the mixture of oil and precipitant should be very close to the onset of precipitation to avoid forming a filter cake at the inlet face of the core plug. Asphaltenes should be able to go inside of the core plug and form deposit layers across the core. To find the proper ratio of the oil and precipitant, a stainless-steel sintered filter with a 0.5 μm mesh size can be installed before the core holder and close to the inlet as shown in Figure 5.10a. The injection of the precipitant starts at a low volume percent, and it is slowly increased until a pressure drop in the filter is detected. This increase in the filter pressure drop shows that asphaltene particles have precipitated out and have been trapped by the filter. Therefore, this volume percentage of the precipitant corresponds to a value slightly higher than the onset of precipitation.

It should be noted that the pressure drop in the filter is monitored only by the pressure transducer that measures the pressure drop in the first segment of the core plug. The pressure drop in the filter is not considered in the overall pressure drop.

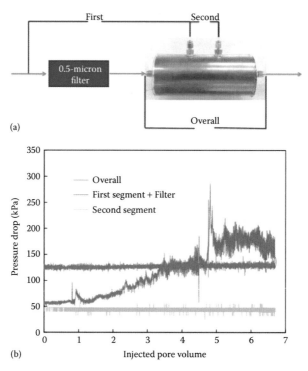

(a)

(b)

FIGURE 5.10 (a) Schematic of design used to detect asphaltene precipitation with a filter. (b) Pressure drop versus injected pore volume for injection with *n*-pentane at 353 K and 0.689 MPa. (Reprinted from Tavakkoli, M. et al., Probing asphaltene deposition tendency in porous media using core flooding experiments, In Preparation, 2017.)

With this design, if asphaltenes deposit in the filter, but not in the core plug, the overall pressure drop does not change significantly. However, the pressure transducer that monitors the pressure drop in the first segment of the core plug shows an increase in the pressure drop values.

Figure 5.10b shows the pressure drop value versus injected pore volume for injection of a crude oil mixed with n-pentane at 353 K and 0.7 MPa, with the filter installed before the inlet of the core holder. Injection velocity is 1.8 m/d and the amount of n-pentane in the mixture is 40 vol%. Based on the results presented in Figure 5.10b, there is a significant pressure drop recorded by the pressure transducer that monitors the pressure drop in the first segment of the core plug as well as the pressure drop in the filter. This increase in the pressure drop value is as a result of the asphaltenes entrapment in the filter, but not asphaltene deposition in the first segment of the core. Asphaltene deposition in all three segments of the core plug is not significant, and therefore, the overall pressure is almost constant during the experiment.

5.1.2.1.2 Capabilities and Limitations

Overall, the new methodology described here to prepare the core holder provides a cheaper, faster, and more modular design in comparison to the conventional metallic core holder without detracting from experimental capabilities. Because of the strong bond the epoxy forms to the core plug, it can withstand the HPHT conditions. Moreover, multiple holders can be prepared and used simultaneously, reducing the duration of the core aging and execution of multiple experiments. Although there is great potential for this new method for preparing core holders, there are some limitations. Currently, the operating pressures for this core holder are limited to below 3.45 MPa. Further modification of the design is currently being investigated to raise the possible operating pressures to around 20.68–27.58 MPa.

5.1.2.2 Microfluidic Device

The so-called *lab-on-a-chip* microfluidics offer unique benefits that facilitate the experimental determination of asphaltene deposition in porous media. Because of their high surface-to-volume ratios and their microscopic length scales, microfluidic systems allow for the controlled study of deposition while minimizing heat and mass transfer resistances. Studying asphaltene deposition in microfluidics offers the potential to analyze the kinetic factors that govern deposition mechanisms while limiting transport shortcomings common in macroscopic systems (Jensen 2001).

5.1.2.2.1 Experimental Setup

Numerous designs of microfluidic devices have been used to study asphaltene deposition in micropores. This section examines these experimental setups and how they are used to observe deposition at the microscale.

Hu et al. (2014) used a transparent packed bed microreactor (μPBR) filled with quartz particles to investigate asphaltene deposition in porous media. The microreactor is formed from 1-mm silicon wafers with 1.1-mm Pyrex wafer caps, with final dimensions of $5.0 \times 1.8 \times 0.21$ cm. The microreactor has a microchannel with

300 μm in depth and 9 mm in width. At the outlet of the microchannel, 30 rows of cylindrical pillars of 20 μm in diameter are etched. Finally, the μPBR is completed after injecting the reactor with an ethanol solution with quartz crystals of a mean diameter of 29 μm (Hu et al. 2014).

The experimental setup requires the use of two high-pressure pumps to inject two solutions into the μPBR (Hu et al. 2014). One pump is filled with model oil containing asphaltenes dissolved in toluene and another with *n*-heptane. Inline check valves are placed to prevent backflow during the operation, and two pressure reducing valves are also added. The lines of *n*-heptane and asphaltenes in toluene are mixed at a T-junction in an ultrasonic bath. Additionally, the μPBR is connected to a heating bath to ensure that the temperature at the outlet of the μPBR at 343 K. A back-pressure regulator at the outlet maintains a constant pressure of 34.5 kPa to minimize microchanneling within the μPBR.

Before testing and quantifying asphaltene deposition in the μPBR, Hu et al. (2014) established potential parameters that could affect deposition on the microscale. They determined that flow rates of *n*-heptane, volume percentage of *n*-heptane to asphaltenes in toluene, Reynolds number, and pore volume would affect the amount of asphaltene deposition in their device. To test these parameters, they used a charge-coupled device (CCD) camera to take photographs of their transparent μPBR. To directly test the effect of pore volume on deposition, Hu et al. (2014) set their Reynolds number to approximately 1.30×10^{-2} and porosity constant to 40.6%. They varied the number pore volume of the μPBR from 0 to 77 and captured photos using the CCD camera under these conditions. This method allows them to view the color changes in the μPBR and the distribution of asphaltene deposition.

Additionally, Hu et al. (2014) varied the flow rates of asphaltenes in toluene (40 wt%) and *n*-heptane from 7.5 to 40.0 μL/min at an *n*-heptane concentration of 60 vol%. The Reynolds number ranged from 0.52×10^{-2} to 2.76×10^{-2}. Ultimately, the mass of asphaltenes deposited in the μPBR was quantified by a mass balance calculation between the mass of asphaltenes flowing into and out of the μPBR. Hu et al. (2014) found that as Reynolds number decreased from 2.76×10^{-2} to 0.52×10^{-2}, the mass of asphaltenes deposited increased from 1.1 to 2.1 mg. From these results, they claimed that Reynolds number played a vital role in the deposition mechanism of asphaltenes in a μPBR.

Hu et al. (2014) proposed that a higher Reynolds number meant more particles could pass along stream lines and plug the pore entrances through hydrodynamic bridging. In this instance, hydrodynamic forces prevent simultaneous asphaltene particles from entering the pore entrance, causing a particle bridge to form (Ramachandran and Fogler 1999). Because these pore entrances are constricted from a higher flow of asphaltenes, fewer asphaltenes can deposit in the μPBR because most of them aggregate toward the pore entrances. For lower Reynolds numbers, asphaltenes are able to deposit deeper and more uniformly in the μPBR, which lead to a larger overall mass of deposited asphaltenes. Moreover, Hu et al. (2014) observed that the pressure drop increased more gradually for lower Reynolds numbers compared to severe increases at higher Reynolds numbers. This result confirms that hydrodynamic bridging is more likely to occur in the μPBR at higher Reynolds numbers.

Lin et al. (2016) also used a microfluidic device to study asphaltene solubility on deposition in model porous media. Their microfluidic device was made with Norland Optical Adhesive (NOA) 81, which was solvent resistant and temperature tolerant. As shown in Figure 5.11b, the microfluidic device has a homogenous network of circular posts that are 125 μm in diameter. This network forms a minimum pore-throat spacing of 125 μm. In total, the microfluidic device has a permeability of approximately 5.23 D and a surface contact angle of about 79.5 degrees.

According to Figure 5.11a, a model oil containing 5 wt% of asphaltene in toluene and *n*-heptane are injected into the microreactor from two syringe pumps. Both lines from the pumps are mixed at a T-junction, and the flow rate of the mixture is maintained at 60 μL/min. The microfluidic device is placed on an inverted microscope (Olympus IX 71) and a high-speed complementary metal-oxide semiconductor (CMOS) camera is used to capture the deposition in the microfluidic device. Additionally, a differential pressure transducer is connected to the system to measure the pressure drop across the inlet and the outlet of the microfluidic device (Figure 5.11c).

To quantify asphaltene deposition in this device, the images are first processed using ImageJ (Schneider et al. 2012) and a Python image processing module (Hunter 2007). Next, locations of the posts are identified. The pixels of asphaltene deposition are differentiated from the image using Otsu's (1979) method. This allows the pixels to be converted into the total area of deposition. Next, the volume of deposition is

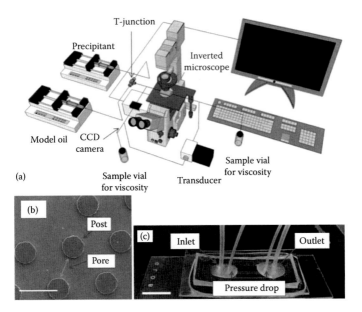

FIGURE 5.11 (a) Experimental setup of the microfluidic device. (b) SEM image of posts and pore in the microfluidic device. (c) Microchannel in the device. CCD, charge-coupled device; SEM, scanning electron microscope. (Reprinted from Lin, Y.-J. et al., *Langmuir*, 32, 8729–8734, 2016.)

calculated by multiplying the deposition area by the channel height ($h = 20$ μm). Finally, the mass of asphaltenes deposition in the microfluidic is calculated by multiplying the volume of deposition and the average asphaltene density, which is assumed to be 1,200 kg/m^3.

By keeping the same amount of injected asphaltenes, Lin et al. (2016) varied the amount of n-heptane from 40 vol% to 80 vol% and the corresponding experimental duration from 10 to 30 minutes. They found that submicrometer-sized asphaltene aggregates began to form at around 40 vol% of n-heptane. However, these aggregates were too small to deposit within the residence time in the device. The deposition rate increased and then decreased between 40 vol% and 65 vol% of n-heptane, possibly because of the competition between deposition and convection. Additionally, the deposition rate increased from 70 vol% to 80 vol% of n-heptane as the aggregates grew in size (Lin et al. 2016). A plot of the mass of deposited asphaltenes as a function of the amount of injected asphaltenes is shown in Figure 5.12.

According to these experiments conducted by Hu et al. (2014) and Lin et al. (2016), it is clear that designing an experiment for determining asphaltene deposition in microfluidics requires several major components. One is the physical device—a clear microfluidic device with a known pore size. Additionally, two pumps are required to inject the precipitant (e.g., n-heptane) and the oil into the microchannels. Next, a CCD camera needs to capture images of the deposition process. A pressure transducer is required to measure changes in pressure drop across the device, which can be useful for detecting hydrodynamic bridging from asphaltene deposition. Finally, there needs to be a method for quantifying the asphaltene deposition. In these examples, Hu et al. (2014) used a mass balance calculation for asphaltenes flowing in and out of the device, whereas Lin et al. (2016) used image processing technology to determine the volume and mass of the deposited asphaltenes.

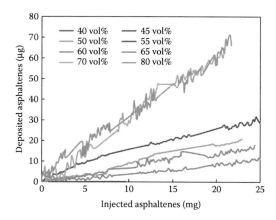

FIGURE 5.12 Asphaltene deposition at different volume ratios of n-heptane. (Reprinted from Lin, Y.-J. et al., *Langmuir*, 32, 8729–8734, 2016.)

5.1.2.2.2   Capabilities and Limitations

The use of microfluidic devices to conduct deposition tests has a number of significant advantages. The volume of fluids within these channels is very small; hence, the amount of crude oil sample required is quite less. Also, the fabrications techniques used to construct microfluidic devices are relatively inexpensive and are amenable both to highly elaborate, multiplexed devices and mass production. From the examples described previously, microfluidic devices are effective in determining asphaltene deposition in controlled porous media while reducing transport limitations found in macroscale experimental setups. These experimental setups offer capabilities that may be useful for future research. For example, different precipitants (e.g., *n*-pentane) may be used to determine the effect of the type of precipitant on asphaltene deposition. Additionally, clear microfluidic devices, along with CCD cameras and microscopes, allow the user to visually capture asphaltene deposition on the microscale. As demonstrated in the setup by Hu et al. (2014), connecting the microreactor to a heating bath can be useful for determining the effect of temperature on deposition.

Some limitations may also be present in these experimental configurations. For example, new microfluidic systems need to be constructed after each experiment because deposition may damage the device or undesirably affect future experiments. Additionally, deposition at the microscale could potentially have factors that do not apply once scaling up to the macroscopic porous reservoirs. So far, these experiments have been conducted at relatively low pressures compared to the core-flooding experiments. Also, the results could overestimate the formation damage in the porous media because the flow rate of the experiment is usually fast. Unlike the asphaltene destabilization caused by pressure depletion in the porous reservoirs, asphaltenes are induced by the addition of *n*-alkane in the current microfluidic devices. Nevertheless, these microfluidic systems provide researchers the unique potential to discover and analyze the kinetics of asphaltene deposition at the microscale.

5.1.2.3 Comparisons of Different Techniques

The experimental determination of asphaltene deposition in the porous media is usually conducted by the core-flooding test or in the microfluidic device. The core-flooding test is able to assess asphaltene deposition in a real rock sample. Also, asphaltene deposition can be investigated at HPHT conditions with the presence of a live oil sample. Thus, the core-flooding test can simulate the operating conditions in the near-wellbore regions, and representative results on asphaltene deposition in porous media can be obtained. However, core-flooding experiments are usually complicated and costly. Depending on the rock samples, the consumption of oil can be relatively large. Also, the core-flooding test is limited to investigate the macroscopic properties such as the permeability reduction resulting from asphaltene deposition. The mechanisms of asphaltene precipitation and deposition within porous media cannot be visualized at the pore scale. Another disadvantage of the core-flooding experiment is that it is not practical to run many pore volumes in a given experiment. Thus, the effect of asphaltene deposition is limited to what happens in the early stage. In comparison, the microfluidic device has been recently adapted to study asphaltene deposition in the microchannels, which represent the porous media of the rock sample. By using this setup, the accumulation of the deposited materials can be visualized by high-speed optical microscopy. The fabrication of the porous media microchannel is relatively cheap and easy. Nevertheless, most current microfluidic devices can only allow the assessment of asphaltene deposition using dead oil samples at ambient conditions. The manually fabricated circular posts and pore-throat spacing may not well represent the permeability in the real rock sample. Moreover, the deposition profile in the two-dimensional structure of the microchannels may not be representative of the actual deposition process in the near-wellbore regions.

5.2 ASSESSMENT OF DIFFERENT FACTORS THAT AFFECT ASPHALTENE DEPOSITION

5.2.1 EFFECT OF WATER

In the laboratory, water is usually removed from crude oil samples before assessing asphaltene deposition issues, leading to no or very low concentrations of water (<0.5 wt%). However, water-in-oil emulsions are virtually unavoidable in the crude oil production process. The injected water from secondary or enhanced oil recovery processes can co-produce with the crude oil. Usually, this water is emulsified with the crude oil, but water can possess significant solubility in crude oil at temperatures greater than 523 K (Tharanivasan et al. 2012). Therefore, understanding the effect of water is a critical factor for assessing and mitigating asphaltene deposition under realistic production conditions.

Different studies have investigated how emulsified water affects crude oil stability and asphaltene behavior. Tavakkoli et al. (2016) determined that asphaltenes were highly attracted to the oil–water interface, which could directly affect the stability of the crude oil–water emulsion compared to crude oil alone. They also found that specific subfractions of asphaltenes (i.e., C_{7+} asphaltenes) had a higher affinity to the oil–water interface. Additionally, Aslan and Firoozabadi (2014) investigated the direct effects of water on asphaltene deposition in steel tubing. Their experimental setup consisted of a 30 cm stainless steel pipe (1 mm ID), three syringe pumps, two manifolds, a T-junction, and a pressure transducer. This experimental setup allowed them to inject DI water, n-heptane, and crude oil simultaneously into the pipe to determine the effect of water on deposition. Aslan and Firoozabadi (2014) observed that delayed deposition occurred at various concentrations of DI water ranging from 500 to 7,000 ppm in a mixture of n-heptane and oil with 2 volume-to-volume ratio. Figure 5.13 shows the pressure drop in the tube at different concentrations of water. Crude oil without water clogs at around 400 pore volumes, whereas the addition of DI water delays the clogging up until 1,300 pore volumes. The clogging is associated with the sharp increase in pressure drop.

According to Figure 5.13, higher dosages of water can effectively delay asphaltene deposition in a stainless steel pipe at room temperature (Aslan and Firoozabadi 2014). One explanation could be that the hydrogen bonding between water and the heteroatoms (O, N, S) on the surface of asphaltene aggregates delays deposition. It is hypothesized that the hydrogen bonding changes the state of asphaltenes from hydrophobic to hydrophilic (Aslan and Firoozabadi 2014). It is likely that the hydrogen

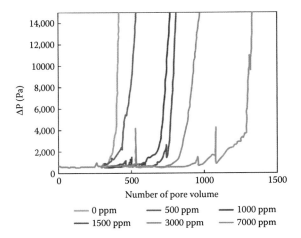

FIGURE 5.13 Pressure drop as a function of pore volume using *n*-heptane and oil with a 2 volume-to-volume ratio. The number for each curve is the concentration of deionized water in ppm. (Reprinted from Aslan, S., and Firoozabadim, A., *Langmuir*, 30, 3658–3664, 2014.)

bonding is the underlying reason why asphaltene molecules are more attracted to the oil–water interface (Tavakkoli et al. 2016). In addition, water–oil interfacial surface is probably much greater than the surface of the production tubing; thus, asphaltenes may ultimately trapped by the water–oil interface (Tavakkoli et al. 2016). Also, water creates discontinuities that make it difficult for asphaltenes to move to the metal surface. Thus, it is possible that interactions between water molecules and asphaltene molecules play a major role in delaying the onset of asphaltene deposition in experimental crude oil and water systems. However, the presence of water–brine could cause other effects such as interaction of electrolytes and asphaltenes, or corrosion, which might in turn increase the tendency of asphaltene deposition in the production tubing. The effects of electrolytes and corrosion on asphaltene deposition will be discussed in Sections 5.2.2 and 5.2.3.

5.2.2 EFFECT OF ELECTROLYTES

As mentioned, water is usually present in the oil production process. In an oil reservoir, however, the water emulsified with oil is not pure. Electrolytes present in brine with concentrations of 180,000–20,000 ppm can cause results to deviate from emulsions with DI water alone (Tavakkoli et al. 2016). Understanding the effect of electrolytes in oil–brine emulsions on asphaltene deposition is a crucial to the understanding and mitigation of flow assurance problems in the field. The effects of metal ions such as ferric ions and aluminum ions in the synthetic brine have been investigated with respect to deposition by Wang et al. (2014).

Previous experiments have been performed to test the effects of electrolytes on asphaltene deposition. For example, Fogden (2017) investigated the effect of sodium chloride (NaCl) and calcium chloride ($CaCl_2$) on asphaltene deposition in porous silica films. In their experiments, the porous silica films were submerged and aged in crude oil and various concentrations of brine with NaCl and $CaCl_2$. They determined that the presence of NaCl (from 16 to 512 mM) had little impact on asphaltene deposition, whereas increasing the amount of $CaCl_2$ up to 10% (mol/mol) reduced and suppressed deposition (Fogden 2017).

Additionally, Wang et al. (2014) found that adding brine with 6.5 wt% of NaCl decreased the deposition in stainless steel capillary tubes by around 40% compared to crude oil without water. However, they also determined that the addition of only DI water reduced asphaltene deposition the most. Compared to the condition without water, Wang et al. (2014) observed 44% reduction in asphaltene deposition after the addition of 2 vol% DI water in the crude oil. They also determined that the deposition rate returned to or exceeded the rate of crude oil without water after adding NaCl with Fe^{3+} or Al^{3+} ions. In fact, they found that the addition of more than 1,000 ppm of Fe^{3+} and Al^{3+} ions could worsen the asphaltene deposition problem, depositing a higher percentage of asphaltenes compared to the crude oil alone (Wang et al. 2014). Figure 5.14 displays the effect of different electrolytes on asphaltene deposition in the capillary flow loop.

Experimental results demonstrate that DI water is more effective to reduce asphaltene deposition than brine. It is likely that the presence of NaCl ions can disrupt the stability of asphaltenes dispersed in water (Demir et al. 2016). $CaCl_2$ may play a role in reducing asphaltene deposition, although its effectiveness compared to DI water alone needs to be further studied (Fogden 2017). Ions such as Fe^{3+} and Al^{3+} at higher concentrations (greater than 1,000 ppm) appear to destabilize crude oil and worsen the asphaltene deposition problem. The effect of iron ions on asphaltene deposition is

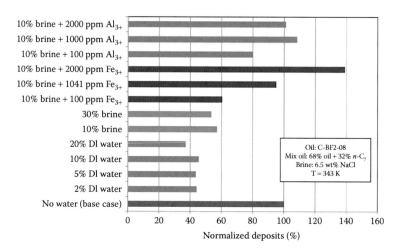

FIGURE 5.14   Deposition tests with crude oil and n-C_7 in 0.762-mm ID capillaries at 343 K, 12 cm^3/h flow rate, and 24 hours run time. Deposits were normalized with respect to the base case of crude oil without water. DI, deionized; ID, inner diameter. (Reprinted from Wang, J. et al., In Offshore Technology Conference. doi:10.4043/25411-MS, 2014.)

also assessed in the packed bed column by Sung et al. (2017), and it will be discussed in Section 5.2.3.1. Nevertheless, further research needs to be done on the mechanism and interactions between electrolytes and asphaltenes to determine mitigation strategies for deposition problems encountered in the field.

5.2.3 EFFECT OF CORROSION

Corrosion of oilfield pipelines is one of most recurring and challenging problems that are commonly found both in upstream and downstream operations. An increased risk of corrosion might occur as upstream operations move to harsher environments and deeper waters. Although the industry relies heavily on carbon and low-alloy steels to meet mechanical and low cost requirements, these materials do not possess a significant corrosion resistance (Kermani and Morshed 2003). As a result, corrosion-related failures contribute to more than 25% of total failures in the oil and gas industry (Kermani and Harrop 1996). Thus, factors such as pipeline materials, the presence of brine and acid gases such as H_2S or CO_2, high temperatures, and high pressures make corrosion an inevitable challenge in the oil production process. Not only does corrosion affect the mechanical integrity of operations, but it has also been shown that corrosion may further aggravate asphaltene deposition in wellbore and surface facilities (Sung et al. 2016). Thus, because of the prevalence of the corrosion problem, it is necessary to understand how corrosion can directly affect asphaltene deposition.

Internal corrosion problems can be caused by CO_2 and H_2S, both of which are present in fluids produced during upstream productions (Kermani and Morshed 2003). CO_2 in water produces carbonic acid, H_2CO_3, which lowers the pH and increases the corrosivity of the fluid. Additionally, as the operating pressures increase to 8.27 MPa, the partial pressure of CO_2 and H_2S can reach 0.62 MPa and 262 Pa, respectively. It should be mentioned that the partial pressures of CO_2 and H_2S also depend on the concentrations of CO_2 and H_2S in the fluid. These chemicals, along with high pressure, low pH, water, and light hydrocarbons, lead to potential corrosion issues in the oilfield pipelines (Palacios et al. 1997). One possible solution to this type of corrosion is to avoid using carbon steel pipelines in presence of acid gases.

Sung et al. (2017) performed experiments to simulate the effect of corroded surfaces on asphaltene deposition in the production tubing. In their experiments, carbon steel spheres were exposed to an acidic solution of hydrochloric acid until corrosion with rust was detected on the spheres through a digital microscope. Corroded spheres without rust and the original carbon steel spheres were also tested. The difference between the corroded spheres with and without rust affects the surface roughness, which is further elaborated in Section 5.2.3.2. The experiments were conducted in the packed bed column deposition test apparatus depicted in Figure 5.5. The aggregated asphaltenes that did not deposit on the spheres were collected from the effluent. Thus, a larger amount of asphaltenes in the effluent indicated a lower rate of deposition in the column.

As presented in Figure 5.15, it appears that the original spheres have the largest amount of aggregated asphaltenes in the outlet. In contrast, much smaller amounts

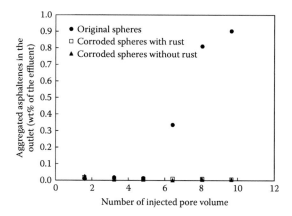

FIGURE 5.15 Aggregated asphaltenes collected from the effluent versus the injected pore volume. (Reprinted from Sung, C.-A. et al., Effect of pipeline corrosion on asphaltene deposition tendency on metallic surfaces, In Preparation, 2017.)

of aggregated asphaltenes are obtained from the effluent for the cases with corroded spheres, providing evidence that asphaltenes are more likely to deposit on corroded spheres, both with and without the presence of rust (Sung et al. 2017). High affinity of asphaltenes to the iron oxide facilitates the deposition of asphaltenes on corroded spheres with the presence of rust. According to a study by Murgich et al. (2001), low hydrogen-to-carbon ratio and heteroatom contents of asphaltenes could promote the adsorption of asphaltenes on the iron oxide mineral. The corroded spheres without the presence of rust have a higher surface roughness because of the deterioration of iron in the carbon steel spheres compared to the original spheres. The higher surface roughness leads to higher surface area for asphaltenes to deposit on (Sung et al. 2017).

5.2.3.1 Effect of Iron Ions

In the oilfield pipelines and wellbores, the corrosion of iron materials can produce iron ions (Schramm 2000). The increased iron concentration in oil causes the oil to become more unstable (Ibrahim and Idem 2004). Thus, a better understanding of the effect of iron ions on asphaltene deposition can reveal valuable information to better understand the interrelation between corrosion and asphaltene deposition. Sung et al. (2017) investigated the effect of Fe^{3+} ions in the packed bed column deposition test apparatus depicted in Figure 5.5. An emulsified solution containing 5 wt% asphaltenes in toluene and 2 wt% Fe^{3+} ions in DI water was injected into the column using a syringe pump. A precipitant of n-heptane was co-injected from another syringe pump at a 70 vol% ratio with the emulsion. As a control, the procedure was repeated with an emulsified solution containing 5 wt% asphaltenes in toluene and DI water. The results showed that 25.0% of the total infused asphaltenes in the system deposited onto the carbon steel spheres in the presence of emulsified DI water. In comparison, the experiment with emulsified Fe^{3+} ion solution caused 36.2% of the total infused asphaltenes deposited onto the spheres. Thus, Sung et al. (2017)

concluded that asphaltenes are more likely to deposit on the surface of spheres in the presence of Fe^{3+} ions, which is in agreement with the results obtained by Wang et al. (2014). However, Sung et al. (2017) reported that the deposited material is not pure asphaltenes and contains both asphaltenes and ferric ions. They concluded that Fe^{3+} ions can form a complex with asphaltenes in solution, and this complex has a high affinity to the carbon steel spheres.

It is necessary to mention that the spheres may be corroded during the experiment with the presence of acidic Fe^{3+} solution. However, Sung et al. (2017) neglected the effect of Fe^{3+} solution acidity on asphaltene deposition compared to the case with DI water.

5.2.3.2 Effect of Surface Roughness

Akbarzadeh et al. (2012) investigated the effect of surface roughness on asphaltene deposition tendency in their RealView deposition cell. They conducted the experiment using a rough cell wall insert with an average roughness of 30 μm and a smooth cell wall insert with roughness smaller than 8 μm. The experiments were compared with the regular cell wall with an average roughness of 8 μm. For each experiment, Akbarzadeh et al. (2012) quantified the mass of deposited asphaltenes from the RealView cell. Figure 5.16 shows the impact of surface roughness on asphaltene deposition.

The results demonstrate that a higher surface roughness further aggravates asphaltene deposition problems. It is likely that the increase in surface area associated with surface roughness allows for more space for the asphaltenes to deposit. Therefore, using a smoother stainless steel surface in pipelines and wellbores can possibly reduce the amount of deposition that occurs. However, this may not be practical in the production lines. For example, deposition in pipelines can induce further deposition because of the associated increase in surface roughness. Additionally,

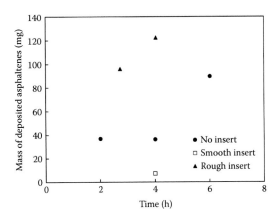

FIGURE 5.16 Effect of surface roughness on the deposition tendency of asphaltenes according to the experiments conducted in the RealView deposition cell. (Reprinted from Akbarzadeh, K. et al., *Energy Fuels*, 26, 495–510, 2012.)

corrosion in pipelines and wellbores has the potential to increase the surface roughness and thus induce or worsen asphaltene deposition (Landolt 2007).

It can be difficult to isolate and control the surface roughness of pipelines during operations. Likely, reducing asphaltene deposition induced by surface roughness requires simultaneously addressing multiple factors including corrosion. Innovations such as new alloys, inhibitors, and coatings demonstrate promise for controlling corrosion and surface roughness in the industry (Tuttle 1987). Moreover, it is valuable to develop integrated approaches that can simultaneously mitigate multiple problems such as corrosion and asphaltene deposition in the production lines.

5.2.4 EFFECT OF COATING

Asphaltene deposition occurs when the attraction between the asphaltene aggregates and the substrate surface is larger than that between the solvent and the surface (Gonzalez 2008). To unveil the mechanisms of asphaltene deposition in the production lines, it is crucial to investigate the interactions between the precipitated asphaltenes in the bulk phase and the substrate surface. The better understanding of these mechanisms can provide valuable insight into the selection of the best coating materials to mitigate asphaltene deposition in the wellbores and pipelines.

The surface properties such as surface free energies between the liquid phase and the solid phase can provide important information on the evaluation of asphaltene deposition tendency at the solid–liquid interface (Gonzalez 2008). As discussed in Section 2.2.2.3, the contact angle measurement has been widely used to characterize the competing interactions between the cohesive energy of the liquid molecules and the adhesion between the liquid and the solid phases (Gonzalez 2008). Gonzalez (2008) determined the surface energies from contact angle measurements for different coating materials and reported that the surface energy of PTFE was 18.4 mN/m, which was much lower than the surface energies of the selected commercial coating materials (30–35 mN/m).

To assess the asphaltene deposition tendency on various coating materials, Gonzalez (2008) used the Hamaker-Lifshitz theory to predict the interactions among the asphaltene-lean phase (solvent or crude oil), the precipitated asphaltene-rich phase, and the substrate surface. By only considering the van der Waals intermolecular interaction between the phases, the Hamaker constant (A) is defined in Equation 5.4 (Israelachvili 2011),

$$A_{12} = \pi^2 C_{12} \rho_1 \rho_2 \qquad (5.4)$$

where:
   C_{12} is the van der Waals constant for a pair of molecule
   ρ_1 and ρ_2 is the corresponding molecular density for molecules 1 and 2

Thus, the Hamaker constant can show the difference between the cohesive interaction with the bulk phase and the adhesive interaction with the solid surface (Gonzalez 2008).

Lifshitz (1956) developed an equation to calculate the Hamaker constant from the measurable refractive indexes and dielectric constants of the interacting bodies. By neglecting the atomic structure, Lifshitz (1956) treated the forces between the interacting bodies as continuous media. According to Gonzalez (2008), the Hamaker constant for the interaction of medium 1 and medium 2 across medium 3 can be derived based on the Lifshitz theory:

$$A_{132} \approx \frac{3}{4} k_b T \left(\frac{\varepsilon_1 - \varepsilon_3}{\varepsilon_1 + \varepsilon_3} \right) \left(\frac{\varepsilon_2 - \varepsilon_3}{\varepsilon_2 + \varepsilon_3} \right)$$

$$+ \frac{3hv_e}{8\sqrt{2}} \frac{\left(n_1^2 - n_3^2 \right)\left(n_2^2 - n_3^2 \right)}{\left(n_1^2 + n_3^2 \right)^{1/2} \left(n_2^2 + n_3^2 \right)^{1/2} \left\{ \left(n_1^2 + n_3^2 \right)^{1/2} + \left(n_2^2 + n_3^2 \right)^{1/2} \right\}} \quad (5.5)$$

where:

ε_1, ε_2, and ε_3 represents the dielectric constants of the three media
n_1, n_2, and n_3 are the refractive indexes extrapolated to zero frequency of materials
T is the temperature
k_b is the Boltzmann constant
h is the Planck constant
v_e is the main adsorption frequency in the UV region

Most of the polymeric materials have relatively small Hamaker constants. For example, the Hamaker constant for PTFE is 3.8×10^{-20} J (Israelachvili 2011; Hough and White 1980). Because metals and metal oxides are highly polarizable, their Hamaker constants are usually an order of magnitude higher than those of non-conducting surfaces (Israelachvili 2011; Gonzalez 2008). The composite Hamaker constants can be calculated to determine the sticking and spreading tendencies of the asphaltene-rich phase on different solid surfaces (Gonzalez 2008). According to Figure 5.17a, the composite Hamaker constant, A_{SLR}, can be calculated to predict the sticking tendency of the precipitated asphaltenes on the substrate, as the asphaltene-lean phase being the medium between the substrate and rich phase. As shown in Figure 5.17b,

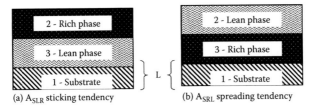

(a) A_{SLR} sticking tendency   (b) A_{SRL} spreading tendency

FIGURE 5.17 Graphical representation of the composite Hamaker constant for (a) A_{SLR} sticking tendency and (b) A_{SRL} spreading tendency. SLR, solid, lean, rich (phases). (Reprinted from Gonzalez, D. L., Modeling of asphaltene precipitation and deposition tendency using the PC-SAFT equation of state, Rice University, Retrieved from https://scholarship.rice.edu/handle/1911/22140, 2008.)

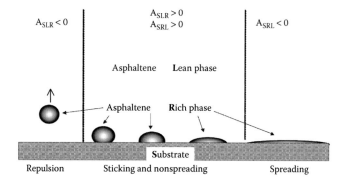

FIGURE 5.18 Illustration for the signs of the composite Hamaker constants on the sticking and spreading tendencies of the asphaltene-rich phase on the solid surface. SLR, solid, lean, rich (phases). (Reprinted from Gonzalez, D. L., Modeling of asphaltene precipitation and deposition tendency using the PC-SAFT equation of state, Rice University, Retrieved from https://scholarship.rice.edu/handle/1911/22140, 2008.)

the other composite Hamaker constant, A_{SRL}, can be used to predict the spreading tendency of the asphaltene-rich phase on the surface, as the asphaltene-rich phase being the medium between the substrate and lean phase (Gonzalez 2008).

The illustration for the signs of the composite Hamaker constants, A_{SLR} and A_{SRL}, on the determination of the asphaltene deposition tendency is presented in Figure 5.18. A positive value of A_{SLR} indicates that the asphaltenes in the rich phase can stick to the solid surface, whereas a negative value represents the repulsion of the precipitated asphaltenes from the surface. Likewise, a positive value of A_{SRL} means that there is no spreading tendency of the asphaltene-rich phase on the surface, whereas a negative value predicts that the rich phase has high spreading tendency over the substrate and the adsorbed layer approaches infinite thickness at saturation (Hirasaki 1993; Gonzalez 2008).

According to Gonzalez (2008), the effect of different coating materials on asphaltene deposition was assessed in the capillary deposition flow loop as shown in Section 5.1.1.1. The result showed that the deposit layer on the stainless steel surface (47 µm) were much thicker than that on the pure phenolic-resin surface (27 µm). This provides evidence that the metallic materials have a higher affinity to asphaltenes compared to the polymeric materials. However, the correlation between the signs of the Hamaker constants and the asphaltene deposition tendency has not been experimentally validated yet. The effect of coatings on the prevention of asphaltene deposition will be discussed in Chapter 7.

5.3 FINAL REMARKS

The common experimental setups for the determination of asphaltene deposition on the metallic surface, including the capillary flow loop, the QCM, the RealView deposition cell, and the packed bed column, were discussed and compared. The packed bed column is a newly developed apparatus to investigate asphaltene deposition based on direct mass balance calculation using a small amount of oil sample. The physical and chemical properties of the deposits can be easily characterized. Also, the packed bed column can be adapted to investigate the effect of material composition, surface area, surface roughness, and water, corrosion on asphaltene deposition. Further effort is needed to enable the feasibility of studying asphaltene deposition induced by pressure depletion in live oil samples by the packed bed column. Besides that, the core-flooding test and the microfluidic device were compared as techniques to assess asphaltene deposition in the porous media. Moreover, the impacts of water, electrolytes, corrosion, and internal surface coatings on asphaltene deposition were presented.

Much effort has been devoted to investigating asphaltene deposition on the metallic surface and porous media experimentally. The development of reliable and cost-effective approaches to probe asphaltene deposition problems will not only improve the accuracy for the predictions of asphaltene deposition in the production tubing and near-wellbore regions, but also provide better understandings on the mitigation and remediation of asphaltene deposition.

REFERENCES

Abudu, A., and L. Goual. 2009. Adsorption of crude oil on surfaces using quartz crystal microbalance with dissipation (QCM-D) under flow conditions. *Energy & Fuels* 23 (3): 1237–1248. doi:10.1021/ef800616x.

Akbarzadeh, K., A. Hammami, A. Kharrat, D. Zhang, S. Allenson, J. Creek, S. Kabir et al. 2007. Asphaltenes—problematic but rich in potential. *Oilfield Review* 19 (2): 22–43.

Akbarzadeh, K., D. Eskin, J. Ratulowski, and S. Taylor. 2012. Asphaltene deposition measurement and modeling for flow assurance of tubings and flow lines. *Energy & Fuels* 26 (1): 495–510. doi:10.1021/ef2009474.

Akbarzadeh, K., J. Ratulowski, T. Lindvig, T. L. Davies, Z. Huo, G. Broze, R. Howe, and K. Lagers. 2009. The importance of asphaltene deposition measurements in the design and operation of subsea pipelines. In Society of Petroleum Engineers. doi:10.2118/124956-MS.

Ali, M. A., and M. R. Islam. 1998. The effect of asphaltene precipitation on carbonate-rock permeability: An experimental and numerical approach. *SPE Production & Facilities* 13 (3): 178–183. doi:10.2118/50963-PA.

Aslan, S., and A. Firoozabadi. 2014. Effect of water on deposition, aggregate size, and viscosity of asphaltenes. *Langmuir* 30 (13): 3658–3664. doi:10.1021/la404064t.

ASTM, D. 1998. Standard test method for characteristic groups in rubber extender and processing oils and other petroleum-derived oils by the clay-gel absorption chromatographic method. *Links*. In Preparation.

Bagheri, M. B., R. Kharrat, and C. Ghotby. 2011. Experimental investigation of the asphaltene deposition process during different production schemes *Oil & Gas Science and Technology IFP* 66 (3): 507–519. doi:10.2516/ogst/2010029.

Broseta, D., M. Robin, T. Savvidis, C. Féjean, M. Durandeau, and H. Zhou. 2000. Detection of asphaltene deposition by capillary flow measurements. In *SPE/DOE Improved Oil Recovery Symposium*. Tulsa, Oklahoma: Society of Petroleum Engineers. Retrieved from https://www.onepetro.org/conference-paper/SPE-59294-MS.

Buckley, J. S. 2012. Asphaltene deposition. *Energy & Fuels* 26 (7): 4086–4090. doi:10.1021/ef300268s.

Chaisoontornyotin, W., N. Haji-Akbari, H. Scott Fogler, and M. Paul Hoepfner. 2016. A combined asphaltene aggregation and deposition investigation. *Energy & Fuels.* doi:10.1021/acs.energyfuels.5b02427.

Demir, A. B., H. Ilkin Bilgesu, and B. Hascakir. 2016. The effect of clay and salinity on asphaltene stability. In Society of Petroleum Engineers. doi:10.2118/180425-MS.

Dudášová, D., A. Silset, and J. Sjöblom. 2008. Quartz crystal microbalance monitoring of asphaltene adsorption/deposition. *Journal of Dispersion Science and Technology* 29 (1): 139–146. doi:10.1080/01932690701688904.

Ekholm, P., E. Blomberg, P. Claesson, I. H. Auflem, J. Sjöblom, and A. Kornfeldt. 2002. A quartz crystal microbalance study of the adsorption of asphaltenes and resins onto a hydrophilic surface. *Journal of Colloid and Interface Science* 247 (2): 342–350.

Eskin, D., J. Ratulowski, K. Akbarzadeh, and S. Pan. 2011. Modelling asphaltene deposition in turbulent pipeline flows. *The Canadian Journal of Chemical Engineering* 89 (3): 421–441.

Fogden, A. 2017. Experimental investigation of deposition of crude oil components in brine-filled pores. Accessed April 24. Retrieved from http://www.scaweb.org/abstracts/977.html.

Ghahfarokhi, A. K., P. Kor, R. Kharrat, and B. S. Soulgani. 2017. Characterization of asphaltene deposition process in flow loop apparatus; An experimental investigation and modeling approach. *Journal of Petroleum Science and Engineering* 151: 330–340. doi:10.1016/j.petrol.2017.01.009.

Gonzalez, D. L. 2008. Modeling of asphaltene precipitation and deposition tendency using the PC-SAFT equation of state. Rice University. Retrieved from https://scholarship.rice.edu/handle/1911/22140.

Haskett, C. E., and M. Tartera. 1965. A practical solution to the problem of asphaltene deposits-Hassi Messaoud Field, Algeria. *Journal of Petroleum Technology* 17 (4): 387–391. doi:10.2118/994-PA.

Hirasaki, G. J. 1993. Structural interactions in the wetting and spreading of van der Waals fluids. *Journal of Adhesion Science and Technology* 7 (3): 285–322.

Hoepfner, M. P., V. Limsakoune, V. Chuenmeechao, T. Maqbool, and H. S. Fogler. 2013. A fundamental study of asphaltene deposition. *Energy & Fuels* 27 (2): 725–735. doi:10.1021/ef3017392.

Hough, D. B., and L. R. White. 1980. The calculation of Hamaker constants from Liftshitz theory with applications to wetting phenomena. *Advances in Colloid and Interface Science* 14 (1): 3–41.

Hu, C., J. E. Morris, and R. L. Hartman. 2014. Microfluidic investigation of the deposition of asphaltenes in porous media 14 (12): 2014–2022. doi:10.1039/C4LC00192C.

Hubbert, M. K. 1956. Darcy's Law and the field equations of the flow of underground fluids. Retrieved from https://www.onepetro.org/general/SPE-749-G.

Hunter, J. D. 2007. Matplotlib: A 2D graphics environment. *Computing in Science & Engineering* 9 (3): 90–95.

Ibrahim, H. H., and R. O. Idem. 2004. Correlations of characteristics of Saskatchewan crude oils/asphaltenes with their asphaltenes precipitation behavior and inhibition mechanisms: Differences between CO_2- and n-heptane-induced asphaltene precipitation. *Energy & Fuels* 18 (5): 1354–1369. doi:10.1021/ef034044f.

Israelachvili, J. N. 2011. *Intermolecular and Surface Forces.* Academic Press.

Jamialahmadi, M., B. Soltani, H. Müller-Steinhagen, and D. Rashtchian. 2009. Measurement and prediction of the rate of deposition of flocculated asphaltene particles from oil. *International Journal of Heat and Mass Transfer* 52 (19): 4624–4634.

Jensen, K. F. 2001. Microreaction engineering—Is small better? *Chemical Engineering Science, 16th International Conference on Chemical Reactor Engineering* 56 (2): 293–303. doi:10.1016/S0009-2509(00)00230-X.

Juyal, P., A. M. McKenna, T. Fan, T. Cao, R. I. Rueda-Velásquez, J. E. Fitzsimmons, A. Yen et al. 2013. Joint industrial case study for asphaltene deposition. *Energy & Fuels* 27 (4): 1899–1908. doi:10.1021/ef301956x.

Kermani, M. B., and D. Harrop. 1996. The impact of corrosion on oil and gas industry. *SPE Production & Facilities* 11 (3): 186–190. doi:10.2118/29784-PA.

Kermani, M. B., and A. Morshed. 2003. Carbon dioxide corrosion in oil and gas production A compendium. *Corrosion* 59 (8). https://www.onepetro.org/journal-paper/NACE-03080659.

Khaleel, A., M. Abutaqiya, M. Tavakkoli, A. A. Melendez-Alvarez, and F. M. Vargas. 2015. On the prediction, prevention and remediation of asphaltene deposition. In Society of Petroleum Engineers. doi:10.2118/177941-MS.

Kord, S., R. Miri, S. Ayatollahi, and M. Escrochi. 2012. Asphaltene deposition in carbonate rocks: Experimental investigation and numerical simulation. *Energy & Fuels* 26 (10): 6186–6199. doi:10.1021/ef300692e.

Kuang, J., A. Melendez-Alvarez, J. Yarbrough, M. Garcia-Bermudes, M. Tavakkoli, D. S. Abdallah, and F. M. Vargas. 2017. Novel way to assess the performance of asphaltene dispersants on the prevention of asphaltene deposition. In Preparation.

Kurup, A. S., F. M. Vargas, J. Wang, J. Buckley, J. L. Creek, J. Subramani, and W. G. Chapman. 2011. Development and application of an asphaltene deposition tool (ADEPT) for well bores. *Energy & Fuels* 25 (10): 4506–4516.

Landolt, D. 2007. *Corrosion and Surface Chemistry of Metals*. Boca Raton, FL: CRC Press.

Lifshitz, E. M. 1956. The theory of molecular attractive forces between solids. *Journal of Experimental and Theoretical Physics* 2: 73–83.

Lin, Y.-J., P. He, M. Tavakkoli, N. T. Mathew, Y. Y. Fatt, J. C. Chai, A. Goharzadeh, F. M. Vargas, and S. L. Biswal. 2016. Examining asphaltene solubility on deposition in model porous media. *Langmuir* 32 (34): 8729–8734. doi:10.1021/acs.langmuir.6b02376.

Manbachi, A., and R. S. C. Cobbold. 2011. Development and application of piezoelectric materials for ultrasound generation and detection. *Ultrasound* 19 (4): 187–196.

Murgich, J., E. Rogel, O. León, and R. Isea. 2001. A molecular mechanics-density functional study of the adsorption of fragments of asphaltenes and resins on the (001) surface of Fe_2O_3. *Petroleum Science and Technology* 19 (3–4): 437–455.

Nabzar, L., and M. E. Aguiléra. 2008. The colloidal approach. A promising route for asphaltene deposition modelling. *Oil & Gas Science and Technology - Revue de l'IFP* 63 (1): 21–35. doi:10.2516/ogst/2007083.

Otsu, N. 1979. A threshold selection method from gray-level histograms. *IEEE Transactions on Systems, Man and Cybernetics* 9 (1): 62–66.

Palacios T, C. A., J. L. Morales, and A. Viloria. 1997. Effect of asphaltene deposition on the internal corrosion in the oil and gas industry. NACE International, Houston, TX (United States). Retrieved from http://www.osti.gov/scitech/biblio/504730.

Ramachandran, V., and H. S. Fogler. 1999. Plugging by hydrodynamic bridging during flow of stable colloidal particles within cylindrical pores. *Journal of Fluid Mechanics* 385: 129–156.

Rodahl, M., F. Höök, A. Krozer, P. Brzezinski, and B. Kasemo. 1995. Quartz crystal microbalance setup for frequency and q-factor measurements in gaseous and liquid environments. *Review of Scientific Instruments* 66 (7): 3924–3930.

Sauerbrey, G. 1959. Verwendung von Schwingquarzen Zur Wägung Dünner Schichten Und Zur Mikrowägung. *Zeitschrift Für Physik A Hadrons and Nuclei* 155 (2): 206–222.

Schneider, C. A., W. S. Rasband, and K. W. Eliceiri. 2012. NIH image to ImageJ: 25 Years of image analysis. *Nature Methods* 9 (7): 671–675. doi:10.1038/nmeth.2089.

Schramm, L. L. 2000. *Surfactants: Fundamentals and Applications in the Petroleum Industry*. Cambridge: Cambridge University Press.

Sim, S. S. K., K. Okatsu, K. Takabayashi, and D. B. Fisher. 2005. Asphaltene-induced formation damage: Effect of asphaltene particle size and core permeability. In Dallas, TX:
Society of Petroleum Engineers. doi:10.2118/95515-MS.

Srivastava, R. K., S. S. Huang, and M. Dong. 1999. Asphaltene deposition during CO_2 flooding. *SPE Production & Facilities* 14 (4). doi:10.2118/59092-PA.

Stockbridge, C. D., and A. W. Warner. 1962. A vacuum system for mass and thermal measurement
with resonating crystalline quartz. In *Vacuum Microbalance Techniques*, pp. 93–114.
Springer. http://link.springer.com/chapter/10.1007/978-1-4899-6285-0_8.

Sung, C.-A., M. Tavakkoli, A. Chen, J. Hu, and F. M. Vargas. 2017. Effect of pipeline corrosion on asphaltene deposition tendency on metallic surfaces. In Preparation.

Sung, C.-A., M. Tavakkoli, A. Chen, and F. M. Vargas. 2016. Prevention and control of
corrosion-induced asphaltene deposition. In Offshore Technology Conference.
doi:10.4043/27008-MS.

Tavakkoli, M., A. Chen, C.-A. Sung, K. M. Kidder, J. J. Lee, S. M. Alhassan, and F. M. Vargas.
2016. Effect of emulsified water on asphaltene instability in crude oils. *Energy & Fuels*
30 (5): 3676–3686. doi:10.1021/acs.energyfuels.5b02180.

Tavakkoli, M., S. R. Panuganti, V. Taghikhani, M. Reza Pishvaie, and W. G. Chapman. 2014a.
Asphaltene deposition in different depositing environments: Part 2. Real oil. *Energy &
Fuels* 28 (6): 3594–3603. doi:10.1021/ef401868d.

Tavakkoli, M., S. R. Panuganti, F. M. Vargas, V. Taghikhani, M. R. Pishvaie, and W. G.
Chapman. 2014b. Asphaltene deposition in different depositing environments: Part 1.
Model oil. *Energy & Fuels* 28 (3): 1617–1628. doi:10.1021/ef401857t.

Tavakkoli, M., M. Puerto, P. He, P.-H. Lin, J. L. Creek, J. Wang, J. Gomes et al. 2017. Probing
asphaltene deposition tendency in porous media using core flooding experiments. In
Preparation.

Tharanivasan, A. K., H. W. Yarranton, and S. D. Taylor. 2012. Asphaltene precipitation from
crude oils in the presence of emulsified water. *Energy & Fuels* 26 (11): 6869–6875.
doi:10.1021/ef301200v.

Tuttle, R. N. 1987. Corrosion in oil and gas Production. *Journal of Petroleum Technology* 39
(7): 756–762. doi:10.2118/17004-PA.

Vargas, F. M. 2009. Modeling of asphaltene precipitation and arterial deposition. Rice
University. Retrieved from https://scholarship.rice.edu/handle/1911/62084.

Vargas, F, M., J. L. Creek, and W. G. Chapman. 2010. On the development of an asphaltene
deposition simulator. *Energy & Fuels* 24 (4): 2294–2299. doi:10.1021/ef900951n.

Vilas Boas Favero, C., A. Hanpan, P. Phichphimok, K. Binabdullah, and H. S. Fogler. 2016.
Mechanistic investigation of asphaltene deposition. *Energy & Fuels* 30 (11): 8915–8921.
doi:10.1021/acs.energyfuels.6b01289.

Wang, J., and J. S Buckley. 2006. *Estimate Thickness of Deposit Layer from Displacement
Test*. Socorro, NM: New Mexico Tech.

Wang, J., J. S. Buckley, and J. L. Creek. 2004. Asphaltene deposition on metallic surfaces.
Journal of Dispersion Science and Technology 25 (3): 287–298.

Wang, J., T. Fan, J. S. Buckley, and J. L. Creek. 2014. Impact of water cut on asphaltene deposition tendency. In Offshore Technology Conference. doi:10.4043/25411-MS.

Zanganeh, P., S. Ayatollahi, A. Alamdari, A. Zolghadr, H. Dashti, and S. Kord. 2012.
Asphaltene deposition during CO_2 injection and pressure depletion: A visual study.
Energy & Fuels 26 (2): 1412–1419. doi:10.1021/ef2012744.

Zougari, M., S. Jacobs, J. Ratulowski, A. Hammami, G. Broze, M. Flannery, A. Stankiewicz,
and K. Karan. 2006. Novel organic solids deposition and control device for live-oils:
Design and applications. *Energy & Fuels* 20 (4): 1656–1663.

Zougari, M., J. Ratulowski, A. Hammami, and A. Kharrat. 2005. Live oils novel organic solid
deposition and control device: Wax and asphaltene deposition validation. In Society of
Petroleum Engineers. doi:10.2118/93558-MS.

6 Modeling Methods for Prediction of Asphaltene Deposition

*N. Rajan Babu, P. Lin, J. Zhang,
M. Tavakkoli, and F. M. Vargas*

CONTENTS

The potential for asphaltene to precipitate and deposit in wellbore and flow lines as a result of changes in pressure, temperature, and composition of crude oil is a major concern for the oil and gas industry. Asphaltene deposit removal from onshore and offshore facilities is an expensive operation. To properly assess the risk of asphaltene deposition, experimental and modeling techniques have been developed to predict the asphaltene deposition profile and rate. There have been only a few established works on modeling asphaltene deposition in wellbores and pipelines as well as in porous media compared with the number of works dedicated to the development of thermodynamic models to investigate asphaltene precipitation. It is important to establish an efficient and reliable modeling technique to understand the mechanism of asphaltene deposition in the oil field based on the various laboratory-scale experimental results.

Several experimental techniques developed to study asphaltene deposition at different pressure and temperature conditions and flow regimes were presented in Chapter 5. In this chapter, modeling methods for the prediction of asphaltene deposition, developed over the years, will be discussed. Asphaltene deposition modeling in wellbores and pipelines has been investigated by researchers for only the last few years. Although asphaltene deposition in porous media has been studied for a relatively longer period of time, this is still an area of active research to achieve a simple, yet comprehensive model that can capture the relevant physics of asphaltene deposition phenomena.

6.1 ASPHALTENE DEPOSITION IN WELLBORE AND PIPELINES

6.1.1 MODELING ASPHALTENE DEPOSITION

6.1.1.1 Initial Works

A multiphase multicomponent hydrodynamic model was proposed by Ramirez-Jaramillo et al. (2006) to represent the asphaltene deposition phenomenon in production wells. Molecular diffusion and shear removal were considered as the two competing mechanisms that define the radial diffusion and later the deposition of asphaltenes. Particle transport toward the wall was considered to be caused by temperature gradient at the wall. This modeling approach was based on the well-developed theory of wax deposition (Burger et al. 1981). But this theory and modeling approach for asphaltene deposition were not supported by experimental results obtained using the organic solid deposition and control (OSDC) device. The experimental method to analyze asphaltene deposition using the OSDC was described in detail in Section 5.1.1.3. Based on the OSDC results, there was no pronounced effect of temperature gradient at the wall on the asphaltene deposition rate (Akbarzadeh et al. 2009).

Soulgani et al. (2009) introduced an asphaltene deposition model on the basis of a series of experiments on an externally heated stainless steel pipe. These experiments were conducted to observe the role of various parameters such as oil flow rate, temperature, and concentration of asphaltene precipitant on the rate of asphaltene deposition. The model for asphaltene deposition was developed based on the results of these experiments. It was assumed that the deposition on the heated pipe surface is controlled by a chemical reaction mechanism based on an Arrhenius exponential term. The asphaltene deposition rate was fitted to a simple correlation, which included parameters such as temperature, velocity, and precipitated asphaltene concentration. It was found that asphaltene deposition increases as temperature and concentration of precipitated asphaltene particles increases. No solid proof supporting the suggested deposition mechanism was provided.

6.1.1.2 Models Based on Capillary Deposition Tests

The model developed by Vargas et al. (2010) is more advanced in comparison with the predecessors. Submodels describing particle precipitation, aggregation, transport, and deposition on the wall were included. The aggregation and the deposition phenomena were modeled using pseudo–first-order reactions. A deposition simulator was proposed based on species conservation equations coupled with thermodynamic modeling of oil using the perturbed chain statistical associating fluid theory (PC-SAFT) equation of state (EOS), which was described in detail in Section 4.2.3.2. A competing phenomenon between aggregation and deposition was identified. The model contained three parameters that were estimated from experiments. The authors demonstrated that their model adequately describes the asphaltene deposition in capillary tubes. The effect of temperature on the deposition rate was also studied. It was showed that in a capillary, as temperature increases, the asphaltene deposition flux increases toward the capillary inlet and rapidly decreases toward the outlet. This behavior was found to be in good agreement with experimental data obtained from capillary experiments.

The work done by Kurup et al. (2011) is a continuation of the deposition simulator developed by Vargas et al. (2010). It involved the development of an asphaltene deposition tool (ADEPT) that can predict the occurrence and calculate the magnitude and profile of asphaltene deposition in a wellbore. The simulator consisted of a thermodynamic module and a deposition module. PC-SAFT EOS was used in the thermodynamic module to describe the phase behavior of oil, which was first characterized by using thermodynamic properties such as saturation points, asphaltene onset pressure data, and physical properties such as the density of the oil (Section 4.2.3.2). The deposition module was then used along with input from the thermodynamic module to calculate the magnitude of asphaltene deposited along the length of the wellbore and pipelines. Similar to the model developed by Vargas et al. (2010), this deposition model also consisted of three tuning parameters, each for precipitation, aggregation, and deposition mechanisms. But instead of a two-dimensional convection-diffusion transport equation as in Vargas et al. (2010), this model was simplified to a one-dimensional axial dispersion equation. The mathematical model used in the deposition module was first benchmarked and validated by comparing the simulation results against the experimentally measured asphaltene deposition flux in a capillary deposition experiment, as illustrated in Section 5.1.1.1 (Wang et al. 2004; Kurup et al. 2012). The simulator was then used to study deposition in two field cases: Kuwait's Marrat oil well and Algeria's Hassi-Messaoud oil field (Kurup et al. 2012). A proper choice of the kinetic parameters helped the deposition simulator in predicting the deposition profile both qualitatively and quantitatively, and the predictions matched very well with the field observations. It should be noted that the simulator does have limitations in capturing the entire physics of the deposition process because it is just a one-dimensional model. A detailed description of ADEPT is included in Section 6.1.2.

As discussed by Vargas et al. (2010) and Kurup et al. (2011) the transport of precipitated asphaltene particles and their subsequent deposition on surfaces are described by the convection-diffusion equations. Because there is a lack of understanding of the rich physics involved in the deposition process, the transport model is frequently of a mechanistic nature, and it relies heavily on experimental data. A more rigorous analysis of the asphaltene deposition model can be done using a computational fluid dynamic (CFD) approach. As asphaltene starts depositing on the wall surface of the wellbore or pipeline, its cross-section is continuously reduced, subsequently changing the flow field. Hence, it is important to track the deposition front while modeling asphaltene deposition. Ge et al. (2013) presented a general framework for modeling the asphaltene deposition process that focuses on a CFD-based transport model with an evolving depositing front coupled to the associated fluid, mass, and energy transport. The deposition front was captured using either a level-set approach or a total-concentration approach. The deposition process occurring at the depositing front was modeled as a first-order reaction. Fluid flow was modeled using the incompressible Navier-Stokes equations.

Along similar lines, much recently, a one-dimensional model for asphaltene deposition in wellbores or pipelines was presented by Guan et al. (2017). This model consisted of the thermodynamic module and the transport module. The thermodynamic module focused on the modeling of asphaltene precipitation using the Peng-Robinson EOS (Section 4.2.3.1). The transport module included the modeling of fluid transport, particle transport, and asphaltene deposition while tracking the deposition front as well. Three parameters, each for precipitation, aggregation, and deposition kinetics, were required in this model. Although a reasonably accurate prediction of the asphaltene deposit layer profile was attained, enhanced performance of the thermodynamic module could be attained by incorporating the PC-SAFT EOS.

6.1.1.3 Models Based on Deposition Tests in New Geometries

Eskin et al. (2011) developed a deposition model, accounting for major mechanisms of particle transport to the wall such as Brownian and turbulent diffusions, turbophoresis, and used particle flux mass transfer expressions for turbulent flows to model the deposition process. The model parameters were obtained by fitting the model predictions to the deposition results obtained from RealView, which is an OSDC (Section 5.1.1.3). The model consisted of a submodel describing the particle size distribution evolution in time in a Couette device and along a pipe and a submodel for calculating the particle transport to the wall. The concept of critical particle size was applied, where only particles that are smaller than the critical size can deposit, which is in agreement with the model proposed by Vargas et al. (2010). A shear removal term was also introduced to account for the increased shear rate at the wall and decreased deposit at higher speeds of the RealView cell. The model contained six parameters that are determined based on experiments in the RealView cell. Akbarzadeh et al. (2012) used this model and showed simulation results for asphaltene deposition with respect to experiments performed on various oils. The impacts of shear, run time, residence time, pressure, chemical inhibitor, and surface roughness on the deposition of asphaltenes were investigated. Also, for the developed asphaltene deposition model, the number of parameters were reduced from six to two (for the oils under their study) by performing sensitivity analyses. Finally, the simulation results for deposition of asphaltenes in vertical production tubings were qualitatively compared with field observations reported by Haskett and Tartera (1965). A population balance model (similar to the one established by Maqbool et al. 2011 and as discussed in Section 6.1.3) was employed for modeling the asphaltene aggregation instead of a simple second-order reaction mechanism as in Kurup et al. (2012).

Recently, asphaltene deposition was measured using a packed bed apparatus by Vilas Boas Fávero et al. (2016). It was showed that a mass transfer-limited one-dimensional deposition model can explain the asphaltene deposition of nanometer-sized unstable asphaltenes in the viscous flow regime. The model does not have any parameters to be tuned against the experimental data, but the mass-transfer coefficient is estimated using a simplified empirical correlation. Though this is the first work on such an apparatus, the model needs to be more comprehensive of the mechanisms taking place in the column. This demands a more rigorous modeling approach and prediction technique for the simulation of asphaltene deposition in such a geometry.

The capabilities and limitations of the significant models reviewed in this section are summarized in Table 6.1.

6.1.2 Development of Asphaltene Deposition Tool

To comprehend the process of asphaltene deposition taking place in the wellbore and pipelines, a simple, yet comprehensive tool developed by Kurup et al. (2011) will be discussed in detail in this section.

At reservoir conditions, the asphaltenes are mostly stable and soluble in oil. However, a change in pressure, temperature, and composition makes the asphaltenes that are originally dissolved in the oil destabilized and leads to phase separation of asphaltenes from oil. These precipitated asphaltenes are known as *primary particles*. These primary particles then combine with each other to form aggregates. It is proposed that these primary particles are susceptible to deposit on the walls of the wellbore. Hence, the three steps that take place in the wellbore, namely, precipitation, aggregation, and deposition of asphaltenes, can be modeled using a transport (convection-diffusion) equation. The mechanism is summarized in Figure 6.1. Based on this notion, Vargas et al. (2010) developed a two-dimensional convection-diffusion deposition model. ADEPT is a modification of the Vargas et al. (2010) model. It is a simple one-dimensional axial dispersion model. The mathematical model for the deposition mechanism of asphaltene is written as the material balance of the asphaltene particles in the transient state over a control volume of the pipeline or wellbore.

TABLE 6.1

Capabilities and Limitations of Wellbore Asphaltene Deposition Models

Model	Capabilities	Limitations	Parameters
Vargas et al. (2010)	• Submodels describing particle precipitation, aggregation, transport, and deposition • Two-dimensional convection-diffusion equation • Precipitation modeled using PC-SAFT EOS • Supported by capillary deposition tests	Aggregation and deposition phenomena modeled using first-order reactions	3
Eskin et al. (2011)	• Submodels describing particle-size distribution evolution in time using PBM and particle transport to the wall using particle flux mass transfer • Supported by experimental data from RealView	• Experiments on RealView are expensive • Few experimental data are available for tuning parameters	6
Kurup et al. (2012)	• Continuation of the deposition simulator developed by Vargas et al. (2010) • One-dimensional axial dispersion model as it is computationally less expensive • Supported by capillary deposition tests	• Aggregation mechanism modeled as simple second-order reaction mechanism • One-dimensional model does not capture the entire physics of deposition	3
Vilas Boas Fávero et al. (2016)	• No tuning parameters, but mass transfer coefficient estimated using a correlation • Supported by experimental data from packed bed column	One-dimensional model does not capture the entire physics of deposition	1
Guan et al. (2017)	• Submodels describing particle precipitation, aggregation, transport, and deposition • Supported by experimental data from capillary deposition tests • Accounts for the variation in the flow cross-sectional area due to the growing deposit layer • Uses a fixed mesh	• Transport module is for single-phase flow • Uses PR EOS which is not accurate for asphaltene-precipitation modelling	3

EOS, equation of state; PC-SAFT, perturbed chain statistical associating fluid theory; PBM, population balance model; PR, Peng-Robinson.

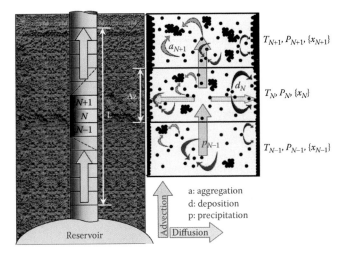

FIGURE 6.1   Mechanism for asphaltene transport in the wellbore. (Reprinted from Vargas, F. M. et al., *Energy & Fuels*, 24, 2294–2299, 2010.)

6.1.2.1   Assumptions

- Only the asphaltene aggregates, which are smaller than the critical particle size, participate in the deposition process. The large aggregated particles are considered to be carried with the flow because of inertia and do not tend to deposit. In ADEPT, 0.2 µm is assumed as the critical particle size.
- The deposition process is not dominated by transport occurring in the core flow but dominated by transport and kinetics occurring in the laminar boundary layer adjacent to the wall of the wellbore or pipeline. The laminar flow inside a capillary tube mimics the laminar boundary layer occurring in the turbulent wellbore and pipeline flows. Hence, the deposition kinetic constant can be obtained from capillary deposition experiments.
- The diffusivity of asphaltene particles in the fluid flow is assumed to be constant.

6.1.2.2   Model Development

Step 1: Precipitation and aggregation kinetics

The precipitation of asphaltene is modeled as a first-order kinetic process. Asphaltene aggregation is assumed as a second-order kinetic process.

$$\frac{dC'}{dt'} = k_p\left(C'_f - C'_{eq}\right) - k_{ag}C'^2 \tag{6.1}$$

$$\frac{dC'_{ag}}{dt'} = k_{ag}C'^2 \tag{6.2}$$

$$\frac{dC'_f}{dt'} = -k_p\left(C'_f - C'_{eq}\right) \tag{6.3}$$

where:

C' is the dimensional concentration of the primary particles

C'_{ag} is the dimensional concentration of aggregated particles

C'_f is the dimensional concentration of dissolved asphaltene in the oil–precipitant mixture

C'_{eq} is the dimensional thermodynamic equilibrium concentration of asphaltene, which can be regarded as the solubility of asphaltene at the given pressure, temperature, and composition

The value of the precipitation kinetic parameter, k_p, and aggregation kinetic parameter, k_{ag}, can be obtained by solving the Equations 6.1 through 6.3 simultaneously, by minimizing the difference between the experimental data and the modeling results, with the initial condition as the initial molar concentration of asphaltene primary particles solubilized in the oil phase, C_0. The thermodynamic modeling to determine C'_{eq} is performed using the PC-SAFT EOS. The experimental data is obtained by adding a certain amount of precipitant (*n*-heptane) to the oil sample, from which the mass of aggregated asphaltenes can be calculated for different aging times (Kurup et al. 2012), as explained in Section 3.4.

Step 2: Mass balance over axial segment of wellbore

Considering Figure 6.2 and taking into account the assumptions already stated in Section 6.1.2.1, a mass balance can be performed for the axial segment δz as,

$$V_{cell}\frac{\partial C'}{\partial t'} = -V_{cell}U_z\frac{\partial C'}{\partial z'} + V_{cell}D_{ax}\frac{\partial^2 C'}{\partial z'^2} + V_{cell}k_p\left(C'_f - C'_{eq}\right) - V_{cell}k_{ag}C'^2 - V_{int}R_{int} \quad (6.4)$$

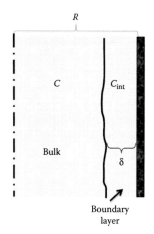

FIGURE 6.2   Laminar boundary near the wall. (Reprinted from Kurup, A. S. et al., *Energy Fuels*, 26, 5702–5710, 2012.)

where:

V_{cell} and V_{int} are the volumes of the axial segment and boundary layer segment

R_{int} is the rate of asphaltene depletion because of the deposition process

U_z is the average axial velocity of the fluid

C', C'_f, and C'_{eq} are the dimensional concentration of asphaltene at instant, t', and position, z', dimensional concentration of asphaltene in oil–precipitant mixture and dimensional thermodynamic concentration of asphaltene, respectively

D_{ax} is the axial dispersion coefficient

For laminar flows, such as in the case of capillary deposition experiments, the dispersion coefficient is calculated using $D_{ax} = D_e + \frac{U_z r^2}{48 D_e}$, provided that the pipe is long enough and the condition that, $D_e \bar{t} / r^2 > 0.125$ is met. D_e is the particle diffusivity, $\bar{t}$ is the residence time, and r is the radius of wellbore.

Step 3: Introduction of nondimensional variables

Introducing nondimensional variables, Equation 6.4 becomes,

$$\frac{\partial C}{\partial t} = -\frac{\partial C}{\partial Z} + \frac{1}{Pe} \frac{\partial^2 C}{\partial Z^2} + Da_p \left(C_f - C_{eq} \right) - Da_{ag} C^2 - \frac{V_{int}}{V_{cell}} \frac{L}{U_z} R_{int} \qquad (6.5)$$

where:

$$\frac{V_{int}}{V_{cell}} = \frac{2\pi R \delta \Delta Z}{\pi R^2 \Delta Z} = \frac{2\delta}{r} \text{ and } R_{int} = k_d C_{int} \qquad (6.6)$$

The dimensionless parameters are given as

$$Pe = \frac{U_z L}{D_{ax}}, \ Da_{ag} = \frac{k_{ag} L C_0}{U_z}, \ Da_p = \frac{k_p L}{U_z}, \ C = \frac{C'}{C_0},$$

$$C_f = \frac{C'_f}{C_0}, \ C_{eq} = \frac{C'_{eq}}{C_0}, \ Z = \frac{z}{L}, \ t = \frac{t' U_z}{L}$$

where:

L is the axial length of the wellbore

δ is the boundary layer thickness

Pe, Da_{ag} and Da_p are the non-dimensional numbers, which are defined as follows:

- Peclet number $\left(Pe = \frac{U_z L}{D_{ax}} \right)$ is the ratio of the convective transport rate to that of the diffusive transport rate.
- Precipitation Damköhler number $\left(Da_p = \frac{k_p L}{U_z} \right)$ is the ratio of the precipitation rate to that of the convective transport rate.
- Aggregation Damköhler number $\left(Da_{ag} = \frac{k_{ag} L C_0}{U_z} \right)$ is the ratio of the aggregation rate to that of the convective transport rate.

Step 4: *Rate of asphaltene deposition*

Two competing processes occur in the boundary layer, namely, transport of asphaltene particles into the boundary layer and depletion of asphaltene because of deposition kinetics. At steady state,

$$R_{mass.tr} = R_{int} \tag{6.7}$$

$$\frac{D_e}{\delta^2}\left(C - C_{int}\right) = k_d C_{int} \tag{6.8}$$

$$\text{If } \frac{C_{int}}{C - C_{int}} = \frac{D_e}{\delta^2 k_d} = \phi, \text{then } C_{int} = \frac{\phi}{\phi + 1} C \tag{6.9}$$

where:

C is the dimensionless concentration of asphaltene in the bulk
C_{int} is the dimensionless concentration in the boundary layer
k_d is the deposition kinetic constant

The diffusion coefficient is calculated using the Stokes–Einstein relation. Substituting the expression for R_{int} in Equation 6.5, gives the mass balance for the deposition of asphaltene in wellbore,

$$\frac{\partial C}{\partial t} = -\frac{\partial C}{\partial Z} + \frac{1}{Pe}\frac{\partial^2 C}{\partial Z^2} + Da_p\left(C_f - C_{eq}\right) - Da_{ag}C^2 - \frac{V_{int}}{V_{cell}}\frac{L}{U_z}k_d C_{int} \tag{6.10}$$

Step 5: *Introduction of Scaling Factor*

The mass balance for the deposition of asphaltene in wellbore is given as,

$$\frac{\partial C}{\partial t} = -\frac{\partial C}{\partial Z} + \frac{1}{Pe}\frac{\partial^2 C}{\partial Z^2} + Da_p\left(C_f - C_{eq}\right) - Da_{ag}C^2 - \frac{L}{U_z}k_d^* C \tag{6.11}$$

$$k_d^* = (k_d)_{cap}\frac{2\delta}{r}\frac{\phi}{\phi + 1} = (k_d)_{cap}ScF \tag{6.12}$$

$$ScF = \frac{2\delta}{r}\frac{\phi}{\phi + 1} \tag{6.13}$$

where *ScF* is the scaling factor used to scale the deposition kinetic constant measured using capillary deposition experiments to that of asphaltene deposition in the wellbore or pipeline.

Step 6: *Initial and boundary conditions*

The required initial and boundary conditions are:

$$C\left(t = 0, Z\right) = 0$$

$$C\left(t, Z = 0\right) = 0$$

$$\frac{\partial C}{\partial Z}(Z=1)=0 \qquad (6.14)$$

Step 7: Experimental results from capillary deposition test

To illustrate the application of ADEPT, a capillary deposition test performed by Kurup et al. (2012) is considered here. The asphaltene deposition study was conducted in a capillary tube with crude oil (stock tank oil) and propane (precipitant). The deposition test was performed at 343.15 K, with a flow rate of 12 cm³/h oil–propane mixture (Kurup et al. 2012). The asphaltene deposition profile along the axial length of the capillary tube obtained from the test is shown in Figure 6.3. The deposition thickness profile shows that the magnitude of asphaltene deposition reaches a maximum at the entrance of the tube because at this point, the driving force for precipitation is maximum. The deposition decreases along the axial length of the tube as the amount of primary particles in the flowing mixture decreases.

Step 8: Deposition simulator predictions

In this example, the values of the precipitation and aggregation kinetic parameters were obtained by solving the equations as given in *Step 1*, to find $k_p = 1.32 \times 10^{-3}\,\mathrm{s}^{-1}$ and $k_{ag} = 7.29 \times 10^{-3}\,\mathrm{m}^3\,\mathrm{mol}^{-1}\,\mathrm{s}^{-1}$. The deposition-kinetic parameter value is obtained by matching the peak of the deposition-flux profile obtained from experimental data to that of the simulator predictions ($(k_d)_{cap} = 1.4 \times 10^{-3}\,\mathrm{s}^{-1}$). The asphaltene deposition flux along the axial length of capillary tube obtained from the both the deposition test and ADEPT simulation are shown in Figure 6.3.

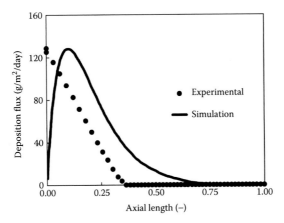

FIGURE 6.3 Comparison of experimental and simulation results for asphaltene deposition flux along the dimensionless axial length of capillary pipe. (Reprinted from Kurup, A. S. et al., *Energy Fuels*, 26, 5702–5710, 2012.)

The comparison between the two curves shown in Figure 6.3 indicates that the simulation results show a delayed deposition compared to the corresponding experiments. This is because of the simulator being modeled in such a way that asphaltene precipitation begins only at the entrance of the pipe. This means that there is no delay between mixing of oil with the precipitant and entrance of this mixture into the pipe. Choosing a higher value of the deposition constant can decrease this discrepancy; however, this can cause an overprediction of the deposition flux peak magnitude compared to experimental measurements

With the simulation of asphaltene deposition in a capillary tube being performed, asphaltene deposition in a wellbore can be predicted now. $(k_d)_{cap}$ value obtained in *Step 8* can be scaled to k_d^* with the help of the scaling factor ScF, introduced in Equation 6.13. One such case study is discussed in detail in Section 8.2.1.

6.1.3 MODELING ASPHALTENE AGGREGATION

Instead of using a simple reaction mechanism to model an asphaltene aggregation process (as seen in ADEPT in Section 6.1.2), a population balance model (PBM) can be implemented. Such a model helps to understand the physics of the process more clearly and obtain a particle size distribution of the asphaltene primary units and aggregates with progress in time. Maqbool et al. (2011) developed a generalized geometric PBM for simulating the growth of asphaltene aggregates from the nanometer scale to micrometer-sized particles. The modeling was performed with respect to results obtained from experiments, where samples were withdrawn from the well-stirred crude oil–heptane mixture at different times and centrifuged. The asphaltene cake obtained as a result of centrifugation was then washed with heptane several times to remove any residual crude oil in the cake, dried in an oven, and weighed to determine the mass of the precipitated asphaltenes (Maqbool et al. 2011). The generation and depletion schemes of the ith aggregate were described by four mechanisms, which are shown in Table 6.2.

In Table 6.2, $\bar{R}$ is the geometric spacing between two aggregates and the kinetic parameter, $K_{i,j}$, which is the collision kernel, and is given as $K_{i,j} = \alpha_{i,j}\beta$, where β is the efficiency of collision and $\alpha_{i,j}$ is the collision frequency, which can be represented using the Brownian aggregation kernel as,

$$\alpha_{i,j} = \frac{2RT\left(d_i + d_j\right)^2}{3\mu d_i d_j} \qquad (6.15)$$

where:
d_i and d_j represent the diameters (m) of colliding aggregates i and j
μ is the viscosity of the medium (kg m^{-1} s^{-1})
R is the universal gas constant ($R = 8.314$ J K^{-1} kmol^{-1})
T is the absolute temperature (K)
β is the single fitting parameter in this geometric PBM and can be estimated from experimental data involving the growth of the aggregates

TABLE 6.2

Mechanism for the Generation and Depletion of the ith Aggregate in the Geometric PBM

Mechanism	Reaction	Rate of Reaction
Generation 1	$\bar{R}A_{i-1} \rightarrow A_i$	$\left(\dfrac{dC_i}{dt}\right)_{GM1} = -\dfrac{K_{i-1,i-1}}{\bar{R}} C_{i-1}^2$
Generation 2	$A_{i-1} + mA_j \rightarrow A_i, j < i-1,$ $m = \left(\bar{R}^{i-1} - \bar{R}^{i-2}\right)/\bar{R}^{j-1}$	$\left(\dfrac{dC_i}{dt}\right)_{GM2} = C_{i-1} \displaystyle\sum_{j=1}^{i-2} K_{i-1,j} \dfrac{\bar{R}^{j-1}}{\bar{R}^{i-1} - \bar{R}^{i-2}} C_j$
Depletion 1	$A_i + mA_j \rightarrow A_{i+1}, j < i, m = \left(\bar{R}^i - \bar{R}^{i-1}\right)/\bar{R}^{j-1}$	$\left(\dfrac{dC_i}{dt}\right)_{DM1} = -C_i \displaystyle\sum_{j=1}^{i-1} K_{i,j} \dfrac{\bar{R}^{j-1}}{\bar{R}^i - \bar{R}^{i-1}} C_j$
Depletion 2	$A_i + A_j \rightarrow A_i, j \geq i$	$\left(\dfrac{dC_i}{dt}\right)_{DM2} = -C_i \displaystyle\sum_{j=1}^{N-1} K_{i,j} C_j$

Source: Adapted from Maqbool, T. et al., *Energy & Fuels*, 25, 1585–1596, 2011.
PBM, Population Balance Model.

The net rate of generation of the ith aggregate is given as sum of the rates of reaction given in Table 6.2, as

$$\frac{dC_i}{dt} = -\frac{K_{i-1,i-1}}{\bar{R}} C_{i-1}^2 + C_{i-1} \sum_{j=1}^{i-2} K_{i-1,j} \frac{\bar{R}^{j-1}}{\bar{R}^{i-1} - \bar{R}^{i-2}} C_j$$

$$-C_i \sum_{j=1}^{i-1} K_{i,j} \frac{\bar{R}^{j-1}}{\bar{R}^i - \bar{R}^{i-1}} C_j - C_i \sum_{j=1}^{N-1} K_{i,j} C_j \tag{6.16}$$

These set of coupled ordinary differential equations need to be solved simultaneously to obtain the molar concentration of the ith aggregate as a function of time, with the initial condition being,

$$C_1(0) = \frac{m_A \varphi_o \rho_o}{M_{w,A} N_{agg}}$$

$$\tag{6.17}$$

$$\text{and} \quad C_i(0) = 0 \qquad \text{for } i > 1$$

where:

m_A (kg kg^{-1}) is the mass fraction of destabilized asphaltene per unit mass of oil resulting from the addition of a specified quantity of heptane

φ_o is the volume fraction of oil in the oil–heptane mixture

ρ_o is the oil density

The total mass of destabilized asphaltene per unit volume of the mixture is $m_A \varphi_o \rho_o$. $M_{w,A}$ is the molecular weight of the asphaltene molecules, and N_{agg} is the number of asphaltene molecules per nanoaggregate ($N_{agg} \approx 8$). And finally, the evolution of the mass of aggregates separated by centrifugation is given by

$$m_A(t) = \frac{\sum_{i=1}^{N} \overline{R}^{i-1} M_{w,A} s_i C_i(t)}{w_o} \tag{6.18}$$

where:

$m_A(t)$ is the mass of separated asphaltene per unit mass of oil (kg kg^{-1})

w_o is the mass of oil per unit volume of mixture (kg m^{-3})

s_i is the separation efficiency of the ith aggregate in the centrifuge

An example from Maqbool et al. (2011) is illustrated here. The experimental results for the mass of asphaltenes collected by centrifugation at different times for 50 vol% heptane addition to an oil sample and the corresponding simulation results are shown in Figure 6.4.

When this result is combined with the molecular weights for each particle size, a mass fraction-based particle size distribution (PSD) can be calculated. The particle size distribution evolution for 50 vol% heptane addition is shown in Figure 6.5. It should be noted that data markers are the only particle sizes allowed by the geometric PBM, and the connecting lines are only a visual guide.

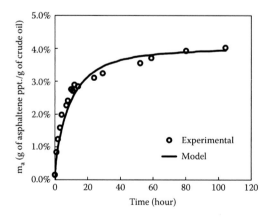

FIGURE 6.4 Experimental and simulated results of the separated aggregates for 50 vol% heptane. (Reprinted from Maqbool, T. et al., *Energy Fuels*, 25, 1585–1596, 2011.)

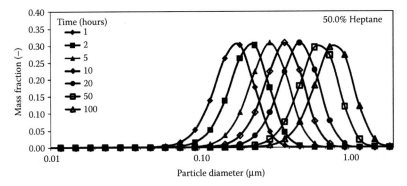

FIGURE 6.5 Particle size distribution of asphaltene particles as a function of time for 50 vol% heptane with oil. (Reprinted from Maqbool, T. et al., *Energy Fuels*, 25, 1585–1596, 2011.)

6.2 ASPHALTENE DEPOSITION IN POROUS MEDIA

To study and comprehend the deposition tendencies of asphaltenes in porous media, it is important to examine the different ways by which asphaltenes could affect its permeability. Permeability is the property of rocks (porous media), that is, an indication of the ability for fluids (gas or liquid) to flow through the rocks. There are several mechanisms that affect the permeability of porous media. The three most common mechanisms are surface deposition, entrainment, and pore-throat plugging. A new mechanism called *pore-throat opening* was recently introduced (Kord et al. 2014).

- Surface deposition: Asphaltene particles begin to deposit on the surface as a result of gravity or intermolecular forces between asphaltene particles and the surface of the rock (grains). Surface deposition is shown in Figure 6.6a. Here, as crude oil flows into the pore space, asphaltene particles start depositing on the surface (black solid circles).
- Entrainment: When more and more asphaltenes get deposited in the pore, the pore volume becomes smaller, which in turn increases the interstitial velocity and erodes away some of the deposited asphaltenes. The phenomenon of entrainment of the deposited asphaltene is shown in Figure 6.6b (grey solid circles).
- Pore-throat plugging: The pore-throat plugging mechanism is shown in Figure 6.6c. This mechanism represents the deposition of asphaltene particles in the pore-throat region (black open circles).
- Pore-throat opening: Pore-throat opening mechanism was proposed by Kord et al. (2014), which is displayed in Figure 6.6d. It is a counterbalance mechanism of pore-throat plugging. When the deposition of asphaltene increases in the pore throat, the cross-sectional area of the pore throat decreases, subsequently increasing the flow rate and thus taking away the deposited asphaltene because of its increased flow rate.

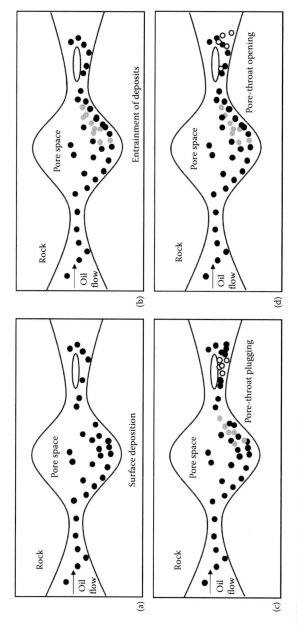

FIGURE 6.6 Mechanisms of deposition: (a) surface deposition, (b) entrainment of deposition, (c) pore-throat plugging, and (d) pore-throat opening. (Reprinted from Kord, S. et al., *Fuel*, 117, 259–268, 2014.)

6.2.1 Deposition Models for Porous Media

In the model proposed by Gruesbeck and Collins (1982), three mechanisms (surface deposition, entrainment, and pore plugging) were considered. Two different pathways were considered inside the porous media: pluggable and nonpluggable pathways. Pluggable pathways represented the small pores, which have the potential to get completely plugged. Nonpluggable pathways denoted the large pores where asphaltene deposition mechanisms, such as surface deposition and entrainment, take place. As the pore-throat cross-sectional area reduces as a result of asphaltene deposition, the local velocity increases rapidly such that it entrains the deposited asphaltenes out of the pores. Therefore, it was assumed that the large pores could not be totally plugged by the asphaltene aggregates.

Ali and Islam (1998) combined two mechanisms for the simulation of asphaltene deposition in porous media. The first mechanism was the asphaltene adsorption using surface excess theory as examined by Sarwar and Islam (1997), and the other one was the deposition model proposed by Gruesbeck and Collins (1982). Three different plugging regimes were defined based on the flow rate, namely, monotonous steady state, quasi–steady state, and continuous plugging, in the order of lower to higher flow rate (Ali and Islam 1998). In this model, the relationship between porosity and permeability was not considered.

Wang and Civan (2001) modified the Gruesbeck and Collins' (1982) model by considering the relationship between porosity and permeability. A polymer solution theory, developed by Hirschberg et al. (1984) simulating the precipitation process, was combined with the deposition model. This theory along with the mass balance calculation of the asphaltenes in the required control volume, and porosity and permeability relationship were incorporated into a three-dimensional, three-phase black-oil simulator. The results were verified using the data from core-flooding experiments (Minssieux 1997).

Almehaideb (2004) developed a model to predict precipitation and deposition of asphaltenes in the near-wellbore region, during primary production. For the deposition model, adsorption mechanism, calculated using Langmuir isotherm equation was added to the modified Wang and Civan's (2001) model.

Lawal et al. (2011) investigated the permeability reduction in porous media by applying the deep-bed filtration (DBF) theory along with the kinetics of asphaltene deposition under dynamic conditions. Filtration coefficient, one of the fitting parameters, was used to represent the overall effect because of asphaltene precipitation, flocculation, and deposition. But the model did not include a dispersive term, and the porous media was assumed to be a homogeneous and isotropic system.

Soulgani et al. (2011) developed a model for asphaltene deposition in porous media, where the rate of deposition was predominantly affected by the velocity of the crude oil, the surface temperature, and the asphaltene concentration of the suspension. The deposition model was coupled with a precipitation model (Hirschberg et al. 1984) and the results were verified using the data obtained from core-flooding experiments (Minssieux 1997).

Jafari Behbahani et al. (2015) developed an asphaltene deposition model that coupled multilayer adsorption and Wang and Civan's model (2001) for mechanical entrapment. The results proved to be in agreement with the experimental data better than the results from Wang and Civan (2001). It was concluded that multilayer kinetic adsorption mechanism plays an important role in asphaltene deposition, wettability alteration, and permeability reduction of the rock.

Kord et al. (2014) modified the surface deposition mechanism in Wang and Civan (2001) by introducing a new mechanism called pore-throat opening. The results from this model matched very well with the core-flooding experimental data.

An analysis of the existing literature reveals that developing an asphaltene deposition model that can describe deposition mechanisms in porous media accurately including the true physics of deposition is still at initial stages of development. Moreover, all the current models are one-dimensional, and some of these models contain a significant number of fitting parameters; hence, they may have limited applicability and predictive capabilities. Table 6.3 presents the capabilities and limitations of the existing models and lists the number of parameters required in each case. The Wang and Civan (2001) model is the most popular model used to simulate asphaltene deposition in porous media because it captures most of the asphaltene deposition mechanisms.

6.2.2 Introduction to Lattice-Boltzmann Modeling

The methods adopted for modeling fluid flow can be classified into three main categories: macroscopic, mesoscopic, and microscopic. In macroscopic-scale modeling, fluid is assumed as a continuous medium. The motion of fluid satisfies the conservation of mass, momentum, and energy. To analyze the fluid flow, nonlinear partial differential Euler equations and Navier-Stokes equations need to be discretized and solved. In microscopic-scale modeling, it is assumed that fluid is composed of many individual molecules. The motion of these molecules is affected by their intermolecular forces and collisions. However, one of the disadvantages of modeling fluid flow based on the microscopic approach includes its excessive computational cost and its inherent limitation to be extended to large-scale systems. Besides macroscopic- and microscopic-flow simulation methods, there is yet another method and is based on mesoscale. In mesoscale modeling, we consider fluid to be a collection of a large number of molecules, instead of focusing on individual molecules as in microscopic approach. Lattice-Boltzmann modeling method is a mesoscale numerical method.

LBM was first introduced by McNamara and Zanetti (1988). Different from the traditional CFD calculations, LBM is based on molecular kinetic theory. It uses a distribution function to represent the behavior of a group of particles, and thus, saving a large amount of computational memory when compared to the molecular dynamic simulation techniques (Mohamad 2011; Guo 2013). Moreover, LBM is comparatively easier to implement and more suitable for parallel computing when compared to other traditional CFD discretization methods such as finite difference, finite element, and finite volume methods. LBM can also be readily implemented on complex geometries, which is important for modeling fluid flow in porous media.

TABLE 6.3
Capabilities and Limitations of Porous Media Asphaltene Deposition Models

Model	Capabilities	Limitations	Parameters
Gruesbeck and Collins (1982)	• First asphaltene deposition model • Combination of surface deposition, entrainment, and pore plugging	• Adsorption term neglected • Many fitting parameters • Did not consider the relationship between porosity and permeability	10
Ali and Islam (1998)	Combination of adsorption term and mechanical entrapment	• Too many fitting parameters • Did not consider the relationship between porosity and permeability	16
Wang and Civan (2001)	• Modified Gruesbeck and Collins model • Considered the relationship between porosity and permeability • Fewer fitting parameters • Extendable to field cases	• Neglected adsorption term • Number of fitting parameters slightly high	7
Almehaideb (2004)	Includes adsorption and mechanical entrapment mechanism	Number of fitting parameters slightly high	7
Lawal (2011)	• Considers deep-bed filtration theory • Only three parameters are needed	• Only one parameter is used to represent the effect of asphaltene precipitation, aggregation, and deposition • Can be applied to only homogenous and isotropic system	3
Jafari Behbahanil et al. (2015)	Considered multilayer adsorption	Too many fitting parameters	10
Kord et al. (2014)	• Added pore-throat opening mechanism. • More accurate than Wang and Civan model.	• Number of fitting parameters slightly high • Neglected adsorption term	8

There are various equations and parameters that comprise the development of the LBM method, which are introduced in Sections 6.2.2.1 through 6.2.2.4. Before the application of LBM for the prediction of asphaltene deposition in porous media, it is of utmost importance to fathom each of these steps carefully.

6.2.2.1   Distribution Function

To account for a large number of molecules introduced in the LBM, a distribution function was proposed to represent the average effect caused by these molecules. The main idea behind this conceptualization is that the velocity and position of each molecule at any instant of time is not important; instead, the distribution function that represents the corresponding property of the collection of particles is important. The distribution function is given as,

$$f(c) = 4\pi \left(\frac{m}{2\pi k_b T} \right)^{\frac{3}{2}} c^2 e^{-\frac{mc^2}{2k_b T}} \tag{6.19}$$

where:
  f is the distribution function
  m is the mass of the particle
  c is the velocity vector of the particle
  T is the temperature
  k_b is the Boltzmann constant

6.2.2.2 Boltzmann Transport Equation

The distribution function is then used in a Boltzmann transport equation. Equation 6.20 represents the Boltzmann transport equation.

$$\frac{\partial f}{\partial t} + \frac{\partial f}{\partial \bar{r}} \cdot c + \frac{F}{m} \cdot \frac{\partial f}{\partial c} = \Omega \tag{6.20}$$

where:
  F represents an external force
  $\bar{r}$ is the position vector of the particle
  Ω is the collision operator

6.2.2.3 The Bhatnagar, Gross, and Krook Model

To overcome the problem of complexity of the collision term in the Boltzmann equation, Bhatnagar, Gross, and Krook (BGK; 1954) used a simplified model to represent the collision operator, which is given as,

$$\Omega = \sigma \left(f^{eq} - f \right) = \frac{1}{\tau} \left(f^{eq} - f \right) \tag{6.21}$$

where:
  σ is the collision frequency
  τ is the relaxation factor
  f^{eq} is the local equilibrium distribution function

Substituting the BGK approximation in the Boltzmann equation,

$$\frac{\partial f_i}{\partial t} + c_i \nabla f_i = \frac{1}{\tau} \left(f_i^{eq} - f_i \right) \tag{6.22}$$

which is the LBM discretized Boltzmann equation. This equation is only valid along some specific directions. To decide the specific directions in which the particles should move, a lattice arrangement is required in LBM.

6.2.2.4 Lattice Arrangement

In LBM, the most common way to represent the lattice arrangement is using the DnQm notation, where n refers to the number of dimensions and m refers to the number of directions that the particles can move. Here are some examples of lattice arrangement normally used in LBM.

- One-dimensional: The most common lattice arrangement for one-dimensional LBM is D1Q2 and D1Q3, which are shown in Figure 6.7a. In D1Q2 arrangement, particles can move to the right or left, whereas, in D1Q3 arrangement, in addition to moving left or right, particles have one more choice to stay at their original position.

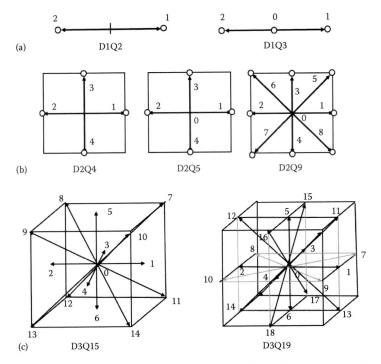

FIGURE 6.7   Lattice arrangements for (a) one-dimensional, (b) two-dimensional, and (c) three-dimensional Lattice-Boltzmann method (LBM). (a and b, Adapted from Guo, Y. and M. Wang, M., *J. Comput. Phys.*, 315, 1–15, 2016; c, Reprinted from Delbosc, N. et al., *Comput. Math. Appl.*, 67, 462–475, 2014.)

- Two-dimensional: The lattice arrangement for two-dimensional LBM are D2Q4, D2Q5, and D2Q9 as shown in Figure 6.7b. In D2Q4, the particles can move in four directions and in D2Q5, additionally, the particles can stay at their original position. In D2Q9 arrangement, particles can move in eight directions or stay at their original position.
- Three-dimensional: There are two kinds of lattice arrangement for three-dimensional LBM, namely, D3Q15 and D3Q19, which are shown in Figure 6.7c. D3Q15 is the most common arrangement used for three-dimensional modeling. (Nit et al. 2013; Premnath et al. 2013). More details on LBM and the implementation of LBM can be found in Mohamad (2011).

6.2.3 APPLICATION OF LATTICE-BOLTZMANN MODELING

LBM has proved to be an effective method to simulate flow in porous media. Succi et al. (1989) first applied LBM to calculate the three-dimensional flow in a porous medium. The relationship between porosity and permeability was also established. Heijs and Lowe (1995) applied LBM to study the permeability reduction using Carman-Kozeny equation (Carman 1939) for porous media composed of a random array of spheres and also for clay soil. Ferréol and Rothman (1995) used three-dimensional tomographic skill to reconstruct the Fontainebleau sandstone. The accuracy of LBM simulation was better for samples of larger size. Simulation results agreed well with the data from laboratory experiments. Hence, LBM can now be used to model asphaltene deposition in porous media as well.

As described in Section 5.1.2.2, asphaltene deposition in porous media is studied at laboratory scale using microfluidic devices (microchannel) and core-flooding experiments. The application of LBM to model asphaltene deposition occurring in a microchannel is illustrated in the following subsections. By incorporating the complexities of channel geometry, fluid flow rates, diffusion coefficients, and possible chemical interactions into a LBM, the behavior of the particular system can be accurately predicted.

6.2.3.1 Assumptions

- The fluid is assumed to be incompressible and Newtonian in nature with constant viscosity.
- The third dimension, depth of the device, is small compared to the other two dimensions such that a shallow planar configuration is created. The governing equation becomes identical to that of the potential flow (which is characterized by an irrotational velocity field) and to the flow of fluid through a porous medium (Darcy's law). It thus permits visualization of this kind of flow in two dimensions.
- The flow is fully developed and laminar.
- The energy dissipation is negligible, and hence, there is no heat transfer to or from the ambient medium.

6.2.3.2 Model Development

Microfluidic devices facilitate in the experimental determination of asphaltene deposition in porous media at laboratory scale. Microchannel experiments and their capabilities were discussed in detail in Section 5.1.2.2. In this section, one such experiment performed by He et al. (2018) will be used to illustrate the application of LBM and further simulate asphaltene deposition in the microchannel geometry.

Step 1: Microchannel geometry

The microchannel geometry for LBM simulation is built as per the device used by He et al. (2018) for their deposition tests. The experiment was performed at ambient pressure and temperature in a microchannel of length 1 cm, width 1.8 mm, and depth 0.02 mm. A number of circular posts of diameter 0.125 mm were present in the device, which in turn acted as an obstacle to the fluid flow and served as a surface for deposition. The porosity of the microchannel was 0.85.

Crude oil and a precipitant (*n*-heptane) were injected into the microchannel. The density and viscosity of the fluid mixture were 0.72 g/cm^3 and 0.5 cP, respectively. The flow rate was maintained at 100 μL/min.

Step 2: Governing equations for fluid flow

The continuity equation is given as,

$$\frac{\partial(\rho v_x)}{\partial x} + \frac{\partial(\rho v_y)}{\partial y} = 0 \qquad (6.23)$$

where:

v_x and v_y represent the velocity fields in x and y directions, respectively

ρ is the density of the fluid in the microchannel

Momentum transfer in x and y directions are given by the following Navier–Stokes equations:

$$\frac{\partial(\rho v_x)}{\partial t} + \frac{\partial(\rho v_x v_x)}{\partial x} + \frac{\partial(\rho v_y v_x)}{\partial y} = -\frac{\partial P}{\partial x} + \frac{\partial}{\partial x}\left(\mu \frac{\partial v_x}{\partial x}\right) + \frac{\partial}{\partial y}\left(\mu \frac{\partial v_x}{\partial y}\right) \qquad (6.24)$$

$$\frac{\partial(\rho v_y)}{\partial t} + \frac{\partial(\rho v_x v_y)}{\partial x} + \frac{\partial(\rho v_y v_y)}{\partial y} = -\frac{\partial P}{\partial y} + \frac{\partial}{\partial x}\left(\mu \frac{\partial v_y}{\partial x}\right) + \frac{\partial}{\partial y}\left(\mu \frac{\partial v_y}{\partial y}\right) \qquad (6.25)$$

where:

P is the pressure

μ is the viscosity of the fluid

Step 3: Fluid flow simulation using LBM

As described in detail in Section 6.2.2, the velocity vector is written in terms of the Boltzmann Equation 6.20, with the collision operator expressed in terms of the BGK model as described in Equation 6.21. Once the BGK model is applied to the Boltzmann equation, a LBM discretized Boltzmann equation is obtained, as shown in Equation 6.22. And as two-dimensional modeling is performed for a microfluidic device, the D2Q9 lattice arrangement is used.

The boundary conditions are applied in such a way that velocity is known at the inlet and pressure (=1 atm) at the outlet. On the wall surfaces and on the surface of the circular posts, no slip boundary condition is applied.

Step 4: Modeling asphaltene deposition

Asphaltene deposition is initiated by the addition of the precipitant (*n*-heptane) in the microchannel. The deposition process is modeled using a two-dimensional convection-diffusion-reaction equation. The mass balance of the suspended asphaltene particles in the required control volume of the microchannel is given as,

$$\frac{\partial m}{\partial t} + \nabla \cdot (\boldsymbol{u}m) = \nabla \cdot \left(D_e \nabla m \right) + R_p - R_{ag} - R_d \tag{6.26}$$

where:

m and m_D represent the mass fraction of the suspended and deposited asphaltene in the crude oil, respectively

D_e is the diffusivity of the asphaltenes in the fluid medium

$\boldsymbol{u}$ is the velocity field, which is obtained by solving the momentum transfer equations using LBM as shown in *Step 3*

R_p, R_{ag}, and R_d represent the rates of asphaltene precipitation, aggregation, and deposition, respectively

In the present example, the rates R_p and R_{ag} are neglected for simplicity. Nevertheless, the rate of asphaltene precipitation can be explicitly calculated if needed, based on the supersaturation degree, which is the difference between the actual concentration of asphaltene in the oil–precipitant mixture and the concentration of asphaltene at thermodynamic equilibrium as shown in Equation 6.3. PC-SAFT EOS (Section 4.2.3.2) can be used to calculate the thermodynamic equilibrium concentration of asphaltene at a particular pressure and temperature. The rate of aggregation can also be modeled using a simple reaction mechanism as shown in Equation 6.2. The rate of asphaltene deposition in the microchannel is modeled as,

$$R_d = -\frac{\partial m_D}{\partial t} = a\,m\,\frac{\partial \boldsymbol{u}}{\partial x} + b\boldsymbol{u} \tag{6.27}$$

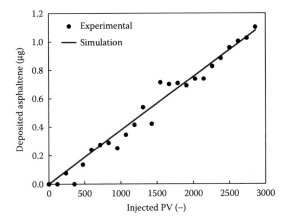

FIGURE 6.8 Amount of deposited asphaltene as a function of injected pore volume for a flow rate 100 μL/min in microchannel. (Experimental results from He, P. et al., Effect of flow rates on the deposition of asphaltenes in porous media, In Preparation, 2018.)

In Equation 6.27, the first term on the right-hand side of the equation demonstrates surface deposition mechanism and the second term indicates the entrainment mechanism. a and b are the phenomenological constants. a is called the surface deposition coefficient, and b is the entrainment coefficient.

Step 5: Estimation of deposition model parameters

The surface deposition and entrainment coefficient values for simulation of asphaltene deposition in the microchannel are obtained by tuning the results from the deposition model against the data obtained from the microfluidic experiments. For this example, the value of a is estimated as 1.69×10^{-5}, and the value of b is 3.69×10^{-3} m^{-1}. The experimental and simulation results for the mass of deposited asphaltene with respect to the injected pore volume (PV) are shown in Figure 6.8. The effect of oil–precipitant mixture flow rate on asphaltene deposition, permeability reduction caused thereby in the microchannel, and many more interesting results are discussed for a case study in Section 8.2.3.

6.3 MODELING OF ASPHALTENE TRANSPORT PROPERTIES

Asphaltene transport properties are important parameters required in the deposition model. While investigating asphaltene deposition in wellbore, transportation pipelines, or porous media, asphaltene transport properties, such as density, viscosity, and diffusion coefficient at that particular pressure and temperature conditions need to be clearly specified. In this section, prediction and calculation of density, viscosity, and diffusion coefficient of asphaltenes will be discussed in detail.

6.3.1 Density

Density is often calculated along with phase-behavior calculation using the same EOS, so that thermodynamic description of phase compositions and properties are consistent. Density, ρ, can be expressed as the following,

$$\rho = \frac{M}{\hat{V}} \tag{6.28}$$

where:
   M is the molecular weight
   $\hat{V}$ is the molar volume

When working with mixtures, the mixture average molecular weight, M_{mix}, and mixture molar volume, $\hat{V}_{mix}$ are used instead of M and $\hat{V}$. M_{mix} can be calculated from the phase composition solved from flash calculation (Section 4.1),

$$M_{mix} = \sum_{i=1}^{N} x_i M_i \tag{6.29}$$

where x_i is the mole fraction of component i. The molar volume for mixture, $\hat{V}_{mix}$ is calculated from the compressibility factor, Z, which is in turn obtained by using an EOS for a given temperature, T, and pressure, P, using Equation 6.30:

$$\hat{V}_{mix} = \frac{ZRT}{P} \tag{6.30}$$

6.3.2 Viscosity

Viscosity is a transport property necessary to simulate fluid flow behavior. The viscosity of hydrocarbon fluids can vary by several orders of magnitude not only because of variation in temperature, pressure, and compositions but also because of the presence of aggregated asphaltene particles when the solution becomes thermodynamically unstable. Therefore, having an accurate viscosity model is important to predict the asphaltene deposition in the wellbore and porous media.

Viscosity models that are applicable to hydrocarbon systems have been thoroughly reviewed (Mehrotra et al. 1996; Vesovic et al. 2014). The gas phase viscosity can be well predicted by models based on the kinetic theory and Chapman-Enskog theory (Mehrotra et al. 1996). The kinetic theory bridges transport properties to molecular properties for dilute gases, where the Boltzmann equation is solved using the Chapman-Enskog theory. On the other hand, predicting the liquid phase viscosity is more challenging. The viscosity models for the liquid phase are usually classified into empirical models and semi-theoretical models. Semi-theoretical models are adopted in industries compared to empirical models because they offer a better performance. The Lohrenz-Bray-Clark (LBC) method, based on the viscosity correlation proposed by Jossi et al. (1962) is an exception, albeit being an empirical model. The LBC method to predict viscosity is often the preferred method in reservoir simulation because of its simplicity and low computational cost compared to the more accurate semi-theoretical models. Although the method is simple, the LBC method is known to provide results of low quality (Pedersen et al. 2015). Hence, while modeling processes that require greater accuracy, application of semi-theoretical models is recommended. Viscosity models for petroleum fluids specifically are often broadly classified either as black-oil approach models or compositional models. Compositional models are usually semi-theoretical. Semi-theoretical models often applied in the industry are the corresponding state principle (CSP) and friction theory (f-theory). Although hard-sphere based models such as the Dymond and Assael approach (Assael et al. 1992) and the Vesovic-Wakeham (VW) model (1989) have found some practical applications, a dependable and consistent model to predict viscosity for petroleum fluids is not always readily available.

6.3.2.1 Lohrenz-Bray-Clark Method

The LBC method (Lohrenz et al. 1964) adopted the viscosity correlation developed by Jossi et al. (1962) for pure components. The viscosity correlation is a fourth-degree polynomial in reduced density ρ_r, which is expressed as,

$$((\mu - \mu^*)\xi + 10^{-4})^{1/4} = a_1 + a_2\rho_r + a_3\rho_r^2 + a_4\rho_r^3 + a_5\rho_r^4 \tag{6.31}$$

$$\xi = \frac{\left(\sum_{i=1}^{nc} z_i T_{ci}\right)^{1/6}}{\left(\sum_{1}^{nc} z_i M_i\right)^{1/2}\left(\sum_{i=1}^{nc} z_i P_{ci}\right)^{2/3}} \tag{6.32}$$

$$\rho_r = \rho/\rho_c \tag{6.33}$$

$$\rho_c = \frac{1}{\hat{V}_c} = \frac{1}{\sum_{i=1}^{nc} z_i \hat{V}_{ci}} \tag{6.34}$$

where:
   μ is the viscosity
   ξ is the viscosity-reducing parameter
   a_i represents the viscosity correlation polynomial coefficients
   ρ_r is the reduced density
   μ^* is the low-pressure gas mixture viscosity
   z_i is the mole fraction of component i
   ρ_c is the critical density
   $\hat{V}_c$ is the critical molar volume

The default values of coefficients a_i are provided in Table 6.4, which were regressed to viscosity data of 11 lighter gases, the heaviest being n-pentane (Jossi et al. 1962).

   Lohrenz et al. (1964) suggested calculating the critical density of a crude oil as follows,

TABLE 6.4

Default Values of Coefficients in the Viscosity Correlation in LBC Method

Parameters in LBC Model	Default Values
a_1	0.10230
a_2	0.023364
a_3	0.058533
a_4	−0.040758
a_5	0.0093324

LBC, Lohrenz-Bray-Clark.

$$\rho_c = \frac{1}{\hat{V}_c} = \frac{1}{\sum_{\substack{i=1 \\ i \neq C_{7+}}}^{nc} z_i \hat{V}_{ci} + z_{C_{7+}} \hat{V}_{cC_{7+}}} \tag{6.35}$$

where nc refers to the number of components, and the critical molar volume of the C_{7+} fraction $\hat{V}_{cC_{7+}}$ can be estimated from,

$$\hat{V}_{cC_{7+}} = 21.573 + 0.015122 M_{C_{7+}} - 27.656 SG_{C_{7+}} + 0.070615 M_{C_{7+}} SG_{C_{7+}} \tag{6.36}$$

Based on the C_{7+} fraction molecular weight $M_{C_{7+}}$ and specific gravity $SG_{C_{7+}}$.

The $\hat{V}_c$ of pseudo-components and the five coefficients a_i can be treated as tuning parameters in the LBC method to obtain better modeling accuracy. The parameter tuning must be undertaken with extreme caution to avoid nonphysical predictions from the fourth-degree polynomial formulation. Because of this concern and the limited viscosity data collected from routine pressure, volume, and temperature experiments, parameter tuning for LBC method can be challenging in the practice.

6.3.2.2 Friction Theory

Since Quiñones-Cisneros et al. (2000) proposed the friction theory, it has been applied to model viscosity for a wide range of fluids including alkanes, natural gas, crude oil, heavy oil, and blends (Quiñones-Cisneros et al. 2000, 2001, 2005; Zeberg-Mikkelsen et al. 2002; Schmidt et al. 2005; Abutaqiya et al. 2018). Friction theory conceptualizes viscosity as a mechanical property instead of a transport property. Viscosity is expressed as a sum of two viscosity contributions,

$$\mu = \mu_0 + \mu_f \tag{6.37}$$

where:
  μ_0 represents the dilute gas contribution
  μ_f represents the friction contribution

The dilute gas contribution, μ_0, can be determined from the model based on the Chapman-Enskog theory (Chung et al. 1988; Quiñones-Cisneros et al. 2001). Quiñones-Cisneros et al. (2000) correlated the friction contribution to the repulsive and attractive pressure terms by analogy with the Amontons-Coulomb friction law (Quiñones-Cisneros et al. 2000). For a cubic EOS, the friction contribution is formulated as follows,

$$\frac{\mu_f}{\mu_c} = \hat{\mu}_f = \hat{k}_a P_a + \hat{k}_r P_r + \hat{k}_{rr} P_r^2 \tag{6.38}$$

where:

μ_c is the characteristic critical viscosity

$\hat{\mu}_f$ is the dimensionless friction viscosity contribution

$\hat{k}$ is the viscosity friction coefficient

The subscript r, a, and rr represent repulsive, attractive, and second-order repulsive term, respectively

P_r and P_a are the repulsive and attractive pressures that can be obtained from a typical cubic EOS

For the Peng-Robinson EOS, P_a and P_r are calculated from the following expression,

$$P_a = -\frac{a}{\hat{V}^2 + 2\hat{V} b - b^2} \tag{6.39}$$

$$P_r = \frac{RT}{\hat{V} - b} \tag{6.40}$$

where the same nomenclature is used as PR EOS which was introduced in Section 4.2.3.1. The characteristic critical viscosity, μ_c, of light alkanes can be found in Quiñones-Cisneros et al. (2001). For undefined components, the characteristic critical viscosity, μ_c, can be estimated from a correlation (Uyehara and Watson 1944),

$$\mu_{c,i} = K_c \frac{M_i^{1/2} P_{c_i}^{2/3}}{T_{c_i}^{1/6}} \tag{6.41}$$

where K_c is the critical viscosity constant, typically tuned against available viscosity data. The expressions and constants to calculate viscous friction coefficients $\hat{k}_a$, $\hat{k}_r$, and $\hat{k}_{rr}$ can be found in Quiñones-Cisneros et al. (2001).

Abutaqiya et al. (2018) have modeled the live oil viscosity of 10 different crudes using the Peng-Robinson EOS version of the friction theory (PR-FT) as seen in Figure 6.9 (Abutaqiya et al. 2018). The properties of the crude oils modeled are listed in Table 6.5. The characterization procedure used is the saturate, aromatic, resin, and asphaltene (SARA)-based method, which has been discussed in

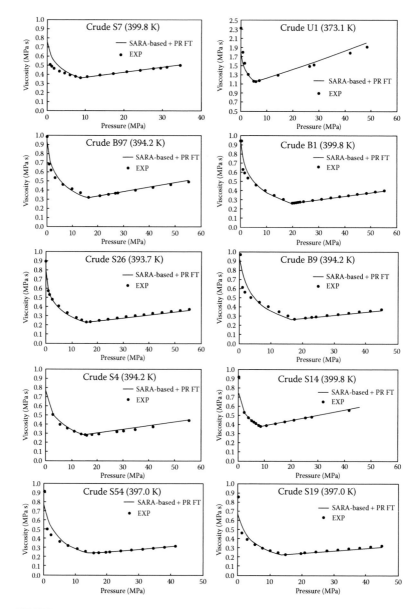

FIGURE 6.9 Live oil viscosity modeling results using SARA-based characterization and Peng-Robinson friction theory. SARA, saturates, aromatic, resin, and asphaltene. (Reprinted from Abutaqiya, M. et al., Viscosity modeling of reservoir fluids using the friction theory with PC-SAFT crude oil characterization method, In Preparation, 2018.)

TABLE 6.5
Properties of Crude Oils Modeled

Property \ Crude	S7	U1	B97	B1	S26	B9	S4	S14	S54	S19
STO M_W (g/mol)	201.5	212.4	201.9	204.0	193.5	208.0	177.5	181.2	193.0	189.7
STO API gravity	37.4	31.7	36.9	36.9	40.5	38.0	40.2	39.2	38.4	39.7
GOR (Sm³/m³)	83.5	38.1	147.8	192.4	189.7	163.7	135.4	63.4	171.4	159.1
ρ^{sat} (kg/m³)	692.0	764.0	654.7	636.2	624.2	647.6	643.0	692.0	632.4	635.5
Reservoir temperature (K)	399.8	373.1	394.2	399.8	393.7	394.2	394.2	399.8	397.0	397.0
P^{sat} (MPa)	8.79	5.76	16.39	19.79	15.79	17.35	15.41	7.63	15.88	14.81
Saturates, wt%	68.6	59.4	70.42	70.2	49.5	—	68.9	68.6	—	—
Aromatics, wt%	22.3	22.57	22.64	22.9	40.2	—	21.8	22.3	—	—
Resins, wt%	6.6	13.97	6.24	4.9	7.2	—	7.1	6.5	—	—
Asphaltene wt%	0.4	1.73	0.18	0.1	3.1	—	0.5	0.5	—	—

Source: Abutaqiya. M. et al., Viscosity modeling of reservoir fluids using the friction theory with PC-SAFT crude oil characterization method. In Preparation, 2018.

GOR, gas-to-oil ratio; M_W, molecular weight; STO, stock tank oil.

Section 4.3.1. The general procedure for applying friction theory for modeling live oil viscosity is,

1. Characterize crude oil based on the compositional analysis and phase-behavior information provided in the PVT report (discussed in Section 4.3)
2. Create pseudo-components and their EOS parameters that predict the correct phase behavior (also discussed in Section 4.3)
3. Tune the critical viscosity constant, K_c, in Equation 6.41 to match the available viscosity data.

Abutaqiya et al. (2018) obtained an overall average absolute percent deviation (AAPD) of 3.7% for the live oil viscosity modeling using the SARA-based characterization and Peng-Robinson EOS version of the friction theory by tuning K_c to viscosity data at saturation point. The average absolute percent deviation for predictions above P^{sat} was 1.8% overall.

6.3.2.3 Corresponding State Principle Models

The corresponding state principle (CSP) has been widely used to model properties of substances (Ely and Hanley 1981; Pedersen et al. 1984). The CSP model for viscosity prediction proposed by Pedersen et al. (1984) was developed from the work of Ely and Hanley (1981) and has found industrial application as well. Ely and Henley expressed reduced viscosity, μ_r, for substances that are members of a homologous series as,

$$\mu_r(\rho,T) = \frac{\mu(\rho,T)}{T_c^{-1/6} P_c^{2/3} M^{1/2}} \tag{6.42}$$

where:
ρ is the density
T is the temperature
T_c is the critical temperature
P_c is the critical pressure
M is the molar mass

The model by Pedersen et al. (1984) uses variables T and P instead of T and ρ as in Ely and Hanley's model (Tham and Gubbins 1970). For a crude oil mixture, Pedersen et al. (1984) proposed the viscosity of mixture μ_{mix} at a given T and P as,

$$\mu_{mix} = \left(\frac{T_{c,mix}}{T_{co}}\right)^{-1/6} \left(\frac{P_{c,mix}}{P_{co}}\right)^{2/3} \left(\frac{M_{mix}}{M_o}\right)^{1/2} \left(\frac{\alpha_{mix}}{\alpha_o}\right) \mu_o\left(P_o, T_o\right) \tag{6.43}$$

$$P_o = \frac{P P_{co} \alpha_o}{P_{c,mix} \alpha_{mix}} \tag{6.44}$$

$$T_o = \frac{T T_{co} \alpha_o}{T_{c,mix} \alpha_{mix}} \tag{6.45}$$

$$\alpha_{mix} = 1 + 7.378 \times 10^{-3} \rho_r^{1.847} M_{mix}^{0.5173} \tag{6.46}$$

where:

The subscript *mix* represents the hydrocarbon mixture
The subscript *o* represents the reference state
α is the rotational coupling coefficient

Pedersen et al. (1984) included the rotational coupling coefficient (α) suggested by Tham and Gubbins to the viscosity CSP model to correct for the deviation from the simple CSP behavior. α can be treated as an adjustable parameter. Pedersen et al. (1984) improved the viscosity prediction by adopting the mixing rule to calculate the mixture molecular weight, M_{mix}

$$M_{mix} = \overline{M_n} + 1.304 \times 10^{-4} \left(\overline{M_w}^{2.303} - \overline{M_n}^{2.303} \right) \tag{6.47}$$

$$\overline{M_w} = \frac{\sum_{i=1}^{nc} z_i M_i^2}{P_{ci}} \tag{6.48}$$

$$\overline{M_n} = \sum_{i=1}^{N} z_i M_i \tag{6.49}$$

6.3.2.4 Effect of Asphaltene Precipitation on Apparent Viscosity of Oil

Asphaltene precipitation is known to increase the apparent viscosity and cause complex viscosity behavior (Barré et al. 2008; Eyssautier et al. 2012; Chávez-Miyauchi et al. 2013; Tavakkoli et al. 2014; Hashmi et al. 2015). Chávez-Miyauchi et al. (2013) summarized from several studies that asphaltene suspension behaves as a dilute system at asphaltene concentration below 10 wt%, and the suspension behaves as a concentrated system at asphaltene concentration above 10 wt%. Properties of the precipitated asphaltenes, such as rate of aggregation, aggregate size distribution, and concentration affect the apparent crude oil viscosity. The viscosity behavior of a dilute suspension of rigid spheres can be modeled using the Einstein equation as,

$$\mu_r = 1 + \left[\mu \right] \phi \tag{6.50}$$

where relative viscosity, μ_r, is expressed as a function of the intrinsic viscosity, $[\mu]$, and the volume fraction of suspended particles, ϕ. To account for the deviation from the ideal behavior predicted using the Einstein equation, researchers have extended the model to reflect the effect of higher particle concentrations, solvation, and non-sphericity (Mooney 1951; Krieger and Dougherty 1959; Pal and Rhodes 1989). The viscosity models for suspensions proposed by Krieger and Dougherty (1959) and that by Pal and Rhodes (1989) have been frequently used to interpret and model asphaltene suspension behaviors. Krieger and Dougherty proposed that the relative viscosity, μ_r, for a suspension can be modeled according to Equation 6.51.

$$\mu_r = \left(1 - \frac{\phi}{\phi_m} \right)^{-[\mu]\phi_m} \tag{6.51}$$

where:
ϕ is the volume fraction of the suspended particles
ϕ_m is the maximum packing volume fraction
$[\mu]$ is the intrinsic viscosity

Pal and Rhodes (1989) suggested that the relative viscosity of a Newtonian emulsion can be expressed as

$$\mu_r = \left(1 - K_0\phi\right)^{-2.5} \tag{6.52}$$

where K_0 is the solvation coefficient. K_0 is intended to correct for the increased effective volume fraction because of the immobilized continuous phase at the surface of the dispersed particles. It was concluded that the Pal and Rhodes (1989) model is unsuitable for modeling asphaltene suspensions because the derivation assumed modeling a suspension of spherical particles and the effect of maximum packing of particles was not considered (Pal and Vargas 2014). Pal and Vargas (2014) reinterpreted 14 sets of viscosity measurements for asphaltene suspensions in various liquid media using the Krieger and Dougherty model. It was found that the maximum packing volume fraction of particles was nearly constant across all data available regardless of their temperature and the nature of asphaltene systems. Hence, Pal and Vargas (2014) proposed that asphaltene suspension can be modeled with the Krieger and Dougherty model, as,

$$\mu_r = \left(1 - \left(\frac{\phi}{0.37}\right)\right)^{-0.37[\mu]} \tag{6.53}$$

With the maximum packing volume fraction $\phi_m = 0.37$.

6.3.3 DIFFUSION COEFFICIENT

A review of the asphaltene deposition models developed over the years shows that the calculation performed to estimate the value of diffusion coefficient, used either in a convection-diffusion transport equation or population balance equations, has been highly influenced by the experimental technique adopted, operating conditions, and properties of the precipitant used to initiate deposition.

In the deposition simulator developed by Vargas et al. (2010), diffusion coefficient of 0.35×10^{-9} m²/s for asphaltenes in toluene is assumed as a rough approximation for its diffusivity in the oil–precipitant mixture used in the capillary deposition test. This value was obtained using fluorescence correlation spectroscopy (FCS) at extremely low concentrations (0.03–3.0 mg/l), where aggregation does not occur corresponding to a hydrodynamic radius of approximately 1 nm (Andrews et al. 2006).

In ADEPT (developed by Kurup et al. 2012), the diffusion coefficient was calculated using the Stokes–Einstein relation, which is given as,

$$D_e = \frac{k_b T}{6\pi v_f r_{cr}} \qquad (6.54)$$

where:
   k_b is the Boltzmann constant
   T is the temperature of the fluid
   v_f is the kinematic viscosity of the fluid
   r_{cr} is the critical size of asphaltene particles in the oil–precipitant mixture

While modeling asphaltene deposition based on experiments carried out in RealView, a cylindrical Couette-Taylor system, asphaltene gets deposited on the outer stationary wall as the inner cylinder rotates at very high speeds (Section 5.1.1.3). For turbulent flows, the eddy diffusion as a result of the fluctuating velocity components and the consequent turbophoresis governs the deposition of aggregated asphaltene on the outer cylinder wall surface. Thus, the particle diffusivity, D_p, in a turbulent flow can be calculated as: $D_p = D_t + D_B$, where D_t is the turbulent particle diffusivity and D_B is the Brownian diffusivity. Because the particles are small (as critical particle size is of the order of nanometers), the particle diffusivity is approximately equal to the eddy diffusivity ($D_p = D_t$), as the Brownian diffusivity plays a significant role for submicron particles only (Eskin et al. 2011). For the sake of convenience, the particle deposition flux only at the laminar boundary sublayer surface can be considered, where the deposition of the asphaltenes to the cylinder surface takes place. The eddy diffusivity at the laminar sublayer surface is given as, $D_t = v_t/Sc_t$, where v_t is the eddy diffusivity (m²/s) and Sc_t is the turbulent Schmidt number. Hence, essentially the calculation of the diffusion coefficient is dependent on the properties of the fluid and flow regimes of the asphaltene deposition experimental techniques.

6.4   FINAL REMARKS

Several significant advancements have been made in developing tools for the investigation and modeling of asphaltene deposition in the wellbore during oil production operations. A detailed review of the modeling of asphaltene deposition in both wellbore and porous media, developed over the years has been presented. The development of ADEPT to predict asphaltene deposition in wellbore and pipelines has been discussed in detail. The application of ADEPT to a real field case will be detailed in Chapter 8. The application of LBM to predict and simulate asphaltene deposition in porous media based on microchannel experiments has also been examined. More applications of LBM to simulate asphaltene deposition in porous media will be discussed in Chapter 8.

The modeling of asphaltene transport properties was also presented. Calculations necessary for obtaining density and diffusion coefficient of asphaltenes were described. Modeling methods adopted to predict asphaltene viscosity at particular pressure and temperature conditions, such as, friction theory and CSP models, have also been investigated.

REFERENCES

Abutaqiya, M., J. Zhang, and F. M. Vargas. 2018. Viscosity modeling of reservoir fluids using the friction theory with PC-SAFT crude oil characterization method. In Preparation.

Akbarzadeh, K., D. Eskin, J. Ratulowski, and S. Taylor. 2012. Asphaltene deposition measurement and modeling for flow assurance of tubings and flow lines. *Energy & Fuels* 26 (1): 495–510. doi:10.1021/ef2009474.

Akbarzadeh, K., J. Ratulowski, T. Lindvig, T. Davis, Z. Huo, and G. Broze. 2009. The importance of asphaltene deposition measurements in the design and operation of subsea pipelines. In *SPE Annual Technical Conference and Exhibition*, pp. 4–7.

Ali, M. A., and M. R. Islam. 1998. The effect of asphaltene precipitation on carbonate-rock permeability: An experimental and numerical approach. *SPE Production & Facilities* 13 (03): 178–183. doi:10.2118/50963-PA.

Almehaideb, R. A. 2004. Asphaltene precipitation and deposition in the near wellbore region: A modeling approach. *Journal of Petroleum Science and Engineering* 42 (2–4): 157–170. doi:10.1016/j.petrol.2003.12.008.

Andrews, A. B., R. E. Guerra, O. C. Mullins, and P. N. Sen. 2006. Diffusivity of asphaltene molecules by fluorescence correlation spectroscopy. *The Journal of Physical Chemistry A* 110 (26): 8093–8097. doi:10.1021/jp062099n.

Assael, M. J., J. H. Dymond, M. Papadaki, and P. M. Patterson. 1992. Correlation and prediction of dense fluid transport coefficients. I. N-Alkanes. *International Journal of Thermophysics* 13 (2): 269–281.

Barré, L., S. Simon, and T. Palermo. 2008. Solution properties of asphaltenes. *Langmuir* 24 (8): 3709–3717. doi:10.1021/la702611s.

Bhatnagar, P. L., E. P. Gross, and M. Krook. 1954. A model for collision processes in gases. I. Small amplitude processes in charged and neutral one-component systems. *Physical Review* 94 (3): 511–525.

Burger, E. D., T. K. Perkins, and J. H. Striegler. 1981. Studies of wax deposition in the trans alaska pipeline. *Journal of Petroleum Technology* 33 (6): 1075–1086. doi:10.2118/8788-PA.

Carman, P. C. 1939. Permeability of saturated sands, soils and clays. *Journal of Agricultural Science* 29: 263–273. doi:10.1017/S0021859600051789.

Chávez-Miyauchi, T. E., L. S. Zamudio-Rivera, and V. Barba-López. 2013. Aromatic polyisobutylene succinimides as viscosity reducers with asphaltene dispersion capability for heavy and extra-heavy crude oils. *Energy & Fuels* 27 (4): 1994–2001. doi:10.1021/ef301748n.

Chung, T. H., M. Ajlan, L. L. Lee, and K. E. Starling. 1988. generalized multiparameter correlation for nonpolar and polar fluid transport properties. *Industrial & Engineering Chemistry Research* 27 (4): 671–679. doi:10.1021/ie00076a024.

Delbosc, N., J. L. Summers, A. I. Khan, N. Kapur, and C. J. Noakes. 2014. Optimized implementation of the Lattice-Boltzmann Method on a graphics processing unit towards real-time fluid simulation. *Computers & Mathematics with Applications*, Mesoscopic Methods for Engineering and Science (Proceedings of ICMMES-2012, Taipei, Taiwan, 23–July 27, 2012), 67 (2): 462–475. doi:10.1016/j.camwa.2013.10.002.

Ely, J. F., and H. J. M. Hanley. 1981. Prediction of transport properties. 1. Viscosity of fluids and mixtures. *Industrial & Engineering Chemistry Fundamentals* 20 (4): 323–332.

Eskin, D., J. Ratulowski, K. Akbarzadeh, and S. Pan. 2011. Modelling asphaltene deposition in turbulent pipeline flows. *The Canadian Journal of Chemical Engineering* 89 (3): 421–441.

Eyssautier, J., I. Hénaut, P. Levitz, D. Espinat, and L. Barré. 2012. Organization of asphaltenes in a vacuum residue: A small-angle X-Ray Scattering (SAXS)–Viscosity approach at high temperatures. *Energy & Fuels* 26 (5): 2696–2704. doi:10.1021/ef201412j.

Ferréol, B., and D. H. Rothman. 1995. Lattice-Boltzmann simulations of flow through Fontainebleau sandstone. *Transport in Porous Media* 20 (1): 3–20. doi:10.1007/BF00616923.

Ge, Q., Y. F. Yap, F. M. Vargas, M. Zhang, and J. C. Chai. 2013. Numerical modeling of asphaltene deposition. *Computational Thermal Sciences* 5 (2): 153–163. doi:10.1615/ComputThermalScien.2013006316.

Gruesbeck, C., and R. E. Collins. 1982. Entrainment and deposition of fine particles in porous media. *Society of Petroleum Engineers Journal* 22 (06): 847–856. doi:10.2118/8430-PA.

Guan, Q., Y. F. Yap, A. Goharzadeh, J. Chai, F. M. Vargas, W. Chapman, and M. Zhang. 2017. Integrated one-dimensional modeling of asphaltene deposition in wellbores/pipelines. 7th International Conference on Modeling, Simulation and Applied Optimization (ICMSA), Sharjah, UAE.

Guo, Y., and M. Wang. 2016. Lattice-Boltzmann modeling of phonon transport. *Journal of Computational Physics* 315: 1–15. doi:10.1016/j.jcp.2016.03.041.

Guo, Z. 2013. *Lattice-Boltzmann Method and Its Applications in Engineering*. Hackensack, NJ: World Scientific Publishing.

Hashmi, S. M., M. Loewenberg, and A. Firoozabadi. 2015. Colloidal asphaltene deposition in laminar pipe flow: Flow rate and parametric effects. *Physics of Fluids* 27 (8): 083302. doi:10.1063/1.4927221.

Haskett, C. E., and M. Tartera. 1965. A practical solution to the problem of asphaltene deposits-Hassi Messaoud Field, Algeria. *Journal of Petroleum Technology* 17 (04): 387–391. doi:10.2118/994-PA.

He, P., Y.-J. Lin, M. Tavakkoli, J. Creek, J. Wang, F. M. Vargas, and S. L. Biswal. 2018. Effect of flow rates on the deposition of asphaltenes in porous media. In Preparation.

Heijs, A. W. J., and C. P. Lowe. 1995. Numerical evaluation of the permeability and the Kozeny constant for two types of porous media. *Physical Review E* 51 (5): 4346–4352.

Hirschberg, A., L. N. J. DeJong, B. A. Schipper, and J. G. Meijer. 1984. Influence of temperature and pressure on asphaltene flocculation. *Society of Petroleum Engineers Journal* 24 (03): 283–293. doi:10.2118/11202-PA.

Jafari Behbahani, T., C. Ghotbi, V. Taghikhani, and A. Shahrabadi. 2015. Experimental study and mathematical modeling of asphaltene deposition mechanism in core samples. *Oil & Gas Science and Technology –Revue d'IFP Energies nouvelles* 70 (6): 1051–1074.

Jossi, J. A., L. I. Stiel, and G. Thodos. 1962. The viscosity of pure substances in the dense gaseous and liquid phases. *AIChE Journal* 8 (1): 59–63.

Kord, S., O. Mohammadzadeh, R. Miri, and B. S. Soulgani. 2014. Further investigation into the mechanisms of asphaltene deposition and permeability impairment in porous media using a modified analytical model. *Fuel* 117, Part A: 259–268. doi:10.1016/j.fuel.2013.09.038.

Krieger, I. M., and T. J. Dougherty. 1959. A mechanism for non-newtonian flow in suspensions of rigid spheres. *Transactions of the Society of Rheology* 3 (1): 137–152.

Kurup, A. S., F. M. Vargas, J. Wang, J. S. Buckley, J. L. Creek, J. Subramani, and W. G. Chapman. 2011. Development and application of an asphaltene deposition tool (ADEPT) for well bores. *Energy & Fuels* 25 (10): 4506–4516.

Kurup, A. S., J. Wang, H. J. Subramani, J. S. Buckley, J. L. Creek, and W. G. Chapman. 2012. Revisiting asphaltene deposition tool (ADEPT): Field application. *Energy & Fuels* 26 (9): 5702–5710.

Lawal, K. A., V. Vesovic, and E. S. Boek. 2011. Modeling permeability impairment in porous media due to asphaltene deposition under dynamic conditions. *Energy & Fuels* 25 (12): 5647–5659. doi:10.1021/ef200764t.

Lohrenz, J., B. G. Bray, and C. R. Clark. 1964. Calculating viscosities of reservoir fluids from their compositions. *Journal of Petroleum Technology* 16 (10). doi:10.2118/915-PA.

Maqbool, T., S. Raha, M. P. Hoepfner, and H. Scott Fogler. 2011. Modeling the aggregation of asphaltene nanoaggregates in crude oil-precipitant systems. *Energy & Fuels* 25 (4): 1585–1596.

McNamara, G. R., and G. Zanetti. 1988. Use of the Boltzmann equation to simulate lattice-gas automata. *Physical Review Letters* 61 (20): 2332–2335.

Mehrotra, A. K., W. D. Monnery, and W. Y. Svrcek. 1996. A review of practical calculation methods for the viscosity of liquid hydrocarbons and their mixtures. *Fluid Phase Equilibria* 117: 344–355.

Minssieux, L. 1997. Core damage from crude asphaltene deposition. In SPE: Society of Petroleum Engineers. doi:10.2118/37250-MS.

Mohamad, A. A. 2011. *Lattice-Boltzmann Method Fundamentals and Engineering Applications with Computer Codes*. London, UK: Springer.

Mooney, M. 1951. The viscosity of a concentrated suspension of spherical particles. *Journal of Colloid Science* 6 (2): 162–170.

Nit, C., L. M. Itu, and C. Suciu. 2013. GPU accelerated blood flow computation using the Lattice-Boltzmann Method. In 1–6. doi:10.1109/HPEC.2013.6670324.

Pal, R., and E. Rhodes. 1989. Viscosity/concentration relationships for emulsions. *Journal of Rheology* 33 (7): 1021. doi:10.1122/1.550044.

Pal, R., and F. Vargas. 2014. On the interpretation of viscosity data of suspensions of asphaltene nano-aggregates. *The Canadian Journal of Chemical Engineering* 92 (3): 573–577. doi:10.1002/cjce.21896.

Pedersen, K. S., P. L. Christensen, and J. A. Shaikh. 2015. *Phase Behavior of Petroleum Reservoir Fluids*, 2nd ed. Boca Raton, FL: CRC Press.

Pedersen, K. S., A. Fredenslund, P. L. Christensen, and P. Thomassen. 1984. Viscosity of crude oils. *Chemical Engineering Science* 39 (6): 1011–1016.

Premnath, K. N., M. J. Pattison, and S. Banerjee. 2013. An investigation of the Lattice-Boltzmann method for large eddy simulation of complex turbulent separated flow. *Journal of Fluids Engineering* 135 (5): 051401. doi:10.1115/1.4023655.

Quiñones-Cisneros, S. E., S. I. Andersen, and J. Creek. 2005. Density and viscosity modeling and characterization of heavy oils. *Energy & Fuels* 19 (4): 1314–1318. doi:10.1021/ef0497715.

Quiñones-Cisneros, S. E., C. K. Zéberg-Mikkelsen, and E. H. Stenby. 2000. The friction theory (F-theory) for viscosity modeling. *Fluid Phase Equilibria* 169 (2): 249–276. doi:10.1016/S0378-3812(00)00310-1.

Quiñones-Cisneros, S. E., C. K. Zéberg-Mikkelsen, and E. H. Stenby. 2001. One parameter friction theory models for viscosity. *Fluid Phase Equilibria* 178 (1–2): 1–16. doi:10.1016/S0378-3812(00)00474-X.

Ramirez-Jaramillo, E., C. Lira-Galeana, and O. Manero. 2006. Modeling asphaltene deposition in production pipelines. *Energy & Fuels* 20 (3): 1184–1196. doi:10.1021/ef050262s.

Sarwar, M., and M. R. Islam. 1997. A non-fickian surface excess model for chemical transport through fractured porous media. *Chemical Engineering Communications* 160 (1): 1–34. doi:10.1080/00986449708936603.

Schmidt, K. A. G., S. E. Quiñones-Cisneros, and B. Kvamme. 2005. Density and viscosity behavior of a north sea crude oil, natural gas liquid, and their mixtures. *Energy & Fuels* 19 (4): 1303–1313. doi:10.1021/ef049774h.

Soulgani, B. S., B. Tohidi, M. Jamialahmadi, and D. Rashtchian. 2011. Modeling formation damage due to asphaltene deposition in the porous media. *Energy & Fuels* 25 (2): 753–761. doi:10.1021/ef101195a.

Soulgani, B. S., D. Rashtchian, B. Tohidi, and M. Jamialahmadi. 2009. Integrated modelling methods for asphaltene deposition in wellstring. *Journal of the Japan Petroleum Institute* 52 (6): 322–331. doi:10.1627/jpi.52.322.

Succi, S., E. Foti, and F. Higuera. 1989. Three-dimensional flows in complex geometries with the Lattice-Boltzmann method. *EPL (Europhysics Letters)* 10 (5): 433.

Tavakkoli, M., V. Taghikhani, M. R. Pishvaie, M. Masihi, S. R. Panuganti, and W. G. Chapman. 2014. Investigation of oil-asphaltene slurry rheological behavior. *Journal of Dispersion Science and Technology* 35 (8): 1155–1162. doi:10.1080/01932691.2013.834421.

Tham, M. J., and K. E. Gubbins. 1970. Correspondence principle for transport properties of dense fluids. Nonpolar polyatomic fluids. *Industrial & Engineering Chemistry Fundamentals* 9 (1): 63–70.

Uyehara, O. A., and K. M. Watson. 1944. A universal viscosity correlation. *National Petroleum News* 36: R-714-722.

Vargas, F. M., J. L. Creek, and W. G. Chapman. 2010. On the development of an asphaltene deposition simulator. *Energy & Fuels* 24 (4): 2294–2299. doi:10.1021/ef900951n.

Vesovic, V., and W. A. Wakeham. 1989. Prediction of the viscosity of fluid mixtures over wide ranges of temperature and pressure. *Chemical Engineering Science* 44: 2181–2189.

Vesovic, V., J. P. M. Trusler, M. J. Assael, N. Riesco, and S. E. Quiñones-Cisneros. 2014. Dense fluids: Viscosity. In *Experimental Thermodynamics Volume IX: Advances in Transport Properties of Fluids*, M. J. Assael, A. R. H Goodwin, V. Vesovic, and W. A. Wakeham (Eds.). Cambridge, UK: The Royal Society of Chemistry.

Vilas Bôas Fávero, C., A. Hanpan, P. Phichphimok, K. Binabdullah, and H. S. Fogler. 2016. Mechanistic investigation of asphaltene deposition. *Energy & Fuels* 30 (11): 8915–8921. doi:10.1021/acs.energyfuels.6b01289.

Wang, J., J. S. Buckley, and J. L. Creek. 2004. Asphaltene deposition on metallic surfaces. *Journal of Dispersion Science and Technology* 25 (3): 287–298. doi:10.1081/DIS-120037697.

Wang, S., and F. Civan. 2001. Productivity decline of vertical and horizontal wells by asphaltene deposition in petroleum reservoirs. In Society of Petroleum Engineers. doi:10.2118/64991-MS.

Zeberg-Mikkelsen, C. K., S. E. Quiñones-Cisneros, and E. H. Stenby. 2002. Viscosity prediction of natural gas using the friction theory. *International Journal of Thermophysics* 23 (2): 437–454.

7 Strategies for Mitigation and Remediation of Asphaltene Deposition

J. Kuang, A. T. Khaleel, J. Yarbrough,
P. Pourreau, M. Tavakkoli, and F. M. Vargas

CONTENTS

As discussed in Chapter 5, asphaltene deposition in the wellbore and near-wellbore regions can lead to restricted flow line pressure and reduced oil field productivity. Problems related to asphaltene deposition cost petroleum industries billions of dollars each year, including the costs of reduced production, wells shut-in, inefficient use of production capacity, prevention and remediation strategies, and so on (Atta and Elsaeed 2011). For oil fields in the Gulf of Mexico, the economic impact associated with this problem is approximately $70 million per well when well shut-ins for ring interventions are required (Gonzalez, 2015). If the deposition occurs in the surface-controlled subsurface safety valve (SCSSV), the cost increases to $100 million per well. Downtime losses based on a 10,000 bbl/day production and an oil price of $50 per barrel can reach up to $500,000 per day. Replacing a lost well with a side track raises the cost to around $150 million (Gonzalez, 2015). To prevent and remediate asphaltene deposition, expensive treatment and clean-up procedures have been widely used. For fields in the Middle East, chemical additive injection costs \$31,000–\$46,000 per well per year. Chemical additive injection for typical Gulf of Mexico production of 10,000 bbl/day costs about \$330,000–\$390,000 per well per year (Gonzalez, 2015). Therefore, the development of a cost-effective approach to controlling asphaltene deposition is of great economical importance to ensure the high efficiency of the oilfield production process.

In the Chapters 5 and 6, current experimental techniques and modeling strategies to investigate the asphaltene deposition phenomenon were discussed and compared. The correct laboratory analyses and modeling predictions can provide important insights into the strategies for prevention and remediation of asphaltene deposition. This chapter will review the most commonly used techniques to control asphaltene deposition in the production lines, including chemical treatment by asphaltene inhibitors, control of operating conditions, internal coatings, physical deposit removal, and solvent washes. The limitations of some current approaches will be addressed, and the best practices for the mitigation of asphaltene deposition will be proposed at the end of the chapter.

7.1 PREVENTION OF ASPHALTENE DEPOSITION

7.1.1 Chemical Inhibition

The most commonly used techniques to control asphaltene deposition include chemical treatment by asphaltene deposition inhibitors (usually called as *asphaltene inhibitors*). Asphaltene dispersants, particular types of asphaltene inhibitors, are chemicals supposedly used to prevent asphaltene deposition in the production tubing. Asphaltene dispersants are believed to peptize asphaltene particles by bridging asphaltenes with their polar groups and stretching the aliphatic chains toward the bulk phase, thereby forming a steric stabilization layer around the asphaltene particles (Sachanen et al. 1945; Leontaritis et al. 1989; Schantz and Stephenson 1991; Barcenas et al. 2008). Chang and Fogler (1994a, 1994b) proved the effectiveness of using alkylbenzene-derived amphiphiles to stabilize asphaltenes in aliphatic solvents. Amphiphilic dispersants are generally composed of an anchoring polar group and a blocking alkyl group (Goual et al. 2014). The polar group usually contains heteroatoms, which could be attached to the asphaltene surface, and the alkyl group refers to the aliphatic tail, which could prevent the aggregation of asphaltene molecules (Goual et al. 2014). Possible interactions between asphaltenes and asphaltene dispersants are π-π interactions between their aromatic rings, hydrogen bonds, acid-base interactions, dipole-dipole interactions, and complex metal ions (Kelland 2009).

Chemical additives, including nonionic amphiphiles, ionic liquids, nanofluids, polymers, and so on show different extents of effectiveness on the prevention of asphaltene precipitation. Extensive research has been conducted to investigate different variables affecting the performance of the chemical additives. It has been reported that asphaltene characteristics (Ibrahim and Idem 2004; Smith et al. 2008; Wang et al. 2009; Juyal et al. 2010), solvent conditions (Ibrahim and Idem

2004; Barcenas et al. 2008), chemical structures of the additives (Chang and Fogler 1994a, 1994b; Hu and Guo 2005; Boukherissa et al. 2009; Goual et al. 2014), and the amount of chemicals adsorbed on asphaltenes (León et al. 2000; Goual et al. 2014; Melendez-Alvarez et al. 2016) play significant roles in the effectiveness of the inhibition of asphaltene precipitation. In this section, the mechanisms of inhibition including dispersion, aging disruption, and electrostatic interactions will be reviewed. Different techniques to assess the performance of chemical additives are summarized and compared.

7.1.1.1 Mechanisms of Inhibition

7.1.1.1.1 Dispersion

During the production process, asphaltene deposition will occur as a result of destabilized asphaltene nano-aggregates precipitating as small clusters, then aggregating into larger clusters (Rogel 2011). These small clusters are often referred to as *primary particles* and generally have a size of a few hundred nanometers (Vargas et al. 2014). The larger asphaltene clusters are commonly referred to as micro-aggregates, and these clusters are generally much larger than primary particles (>1 micrometer). As shown in Figure 1.6, asphaltenes undergo three phases on the way to becoming aged deposits. When asphaltenes are destabilized, nano-aggregates precipitate out of the liquid phase into primary particles. The onset of asphaltene precipitation can be reached by changes in pressure, temperature, or composition of the reservoir fluids (Vargas et al. 2010).

One of the most widely used methods to attempt to mitigate asphaltene deposition is through the use of chemical dispersants. Chemical dispersants function as inhibitors by delaying asphaltene aggregation (Melendez-Alvarez et al. 2016). Because asphaltenes represent the heaviest and most polarizable fraction of crude oils, dispersants are developed to peptize and stabilize these compounds in solution (Karambeigi et al. 2016). The primary mechanism for stabilization is found in the π-π interactions between the chemical dispersants and the asphaltene molecules (Kelland 2009). The specific structure of asphaltene molecules can vary depending on the crude oils, with different asphaltenes consisting of varied heteroatoms, aromatic groups, and hydrocarbon chain lengths. The ideal dispersants are believed to function like resins that can stabilize asphaltene particles. The injection of asphaltene dispersants is motivated for a general belief that they may prevent the aggregation and deposition of asphaltenes (Kraiwattanawong et al. 2009).

A problem with using chemical dispersants as deposition inhibitors is that the reduction in the sizes of asphaltene aggregates has not been shown to always reduce the overall asphaltene deposition. This phenomenon occurs because smaller particles present a higher surface area, which in turn can increase the deposition tendency (Vargas et al. 2010). Furthermore, it has been reported that many conventional dispersants may actually worsen asphaltene deposition in the field (Melendez-Alvarez et al. 2016). The inadequate commercial methods currently used to assess asphaltene inhibitors are responsible for the poor performance of some of these chemicals in the field (Melendez-Alvarez et al. 2016; Kuang et al. 2017). This idea is further discussed in Section 7.3.1.

7.1.1.1.2   Aging Disruption

Another approach that might be useful for preventing asphaltene deposition is through the disruption of the aging process of asphaltenes. When asphaltenes molecules begin to aggregate, they start sticking and stacking on top of each other, as shown in Figure 7.1. When the stacking process begins between different asphaltene molecules, the intermolecular interactions between both molecules allow for a state of increased stability by sticking together to form nano-aggregates. Individual asphaltene molecules are composed of polyaromatic compounds bridged together by flexible aliphatic chains and include substituent aliphatic chains, varied heteroatoms, and one or more polar groups (Zhang et al. 2014). The polarity of asphaltene molecules can vary greatly depending on the different functional groups connected to it (Brandt et al. 1995). In addition, the type of asphaltene can affect how ordered the resultant stacking is. For instance, C_{7+} asphaltenes tend to display more ordered stacking than C_{5+} asphaltenes under the same thermodynamic conditions (Zhang et al. 2014).

When a system experiences a change in equilibrium that encourages precipitation, stacks of asphaltene nano-aggregates begin to grow larger. As these asphaltenes phase-split from the oil-rich liquid phase into an asphaltene-rich liquid phase, larger clusters are formed with other nano-aggregate stacks, represented by the rightmost group in Figure 7.1 as primary particles (Hoepfner 2013).

Asphaltenes clusters experience a microscale rearrangement of their structure as they begin the aging process. During the precipitation process, components like saturates, aromatics, and resins also precipitate out of solution with asphaltenes, giving clusters a physical softness that keeps them in a viscous, liquid-like phase. When the clusters age, the interlocking of the asphaltene cores and aliphatic chains can form a metastable solid phase. This occurs because the dense interlocking of the asphaltene cores often pushes out the other components that coprecipitated out of solution with the nano-aggregates.

This solid state of asphaltenes makes redissolution difficult (Brandt et al. 1995). These asphaltenes are said to be aged, and conditions such as increased temperature

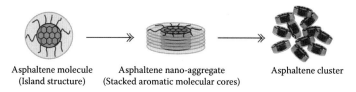

Asphaltene molecule   Asphaltene nano-aggregate   Asphaltene cluster
(Island structure)   (Stacked aromatic molecular cores)

FIGURE 7.1   Progression of asphaltene aggregation, with most aging being seen in larger asphaltene clusters (or primary particles). (Reprinted from Hoepfner, M. P., Investigations into asphaltene deposition, stability, and structure, 231, 2013. With permission.)

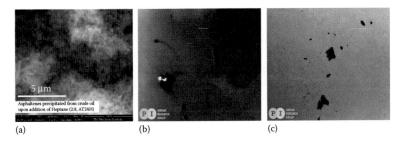

(a)   (b)   (c)

FIGURE 7.2 (a) Scanning electron microscope (SEM) of the precipitated asphaltenes, (b) the precipitated asphaltenes dried at ambient temperature and then put into contact with the original oil, and (c) the precipitated asphaltenes dried at 393 K and then put into contact with the original oil. (Reprinted from Khaleel, A. et al., Society of Petroleum Engineers, doi:10.2118/177941-MS, 2015. With permission.)

(Figure 7.2) can exacerbate this issue. The formation of the aged asphaltenes is the major reason for the asphaltene deposition problem in the production lines, and the transition from the aggregated phase to the aged phase is irreversible (Vargas et al. 2014). Thus, new inhibitors designed based on the mechanism of aging disruption should possess the ability to prevent the formation of compact and solid-like asphaltenes.

7.1.1.1.3 Electrostatic Interactions

Besides the mechanisms of dispersion and aging disruption to inhibit asphaltene deposition, the control of the electrostatic interactions between asphaltenes and chemical additives might be another feasible mechanism of inhibition. Studies have shown that exposing asphaltene solutions to electric fields can alter the adsorptive behavior of asphaltenes, allowing for a new way to prevent deposition (Lichaa and Herrera 1975; Hosseini et al. 2016). Lichaa and Herrera (1975) showed this phenomenon with their electrode experiment. In Licha and Herra's setup (1975), steel electrodes were placed 2 mm apart in a 50:50 solution of crude oil and electroresistant diluting solvent. The solution was then exposed to a voltage of 380 V over a period of 3 to 5 days. Upon final inspection, Lichaa and Herrera (1975) found that the negative electrode was always coated with asphaltene-rich oil, whereas the positive electrode only had a thin film of oil. In separate experiments, Lichaa and Herrera (1975) sought to study the capacity for electric fields to move asphaltenes in solutions. By gradually increasing the voltage across a solution under a microscope, they found an optimal voltage of around 120 volts that was needed to induce significant movement of asphaltene particles in solvents with negligible reactions to electric field exposure (Lichaa and Herrera 1975). In addition, Khvostichenko and Anderson (2009) showed that stable asphaltic materials in oil were negatively charged, whereas asphaltenes destabilized by n-heptane had a net positive charge. Recently, Hosseini et al. (2016) investigated the electrokinetic behavior of asphaltene particles in an electrode-embedded glass micromodel. They observed faster aggregation rate for the asphaltene molecules with larger chromophore, higher structural complexity, and more heteroatoms under an electrostatic field.

These studies are promising because they present a unique way to consider controlling asphaltene deposition. As discussed, surface interactions and varied thermodynamic conditions promote the sticking, layering, and eventual clustering of nanoscale asphaltene groups (Vargas et al. 2010). Although traditional dispersants are developed with the intention of stabilizing asphaltene solutions and maintaining small particle sizes, an electrostatically controlled chemical additive may instead induce flocculation until a specific, average, asphaltene particle size is reached. This corroborates with the conventional idea that the majority of asphaltene deposition is caused by smaller particle sizes (Vargas et al. 2010), whereas larger aggregates are less susceptible to deposit. Thus, controlling the electrostatic interactions involved in deposition mechanisms allows for the manipulation of inherent surface charges on asphaltenes, influenced entirely by the unique structure of different asphaltenes. With more development, an inhibition method that noninvasively controls the movement of asphaltenes in solution could provide major benefits for production lines.

7.1.1.2 Testing Procedures

7.1.1.2.1 Ambient Conditions

7.1.1.2.1.1 Asphaltene Dispersion Test The Asphaltene Dispersion Test (ADT) is one of the most commonly used techniques to assess the efficacy of asphaltene inhibitors to disperse asphaltenes clusters. In short, the ADT is developed as a simple way to measure dispersion efficiency by analyzing asphaltene sedimentation. Compared to actual deposition tests, the ADT is a simple tool for screening multiple chemicals at different dosages.

The ADT is used to investigate the dispersion performance of chemical additives, so the concentration of a dispersant is often varied within a test to study how performance changes with increased dispersant amount (Melendez-Alvarez et al. 2016). First, the oil samples are placed in graduated containers, where each sample unit is then treated with a dispersant that is to be studied. Then, an asphaltene precipitant such as *n*-heptane is added to the well-mixed sample at precipitant to oil ratio of 40:1. The level of asphaltene sedimentation is compared after samples are being aged for a predetermined amount of time, usually 24 hours.

As shown in Figure 7.3, the ADT results for a crude oil treated with 70 and 500 ppm of a dispersant are displayed at two different aging times. After the test, the effect of chemicals and dosages on the reduction of asphaltene sedimentation can be visualized by inspecting the level of sedimentation. A lower amount of asphaltene sedimentation is an indication of the high dispersive ability of the chemical additive. A perfect dispersant would be able to keep asphaltenes suspended in solution.

Because of the qualitative nature of the ADT, making predictions based on similar data can be difficult. For instance, the ADT measures the height of the sediment but compacting of the sediment can cause a great deal of error, especially for more aged sample sets (Melendez-Alvarez et al. 2016). Furthermore, most of the ADT experiments are conducted at ambient temperatures and pressures, which fail to represent reservoir conditions (Khaleel et al. 2015). Equally unrepresentative are the relatively high amounts of precipitant used to dilute the sample sets (Melendez-Alvarez et al. 2016).

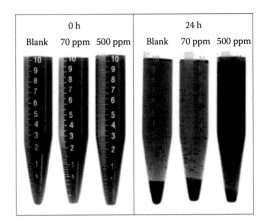

FIGURE 7.3   Asphaltene dispersion test results for a crude oil treated with a dispersant at 70 ppm and 500 ppm for aging times of 0 h and 24 h. (Reprinted from Melendez-Alvarez, A. A. et al., *Fuel*, 179, 210–220, 2016. With permission.)

Although the ADT is widely used to compare the dispersive performance of chemicals, it is not necessarily true that a reduction in the level of sedimentation indicates better prevention of asphaltene deposition on the metallic surfaces.

Juyal et al. (2012) attempted to explore the dispersive capabilities of chemical additives by developing the modified ADT, which assesses a chemical's ability to redisperse asphaltene aggregates in solution. The only difference is that instead of adding the chemical dispersant before the precipitant, the oil is first destabilized by adding the precipitant, and this solution is aged for a set amount of time before applying the dispersant. The same process is followed for assessing the performance of chemical dispersants. Although this method serves as a simple test for qualitatively ranking the dispersive performance and studying the reversibility of asphaltene flocculation, it holds the same weakness as the normal ADT for the determination of actual reduction of asphaltene deposition.

7.1.1.2.1.2 Turbidity Measurement   Turbidity is defined as the cloudiness of a liquid resulting from the suspension of solid particles. Turbidity measurements are usually used to test water quality in water treatment plants ("Turbidity" n.d.). The measurements are performed using a turbidimeter that quantifies the clarity in Nephelometric Turbidity Units (NTU) (Abrahamsen 2012). The measurements are based on light scattering technique, where the suspended particles scatter light passing through the sample and thus reduce the intensity of transmitted light. Turbidimeters include a tungsten-filament lamp and a filter to create light with a known wavelength. The light passes through the sample cell, and the particles in the sample can absorb or scatter light. The scattered light is detected by a detector placed at 90 degrees from the incident beam, and unscattered light is detected by a transmitted light sensor. The absorption of light by the material is corrected for by the ratio of signals from the two detectors (Abrahamsen 2012). The turbidity test provides insight into the optical characteristics of the detected particles. High turbidity value indicates a significant number of suspended particles and thus higher cloudiness (Abrahamsen 2012).

The turbidity test was used to determine the onset of asphaltene flocculation by titrating the sample with a precipitant (Chaogang et al. 2013). This method was also used to identify the effectiveness of different inhibitors to disperse precipitated asphaltenes (Kraiwattanawong et al. 2009; Abrahamsen 2012). It is believed that high turbidity values are associated with good inhibitors that can prevent asphaltene deposition by keeping asphaltenes dispersed.

Kraiwattanawong et al. (2009) used the turbidity test to rate the effectiveness of different chemical dispersants using crude oil N2 from the north slope of Alaska. To perform the turbidity measurements, asphaltene precipitation was induced by adding 7.5 cm^3 of precipitant, such as n-pentane, n-heptane, and n-decane, to 150 µL of crude oil. The sample was mixed to disperse the precipitated asphaltenes. The transmittance of the sample was measured using a Turbiscan MA2000, from Formulaction, as a function of time for 1 hour with a time step of 1 minute. The measurements were performed using a pulsed near-infrared light source with a wavelength of 850 nm. Initially, the transmittance was around zero because all particles got dispersed because of the mixing. As the sample was left undisturbed during the measurements, the upper section of the solution became clearer, allowing more light to be transmitted, as shown in Figure 7.4a. In the case of good asphaltene dispersants, the transmitted light is not expected to change with time and should remain close to zero, indicating efficient dispersion of asphaltenes. From stokes' law, the time needed for the particle to settle depends on its size. For a crude oil tested by Kraiwattanawong et al. (2009), a 1-µm asphaltene particle suspended in n-heptane will need 19 h to travel 5 cm with a settling velocity of 7.25×10^{-5} cm/s. So, if asphaltenes aggregation can be hindered to maintain a particle size less than 1 µm using dispersants, this will prevent deposition by allowing the particles to be transported and removed.

Figure 7.4b presents the mean transmittance (average transmittance along the tube height) of crude N2 treated with 500 ppm of various chemicals, as a function of time. Efficient chemical additives will keep asphaltene particles less than 1 µm such that they stay dispersed in solution. This means that the measured average transmittance will remain constant during the 1 h measurement time. On the other hand, ineffective chemical additives will not be able to reduce asphaltene aggregation, and thus the particles will settle out of solution because their large size and high settling velocity, causing an increase of average transmittance. From Figure 7.4b, DR, DP, and DBSA are classified as ineffective chemicals, whereas the proprietary blends were classified as effective chemicals (Kraiwattanawong et al. 2009).

Despite its simplicity and practicality, the increase in dispersion of asphaltenes detected by the turbidity measurement may not provide accurate indications on the reduction in asphaltene deposition. Based on the mechanism proposed by Vargas et al. (2010), aggregation and deposition are two competing phenomena. Thus, decreasing the rate of aggregation may result in an increase in the rate of deposition.

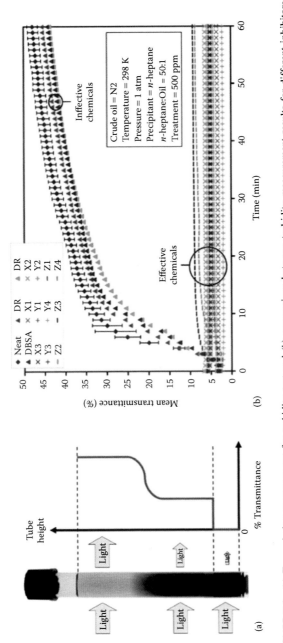

FIGURE 7.4 (a) Transmission measurement for turbidity test and (b) comparison between turbidity measurement results for different inhibitors. (Reprinted from Kraiwattanawong, K. et al., *Energy Fuels*, 23, 1575–1582, 2009. With permission.)

7.1.1.2.1.3 Direct Spectroscopy Recently, Melendez-Alvarez et al. (2016) developed a technique based on near-infrared (NIR) spectroscopy to evaluate the performance of asphaltene dispersants. In comparison to the ADT, the direct spectroscopy method can assess the effectiveness of the dispersants in a wide range of temperatures and compositions. It also serves as a faster and more reproducible approach to quantitatively investigate the performance of asphaltene dispersants (Melendez-Alvarez et al. 2016).

The methodology of the direct spectroscopy method is as follows. Stock solutions are prepared by mixing a specified crude oil with a certain amount of standard dispersant solution. Then, different ratios of stock solution (with and without dispersant) and *n*-heptane are prepared ranging from 0 vol% to 90 vol% of *n*-heptane (Figure 7.5). The masses of the added oil and of the precipitant are measured, and the actual volumes are back-calculated, assuming an ideal solution. The test tubes are then shaken by hand, and the samples are allowed to rest undisturbed for the specified aging time, which in this case is the period of time allotted between sample preparation and the measurement of the NIR transmittance. The sample is transferred to a 10-mm path length quartz cuvette, where each sample is mixed via magnetic stirring bars to keep particles from settling. The NIR transmittance is then measured using a Shimadzu UV–Vis–NIR spectrophotometer (Melendez-Alvarez et al. 2016).

The absorbance of each sample is measured for wavelengths between 1,500 and 1,600 nm. The absorbance values are adjusted by subtracting the absorbance of *n*-heptane, and then mathematically correcting for the effect of dilution. The details for the correction of the dilution effect have been explained by Tavakkoli et al. (2015). The corrected absorbance values are then plotted as a function of the volume fraction of *n*-heptane or crude oil. The same results can also be presented in terms of light intensity (i.e., light transmittance) instead of absorbance, which is how the results are usually reported by direct spectroscopy. In the latter case, asphaltene precipitation is determined by the decrease in the light intensity caused by asphaltene aggregates that block the path of light. The first deviation from the linear trend indicates the precipitation of asphaltenes.

Figure 7.6 shows the results for crude oil S with and without Dispersant 8 at 70 ppm. With lower amounts of precipitant and higher crude oil concentrations, the light intensities for both solutions display consistent linear trends. Then, the light intensity

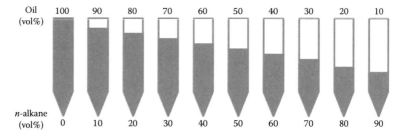

FIGURE 7.5 Sample preparation for the direct spectroscopy method. (Reprinted from Kuang, J. et al., Novel way to assess the performance of asphaltene dispersants on the prevention of asphaltene deposition, 2017.)

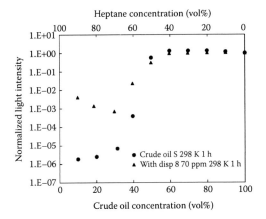

FIGURE 7.6 Plot of crude oil concentration and heptane concentration versus light intensity measured by the spectrophotometer. "Disp 8" refers to crude oil mixed with a conventional industry dispersant. (Reprinted from Melendez-Alvarez, A. A. et al., *Fuel*, 179, 210–220, 2016. With permission.)

values deviate from the horizontal trend between 40 vol% and 45 vol% of crude oil. This point of deviation represents the minimum amount of precipitant required for the detection of asphaltene precipitation (Melendez-Alvarez et al. 2016). The decrease in light intensity after the point of deviation corresponds with an increase in the size and number of aggregates forming. Thus, the solution treated with Dispersant 8 showing a slower rate of decreased light intensity implies slower rates of precipitation and aggregation. Furthermore, the inflection point at 70 vol% of precipitant for the treated solution implies a gradual return to stable solution with fewer and smaller asphaltene aggregates as the concentration of the precipitant approaches 100 vol%.

Melendez-Alvarez et al. (2016) developed an approach to quantify and rank the dispersing power of different chemicals. The dispersive performance efficiency (DPE) can be determined as follows:

$$\text{DPE} = 100 * \left(1 - \frac{A^{Disp}}{A^{Blank}} \right) \tag{7.1}$$

$$A = \int_{10}^{100} -\text{Log}_{10}\left(\text{Light intensity} \right) dC_{\text{crude oil}} \tag{7.2}$$

where A^{Blank} represents the area between the control and the horizontal base line, and A^{Disp} represents the area between the curve of the treated sample set and the horizontal base line, when light intensity is modeled as a function of crude oil concentration. The graphical illustration is presented is Figure 7.7. Numerical method such as trapezoidal rule can be used to approximate the area below the horizontal line.

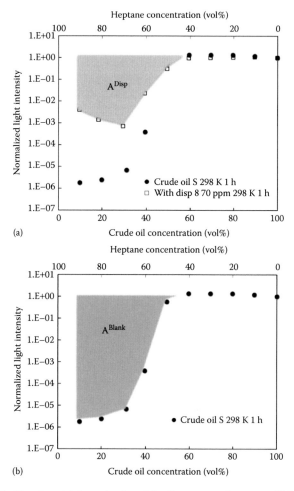

FIGURE 7.7 Experimental determination of the dispersant performance efficiency (DPE).
(a) The shaded area is related to the performance of a given dispersant. (b) The shaded area
is for the blank (without dispersant). The smaller the shaded area in (a), the better the dispersion efficiency.

With this method, dispersion strength can be compared between samples under
the same conditions, with a higher DPE indicating higher dispersion strength. It is
important to point out that this model gives a measure of dispersion strength, but not
of how well a chemical additive will inhibit deposition.

Melendez-Alvarez et al. (2016) investigated the dosage effect of the commercial Dispersants 8, 9, and 15 at 298 K and 1 h aging time. Figure 7.8 shows that
Dispersant 8 performed relatively similarly at both 70 and 500 ppm. These results

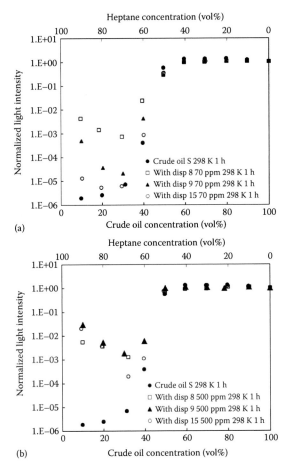

FIGURE 7.8   (a) Crude oil S and crude oil S treated with three different conventional inhibitors at 70 ppm and (b) crude oil S and crude oil S treated with three different conventional inhibitors at 500 ppm. (Reprinted from Melendez-Alvarez, A. A. et al., *Fuel*, 179, 210–220, 2016. With permission.)

support the idea that an optimal concentration exists for dispersants within a given system, beyond which the size of asphaltene aggregates remain approximately the same size (Rogel 2011). On the other hand, Figure 7.8 shows that both Dispersants 9 and 15 significantly reduced asphaltene aggregation at 500 ppm compared to 70 ppm. Despite this, Figure 7.8b indicates that all three dispersants displayed similar dispersive strengths at 500 ppm, whereas Dispersant 8 shows more impressive strength than the others at 70 ppm. Thus, Melendez-Alvarez et al. (2016) determined that Dispersant 8 at 70 ppm is the most effective dispersant for this system.

Direct spectroscopy can also allow for trends to be seen related to aging time and temperature. In their work, Melendez-Alvarez et al. (2016) showed that at constant concentrations, dispersive performance had an inverse relationship with aging time, where aging time is defined as the time between sample preparation and light intensity measurement. The decrease in dispersing power at longer aging time provides indication that these dispersants are kinetic inhibitors. Thus, there should not be a difference between samples with and without inhibitors at infinite time. Furthermore, dispersive performance was also shown to have an inverse relationship with temperature. This effect results from the crude oil acting as a better solvent for asphaltenes at higher temperatures. The addition of the chemicals may have lesser effect because the asphaltenes are more stable in the oil at higher temperatures. Thus, better performance of the inhibitor was observed when the temperature was lower (Melendez-Alvarez et al. 2016).

7.1.1.2.1.4 Capillary Deposition Flow Loop Asphaltene capillary deposition test, also referred to as flow loop test or dynamic loop test was first developed by Broseta et al. (2000), who tested the tendency of asphaltene deposition based on asphaltene stability. Basically, in this technique, a mixture of oil and precipitant is pumped through a metallic capillary tube, to induce asphaltene precipitation and deposition. Asphaltene deposition is monitored by measuring the pressure drop across the capillary tube. Some studies have been performed using this technique to investigate the effect of inhibitors on the deposition of asphaltenes (De Boer et al. 1995; Gon and Fouchard 2016) and scale (Liu et al. 2012) in the capillary flow loop.

The theory behind this technique seems to measure the effectiveness of different inhibitors to prevent asphaltene deposition. However, this may not be true, for several reasons. Firstly, inhibitors will operate differently at high-pressure and high-temperature (HPHT). This was highlighted by De Boer et al. (1995), where they performed the capillary deposition tests to screen the performance of several inhibitors at ambient conditions and translated the results to reservoir conditions. It was found that a lower concentration of inhibitor is required at high pressure and temperature. The addition of a chemical has lesser benefit to slow down asphaltene aggregation because of the increasing stability of oil at high temperature. The concentration of inhibitors needed to mitigate asphaltene deposition at ambient conditions using the capillary deposition test was typically 2 to 3 times greater than the amount required at reservoir conditions. Additionally, Bouts et al. (1995) showed that 1,300 ppm is the optimum concentration for their tested inhibitor (Surdyne A237) to mitigate asphaltene deposition. Their decision was based on *n*-heptane titration experiments performed at ambient conditions. Field tests were performed using the same inhibitor, in an oil field in Alberta, Canada (Bouts et al. 1995). It was found that 66 ppm concentration of the inhibitor was effective at the reservoir conditions, this is 20 times less than the value needed in the same crude oil ambient conditions (Bouts et al. 1995).

On the other hand, Gon and Fouchard (2016) tested the effectiveness of different inhibitors using the capillary deposition test at elevated temperature. The results were highly reproducible for untreated oil from the Gulf of Mexico. However, they pointed out that the technique needs to be improved when used to rank inhibitors because of the inconsistency and irreproducibility of the results. By performing the elemental analysis, some residual oil was found in the deposits. Thus, the extent of deposition cannot be inferred from the pressure drop across the capillary tube. Instead, they decided to quantify the mass of deposited asphaltenes by washing the capillary tube with toluene and then allowing it to evaporate. From contact angle measurements, it was found that inhibitors alter the wetting properties of oil in the capillary tube. The change in contact angle was different for the selected inhibitors. Thus, different amount of residual oil will be observed in the deposits. This issue was addressed by quantifying the amount of residual oil for each experiment. This was obtained by plotting the total amount deposit as a function of run time. The deposition profile was linear for both treated and untreated crude oils. By a simple linear fit, the slope of the straight line is the rate of deposition, and the intercept is the amount of residual oil. At zero time, the amount of residual oil should be zero because it is assumed that there is no deposition at the start of the experiment. Thus, the total amount of asphaltene deposited was corrected by subtracting the value of the intercept. The corrected plot with zero intercepts is presented in Figure 7.9. The effectiveness of the inhibitor can be evaluated by the slope of the straight line (i.e., the rate of deposition). According to Gon and Fouchard (2016), inhibitor C was most effective on the prevention of asphaltene deposition among the three chemicals injected at the same dosage (500 ppm).

Gon and Fouchard (2016) concluded that asphaltene deposition profile is more reliable in quantifying inhibitor performance compared to single data test. However, this technique requires more time, crude oil, and labor than other methods. Despite the

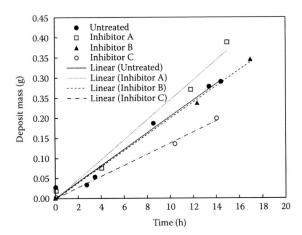

FIGURE 7.9 Asphaltene deposition profile after correction for the mass of residual oil. (Reprinted from Gon, S. and Fouchard, D.M., *Energy Fuels*, 30, 3687–3692, 2016. With permission.)

reproducibility of their results, they concluded that capillary deposition test is not the most appropriate technique to evaluate the performance of different inhibitors.

7.1.1.2.1.5   Packed Bed Column   The packed bed column is a newly developed apparatus to assess asphaltene deposition on the metallic surface (Vilas Boas Favero et al. 2016; Kuang et al. 2017). The details of the design were introduced in Section 5.1.1.4. Compared to other deposition setups, it requires much less oil samples to achieve reproducible results. The amount of deposition can be quantified directly from a mass balance calculation, and the deposition profile can be obtained from the multi-section design. Additionally, the packed bed column allows the determination of the impacts of surface properties such as surface roughness, surface area, and coating materials on the deposition of asphaltenes. Besides its ability to investigate asphaltene deposition under various experimental conditions, the packed bed column is an excellent setup for the evaluation of asphaltene inhibitors targeting the asphaltene deposition process. After each experiment, the deposited materials can be easily obtained and characterized. Thus, valuable information on inhibitor-asphaltene interactions and inhibitor-surface interactions can be determined. The characterization of the asphaltene deposits might provide valuable insight on the development of the next generation asphaltene deposition inhibitors.

The effects of three conventional asphaltene inhibitors (I-8, I-9, and I-15) on the mitigation of asphaltene deposition were assessed by the multi-section column packed with carbon steel spheres at ambient conditions (Kuang et al. 2017). Stock solutions containing 60 ppm of the inhibitor solution and crude oil S9 were prepared. The volume ratio of stock solution and *n*-heptane was maintained at 30:70 with a total flow rate of 9 cm^3/h. The amount of deposits was quantified after 6 hours of experimental duration. Table 7.1 shows the comparisons of the amount of deposition with and without the treatment of three different inhibitors. Good reproducibility was obtained with an average percentage error of 6.8% based on eight experiments being performed. In general, the mass of deposits collected in the bottom section (close to the inlet of flow) of the column is more than that from the top section (close to the outlet of the flow). This provides evidence that asphaltene deposition

---

TABLE 7.1

Results of the Amount of Deposition for the Crude Oil S9 with and without the Addition of 60 ppm of Commercial Dispersant 8, 9, and 15 at Ambient Conditions

	Deposition on Spheres (mg)			
	Conditions			
Surface for Deposition	Crude Oil S9 (Blank)	I-8	I-9	I-15
Spheres from top section	26 ± 12	28 ± 1	28 ± 5	22 ± 8
Spheres from bottom section	52 ± 15	44 ± 4	62 ± 4	40 ± 2
Total	78 ± 3	72 ± 5	91 ± 2	62 ± 10

is more likely to occur instantaneously when oil is destabilized. The effectiveness of the inhibitors on the prevention of asphaltene deposition can be ranked based on the total reduction in the amount of deposition compared to the condition with no chemical treatment. I-15 is the most effective asphaltene deposition inhibitor, which decreases the amount of deposition on spheres by 21%, whereas I-9 increases the deposition by 16%. Thus, the performance of chemical inhibition by different chemicals varies, and the injection of the wrong inhibitor might even worsen the asphaltene deposition problem. Similar observations are obtained in the RealView deposition cell (Akbarzadeh et al. 2012) and capillary deposition flow loop (Gon and Fouchard 2016).

The conventional asphaltene inhibitors are developed based on their performance to disperse asphaltene aggregates; however, it is not necessarily true that the smaller size of asphaltene particles could reduce asphaltene deposition in the production lines. A detailed discussion on the improper assessments of asphaltene inhibitors will be given at the end of this chapter, and a case study showing the discrepancies between the dispersive performance of the asphaltene inhibitor and its ability to reduce asphaltene deposition will be presented in Chapter 8 (Case Study #10).

The effectiveness of alkylphenols on the prevention of asphaltene precipitation and aggregation has been investigated by different researchers (Chang and Fogler 1994a, 1994b; Goual et al. 2014). Also, deposition experiments in the multisection packed bed column have been conducted to assess the impacts of these amphiphiles on the mitigation of asphaltene deposition (Kuang et al. 2017). Three alkylphenols—4-Hexylphenol (HP), 4-Octylphenol (OP), and 4-Dodecylphenol (DP) —were selected to investigate the effect of aliphatic chain length on the prevention of asphaltene deposition in crude oil C. According to Table 7.2, 4-Octylphenol is the best asphaltene deposition inhibitor, which reduces 27% of deposition compared to the condition without chemical treatment. On the other hand, 4-Hexylphenol is the worst asphaltene deposition inhibitor, which increases asphaltene deposition by 18%. Thus, the aliphatic chain length could not only influence the dispersive performance of the amphiphiles, but it may also affect their ability to reduce asphaltene deposition. The deposition tests in the packed bed column provide insights on the existence

TABLE 7.2

Results of the Amount of Deposition for the Crude Oil C with and without the Addition of 30 ppm of 4-Hexylphenol (HP), 4-Octylphenol (OP), and 4-Dodecylphenol (DP) at Ambient Conditions

	Deposition on Spheres (mg)			
	Conditions			
Surface for Deposition	Crude Oil C (Blank)	HP	OP	DP
Spheres from top section	59	56 ± 6	38 ± 4	38 ± 9
Spheres from bottom section	106	138 ± 9	83 ± 15	116 ± 13
Total	165	194 ± 15	121 ± 11	154 ± 4

of an optimum chain length, which can generate the minimum amount of asphaltene deposition. Potential mechanisms of action for the development of future asphaltene deposition inhibitors will be discussed in Section 7.3.2.

7.1.1.2.2 High-Pressure and High-Temperature Conditions

7.1.1.2.2.1 Solid Detection System In an attempt to accurately represent reservoir conditions, techniques have been developed that allow for high-pressure and high-temperature while determining the onset of asphaltene precipitation (as discussed in Section 3.2). Of these methods, the solid detection systems (SDS) have been used by many laboratories to study the precipitation of asphaltenes in live oil. The SDS measures the light transmission through a given sample to determine the onset of precipitation. The SDS can measure the onset of precipitation caused by the changes in temperature, pressure, and composition. The pressure control is done by adjusting the compression of a variable-volume high-pressure, high-temperature pressure-volume-temperature (PVT) cell (Hammami et al. 2000). A central heating source is used to control the temperature of the system. Furthermore, a magnetic impeller is often used to mix the fluid. In addition, the impeller assists in restoring equilibrium in the fluid after a pressure or composition change (Hammami et al. 2000). With this, the light transmittance can be directly measured as a function of either pressure, temperature, or composition of the fluids by the detector probe.

One of the applications of the SDS cell is the evaluation of the performance of chemical inhibitors by determining the shift in the asphaltene onset pressure (AOP). Live oil samples, which contain dissolved gas in solution, are mixed with predetermined concentrations of chemical inhibitors, and then injected into the PVT cell.

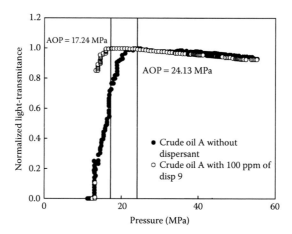

FIGURE 7.10 Light transmittance shown as a function of fluid pressure. For these plots, the onset of precipitation is recognized as the point at which there is a sudden decrease in light transmittance. AOP, asphaltene onset pressure. (Reprinted from Melendez-Alvarez, A. A. et al., *Fuel*, 179, 210–220, 2016. With permission.)

Figure 7.10 shows the results of an SDS experiment performed on crude oil A with and without the injection of commercial Dispersant 9 (Melendez-Alvarez et al. 2016). For the untreated crude oil, the light transmittance increases as pressure decreases from 55.16 to 24.13 MPa. The light transmittance begins to decrease at a pressure below 24.13 MPa. This critical point of 24.13 MPa is known as the *asphaltene onset pressure*. In comparison, the AOP drops to 17.24 MPa for the crude oil treated with Dispersant 9.

Although the SDS tests can be conducted in live oil samples at high-pressure and high-temperature conditions, it is important to highlight that the SDS has a major limitation in the detection of the AOP because of the lack of sensitivity of the instrument (Melendez-Alvarez et al. 2016). When the oil is treated with the dispersant, the precipitation and aggregation of asphaltenes are slowed down by the effect of inhibition (Melendez-Alvarez et al. 2016). Thus, the time needed for the asphaltene particles to grow to a size that can be detected by the instrument is significantly increased. Theoretically, no difference in AOP would be observed between the cases with and without the dispersant when the measurement is performed at infinite time. Therefore, this is an artifact of the experiment that the instrument is not capable of detecting the true onset of asphaltene precipitation.

7.1.1.2.2.2   RealView Deposition Cell   The RealView Deposition Cell is commonly used to assess the performance of asphaltene inhibitors in live oil or combined samples at HPHT conditions. The details of the design were introduced in Section 5.1.1.3. The RealView Deposition Cell aims to represent the thermal and hydrodynamic conditions in the production tubing, and results can be directly obtained from the mass balance calculation after the live oil is destabilized by pressure depletion. Unlike the techniques operating at ambient conditions, the RealView Deposition Cell can provide more representative results on the performance of the inhibitors at realistic conditions. Instead of investigating chemicals' ability to disperse and stabilize asphaltenes, the RealView Deposition Cell can evaluate their effectiveness on the prevention of actual asphaltene deposition. However, the RealView Deposition Cell requires a relatively large amount of oil, close to 1 L, especially for oils with low asphaltene content.

Juyal et al. (2010) reported a study on the assessment of two commercial inhibitors in the RealView Deposition Cell. In a series of experiments, live oil samples with and without 500 ppm of inhibitors were equilibrated in the cell to the temperature and pressure similar to those in the wellbore. The inner cylinder started rotating to achieve the same sheer stress as that in the production lines. Then, the system was depressurized isothermally to 44.82 MPa, which was below the onset pressure of asphaltene precipitation. Finally, the amount of deposits was quantified after 2 hours (Juyal et al. 2010). By comparing the deposition on the outer cylinder of the RealView Deposition Cell for the experiments with and without chemicals treatment, the layer of deposition in the treated system looks darker and thicker (Juyal et al. 2010). The deposits were collected by dichloromethane, and the mass of the deposits was then quantified after evaporating the solvent. Juyal et al. (2010) observed that the injection of the two different inhibitors increased the amount of deposition by 16.7% and 99.0%, respectively. Instead of preventing asphaltene deposition, chemical treatment seems to increase the tendency of asphaltene deposition. They also performed a compositional analysis on the deposits by the electrospray ionization Fourier transform ion cyclotron resonance mass spectrometry (ESI FT-ICR MS). The deposits obtained from the experiments with chemical treatment have lower molecular weight compared to those without chemical treatment, providing evidence that the low molecular weight asphaltenes dispersed by the inhibitors have higher tendency to deposit on the metallic surface (Juyal et al. 2010). Thus, it can be deduced that these chemicals are developed based on their ability to disperse asphaltene aggregates. Experimental results show that it is not necessarily true that smaller asphaltene aggregates can reduce asphaltene deposition in the RealView Deposition Cell. Instead, the generation of more micro-aggregates after the chemical treatment might induce more asphaltene deposition (Vargas et al. 2010). Juyal et al. (2010) also emphasized the necessity of characterizing the deposits for the sake of developing the next generation asphaltene inhibitors targeting the actual deposition process. The ineffectiveness of the conventional asphaltene dispersants on the prevention of asphaltene deposition will be discussed in detail in Section 7.3.1.

Akbarzadeh et al. (2007) assessed the performance of an asphaltene inhibitor on the prevention of deposition in one oil sample from the Gulf of Mexico using the RealView Deposition Cell operating at reservoir temperature and pressure. They reported that the injection of 200 ppm of the inhibitor reduced the deposition rate of asphaltenes by 41.7%. In another case, Akbarzadeh et al. (2012) reported that the mass of asphaltene deposition was reduced from 80 to 26 mg after the treatment with 150 ppm of an inhibitor in the RealView flow-through system. However, a higher dosage of inhibitor (500 ppm) did not further reduce asphaltene deposition. This phenomenon might be explained by the increasing inhibitor-inhibitor interactions and inhibitor-surface interactions after excess chemical treatment (Barcenas et al. 2008; Rogel 2011; Akbarzadeh et al. 2012). In contrast to the experiments conducted in the flow-through system, the same inhibitor did not seem as effective in the batch system because of the limited amount of asphaltene deposition onto the metallic surface (Akbarzadeh et al. 2012).

7.1.2 CONTROL OF OPERATING CONDITIONS

From the conceptual mechanism presented in Figure 1.6, one can conclude that asphaltene solid deposits can be mitigated by either preventing asphaltene precipitation, aggregation, or aging. The chemical additive technique has focused on preventing aggregation through dispersion. Some other efforts have been devoted to developing new inhibitors based on the mechanisms of aging disruption and electrostatic interactions, which will be further elaborated in Section 7.3.2. One of the greatest challenges of the chemical additive technique is to find a chemical with broad spectrum of effectiveness, considering the variability of asphaltenes structure and oil properties in different regions of the world. Thus, one potential way to prevent deposition is through stabilization of asphaltenes and prevention of its precipitation. This is not readily achieved because asphaltene precipitation is governed by the thermodynamics of the system. The stability of asphaltenes is a function of pressure, temperature, and composition. Pressure and temperature are usually dictated by the reservoir condition,

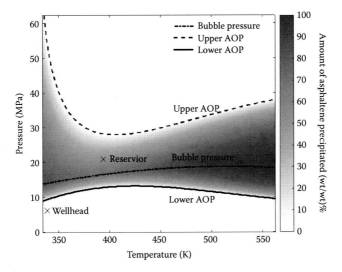

FIGURE 7.11 Phase behavior contour plot for crude oil S26 from the Middle East. AOP, asphaltene onset pressure. (Reprinted from Khaleel, A. and Vargas, F.M., Mitigation of asphaltene deposition by dead oil/maltene reinjection, In Preparation, 2017. With permission.)

wellhead condition, and production profile. However theoretically speaking, from a thermodynamic point of view, controlling the operating conditions can prevent or limit asphaltene precipitation and thus prevent deposition. This section will focus on possible strategies to stabilize or minimize the amount of asphaltene precipitation.

One can determine the effects of pressure and temperature by looking at the phase behavior of asphaltenes. Figure 7.11 shows an asphaltene phase diagram for crude oil S26 from the Middle East. This plot includes the upper AOP, lower AOP, and bubble pressure curves. The color code (contour plot) represents the amount of precipitated asphaltenes in percentage. It is important for the reader to understand that predicting asphaltene precipitation by the AOP alone might be misleading. According to Figure 7.11, asphaltenes start to precipitate around 3.4 MPa below the AOP curve. This is indicated by the white color which stands for 0% of precipitated asphaltenes. Additionally, for modeling asphaltene deposition, the amount of precipitated asphaltenes is a direct input to the model, as discussed in Section 6.1.1, which makes the amount of precipitated asphaltenes more important than the onset point itself.

7.1.2.1 Effect of Pressure and Temperature

Pressure and temperature along the wellbore have a direct impact on the asphaltene precipitation tendency. For the case of crude oil S26, whose phase diagram is presented in Figure 7.11, there is a region where the amount of precipitated asphaltenes is minimum. This is between 366 and 394 K, the region between the minimum of the upper AOP and maximum of the lower AOP. As a rule of thumb for typical reservoir temperature, the higher the temperature in the wellbore the lower the amount of precipitated asphaltenes. If it is economically feasible, thermal insulation of the wellbore could be beneficial not only to reduce wax but also to prevent asphaltene deposition. Another way to increase the wellbore temperature is by steam injection, which has been used to reduce oil viscosity for heavy and viscous oil fields.

If economically and technically feasible, the manipulation of pressure or pressure change in the wellbore could also mitigate asphaltene deposition. Through the production profile, the system goes from a stable state to another stable state by passing through a region of instability. Within this region, precipitation, aggregation, and aging phenomena occur. Precipitation is controlled by thermodynamics. Aggregation and aging are governed by kinetics. If one can increase the rate of pressure drop between the two stable states, such that it overcomes the rate of aggregation and aging, this may result in less and softer asphaltenes. According to the conceptual mechanism discussed in Section 7.1.2, small liquid-like aggregates can potentially be redissolved at the final stable state. The rate of pressure drop can be increased by using the concept of venturi effect through the installation of a choke valve. The location of the choke valve needs to be designed based on each well. This strategy may not be applicable to every wellbore, and it will highly depend on the reservoir pressure and the pressure drop across the wellbore. It may be best used in cases where the reservoir pressure is very high.

7.1.2.2 Effect of Composition

The poor performance of some commercially available mitigation strategies and the lack of understanding of asphaltenes behavior with chemical additives have urged the need to investigate alternative methods to prevent asphaltene deposition. Khaleel and Vargas (2017) proposed a novel technique that revolves around reducing the amount of asphaltene precipitation by changing the composition of the system. The idea is to recycle maltenes or dead oil, which are good asphaltene solvents, into the wellbore region to alter the composition of the system as shown in Figure 7.12. In this technique, the injected chemical is tailored for each well, which makes it universal, unlike the currently used chemical injection mitigation techniques. The decision on whether to recycle dead oil or maltenes will be based on each well, the saturates, aromatics, resins, and asphaltenes (SARA) analysis of the oil, viscosity, and techno economic analysis.

To verify the stabilization of asphaltenes by maltenes/dead oil recycle, the Hildebrand solubility parameter was calculated using perturbed chain form of the statistical associating fluid theory (PC-SAFT) equation of state (EOS). Details on PC-SAFT EOS are discussed in Section 4.2.3.2. Figure 7.13 presents the solubility parameter for crude oil S7 with maltenes and dead oil recycle as a function of pressure. As discussed in Section 4.2.2, the difference between the solubility parameter of two substances gives an indication of its miscibility. In general, the closer the values of the two solubility parameters, the more miscible the substances will be. This proposed technique is feasible at pressures higher than the bubble point pressure, where there is an increase in the solubility parameter of the oil mixture making it a better solvent for asphaltenes. According to Figure 7.13, the reinjection of maltenes and dead oil have a similar effect for this particular crude oil. This is probably because of the low asphaltene content (0.4 wt%) in this crude oil.

To further validate the effectiveness of recycling maltenes on the mitigation of asphaltene deposition, asphaltenes phase diagram with the amount of asphaltene

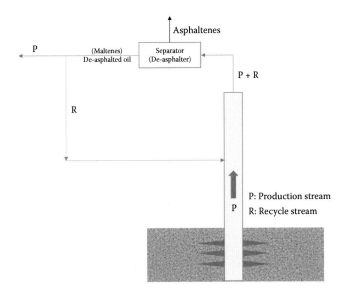

FIGURE 7.12 Schematic of the proposed asphaltene deposition mitigation technique. (Reprinted from Khaleel, A. and Vargas, F. M., Mitigation of asphaltene deposition by dead oil/maltene reinjection, In Preparation, 2017. With permission.)

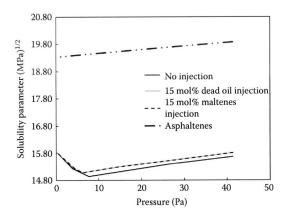

FIGURE 7.13 Solubility parameter with respect to pressure for different recycle fluids. (Reprinted from Khaleel, A. and Vargas, F. M., Mitigation of asphaltene deposition by dead oil/maltene reinjection, In Preparation, 2017. With permission.)

precipitation was plotted and shown in Figure 7.14a. It is important to note that Figure 7.14 was predicted using the PC-SAFT EOS and the PC-SAFT parameters for this crude oil were tuned to AOP data obtained by HPHT conditions. Additionally, the recycle fraction is in mole basis and is the fraction of the overall production, that is, $R = (\text{fraction}) \times (P + R)$. A visible reduction in the amount of asphaltene precipitation

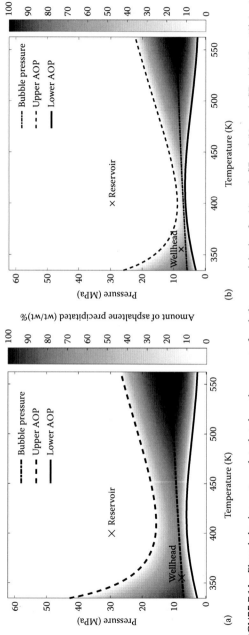

FIGURE 7.14 Phase behavior contour plots that show the amount of asphaltene precipitation for (a) crude S7 and (b) crude S7 with 15 mol% dead oil recycle. The scale bar is in percentage. AOP, asphaltene onset pressure. (Reprinted from Khaleel, A. and Vargas, F. M., Mitigation of asphaltene deposition by dead oil/maltene reinjection, In Preparation, 2017. With permission.)

is observed with a 15% of dead oil recycle as seen in Figure 7.14b. In the practice, even if it is not feasible to reinject this amount of dead oil, any amount will contribute positively to the reduction of asphaltene precipitation and hence deposition.

The validity of this mitigation technique was tested by computer simulation of six different crude oils from the Middle East. This includes crude oils from oil fields that originally had asphaltene deposition problem and fields that showed asphaltene deposition problem after gas injection as part of enhanced oil recovery (EOR) operations. From the screening process, it was concluded that this technique may be more effective in problematic wells with high gas-to-oil ratio (GOR). In all the case studies, light crude oils (0.2 wt%–3 wt% asphaltenes) were tested. It seems that for light crude oils, the effect of recycling dead oil is as good as maltenes. Because the asphaltene deposition problem is usually observed in light crude oils with low asphaltene content, the proposed mitigation technique is promising. The decision on whether to recycle maltenes or dead oils will depend on the asphaltene content, phase behavior, the viscosity of the final mixture, and ultimately on the economics of this process. This section provides promising results on the effectiveness of the recycling technique to reduce the amount of asphaltene precipitation, which can further mitigate the asphaltene deposition problem. It also provides a new tool to investigate different recycle fractions depending on the situation and how severe the asphaltene deposition problem is expected to be.

7.1.3   COATINGS

The techniques of internal coating have been widely used for corrosion protection, hydraulic improvement, and deposition mitigation in the oil and gas industry (Lauer 2007). The effectiveness of coating on the prevention of paraffin and scale deposition has been investigated extensively. Jorda (1966) proved that smooth and nonparaffinic plastic materials could mitigate paraffin deposition successfully. These internal plastic coatings can not only minimize the area of the rough surfaces which trap paraffin mechanically but also provide insulation to the surface of the pipelines (Lauer 2007). However, the application of internal coatings to prevent asphaltene deposition has not been fully developed because of the complexity of the asphaltene deposition problem. The effectiveness of polymer-based coatings has been experimentally determined to mitigate asphaltene deposition.

Gonzalez (2008) evaluated the asphaltene deposition tendency on various coating materials by using the Hamaker-Lifshitz theory to predict the interactions among the asphaltene lean phase (solvent or crude oil), the precipitated asphaltene rich phase, and the substrate surface. The determined Hamaker constants can be used to select the most appropriate coating materials to prevent asphaltene deposition by minimizing the sticking and spreading tendencies of the destabilized asphaltenes. As discussed in Section 5.2.4, a negative value of A_{SLR} corresponds to the repulsion of the precipitated asphaltenes from the surface, thereby reducing the sticking tendency of the asphaltenes on the surface. Thus, according to Figure 7.15, polymeric materials may have less asphaltene deposition than metals. Asphaltenes are least likely to stick on the materials made of polytetrafluoroethylene (PTFE) and polyvinylidene fluoride (PVDF).

On the other hand, a positive value of A_{SRL} means that there is no spreading or wetting tendency of the asphaltene rich phase on the surface. As shown in Figure 7.16,

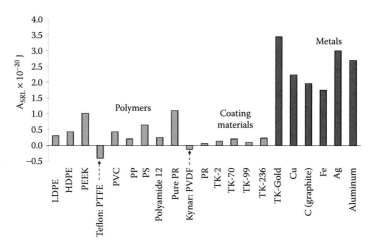

FIGURE 7.15 The Hamaker constant (A_{SLR}) predicting the sticking tendency of asphaltenes on different materials. (Reprinted from Gonzalez, D. L., Modeling of asphaltene precipitation and deposition tendency using the PC-SAFT equation of state, Rice University, Retrieved from https://scholarship.rice.edu/handle/1911/22140, 2008. With permission.)

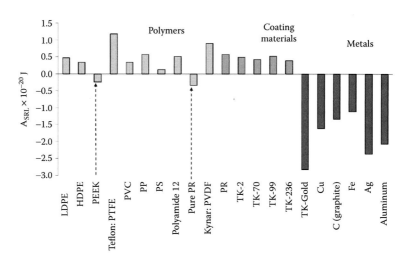

FIGURE 7.16 The Hamaker constant (A_{SRL}) predicting the spreading tendency of asphaltenes on different materials. (Reprinted from Gonzalez, D. L., Modeling of asphaltene precipitation and deposition tendency using the PC-SAFT equation of state, Rice University, Retrieved from https://scholarship.rice.edu/handle/1911/22140, 2008. With permission.)

most polymers are good materials to prevent the adsorption of asphaltenes from solution whereas metal surfaces are usually coated with a layer of asphaltene-rich phase (Gonzalez 2008).

Based on the values of the composite Hamaker constants, polymeric coatings can not only prevent the sticking of asphaltenes but also eliminate the growth of an adsorbed asphaltene-rich phase onto the oilfield pipelines. However, difficulties have been reported of using fluoropolymer coatings to prevent internal deposition in the oilfield conditions. Such coatings can be degraded by the abrasive production conditions (Lauer 2007; Gonzalez 2008). The detachment of the coated materials can not only cause coating failure, but also lead to additional flow assurance problems. Therefore, the development of an effective coating material to prevent asphaltene deposition still remains an objective of current investigations, and the technique to maintain an integrated layer of coating under harsh flowline conditions needs to be considered.

7.2 REMEDIATION OF ASPHALTENE DEPOSITION

The decrease in production because of the deposition of organic solids in various steps of the production of oil has proven to be a significant economic sink for the oil industry. These solids contain a significant amount of asphaltenes, so as one might expect, effective cleaning techniques are centered on the remediation of asphaltene deposition. The most severe deposition problems occur in regions of the reservoir or well that has large driving force for asphaltene precipitation and deposition caused by the changes in temperature, pressure, or composition. Due to the structural complexity and variability of this class of molecules, a universal mitigation approach has yet to be discovered, resulting in the widespread need for effective remediation techniques. Generally, common remediation approaches can be classified into three categories: thermal, mechanical, and chemical. Asphaltenes do not liquefy, but decompose to refractory carbonaceous deposits upon heating, disqualifying thermal treatments as a viable remediation option for asphaltic deposits. We are then left with mechanical and chemical remediation methods to combat asphaltene deposition (Kokal et al. 1995; Straub et al. 1989; Carbognani 2001).

7.2.1 SOLVENT WASH

The categorical definition of asphaltenes is based on solubility: insoluble in n-alkanes, but soluble in aromatic solvents (Kokal et al. 1995). The aromatic nuclei of the characteristic asphaltene structure further affirm the role of aromaticity in the stabilization of asphaltenes in solution. It is important to select the best solvent that can readily dissolve the deposited asphaltenes in the wellbore and near-wellbore regions.

Determination of effective chemical solvents can be facilitated by understanding solubility theory as described by Hansen (2007). Building on the work of Hildebrand

and Scott (1964), Hansen proposed that the energy needed to vaporize a liquid solvent is related to the intermolecular interactions present. He postulated that energies of dispersion (E_D), permanent-dipole (E_P), and electron-exchange (E_H) make up the total cohesive energy of a solvent, such that:

$$E = E_D + E_P + E_H \tag{7.3}$$

$$E/V = E_D/\widehat{V} + E_P/\widehat{V} + E_H/\widehat{V} \tag{7.4}$$

$$\delta = (E/\widehat{V})^{1/2} \tag{7.5}$$

$$\delta = \left(\delta_D^2 + \delta_P^2 + \delta_H^2\right)^{1/2} \tag{7.6}$$

where:
   δ is the solubility parameter,
   E is energy of vaporization, and
   $\widehat{V}$ is the molar volume of the pure solvent.

Partial solubility parameters have been calculated for approximately 1,200 chemicals. Chemicals with similar Hansen solubility parameters (HSP) form stable solutions with one another. Reported values for the overall asphaltene solubility parameter are approximately 20 MPa$^{1/2}$. Table 7.3 is an inexhaustive list of the solubility parameters for a number of important chemicals and confirms relative similarity of the aromatic compounds, notably toluene and xylene, to benzene in contrast to the paraffins and other common compounds listed (Hansen 2007). Light aromatics, such as the ones well-suited for asphaltene dissolution, have unfavorable environmental, health, and safety risks. Exposure has been associated with respiratory and cognitive impairments, and water contamination poses a risk to aquatic life (Thermo Fisher Scientific 2017). Additionally, the low flash points of these solvents may generate combustible vapor to the worksite. As a result, when possible, xylene is chosen because of its higher flash point and is used sparingly unless doped with less hazardous chemicals, known as *cosolvents*. In some companies, the use of xylene is banned because it is considered as a volatile low flash-point component. In that case, the use of aromatic naphtha (commercially known as A150) is preferred. Although cosolvents inevitably diminish the solubility of asphaltenes in the formulation, selection of the right diluent can aid in the dissolution of deposits within a wellbore. Xylene and other aromatics are immiscible in water, so the water-wet asphaltene deposits found within a well prevent a pure aromatic compound from reaching its full dissolution capacity. Blending a water penetrant with xylene results in improved asphaltene stabilization and solid removal under test conditions. The significant difference in HSP values in Table 7.3 between xylene (17.9 MPa$^{1/2}$) and water (47.8 MPa$^{1/2}$) demonstrates the incompatibility of these solvents. To facilitate emulsion of the two chemicals, a cosolvent, whose solubility parameter is between that of xylene and water, may be used. Cosolvent formulations are also known to further benefit the system by

TABLE 7.3
Hansen Solubility Parameters (MPa$^{1/2}$)

Name	δ_D	δ_P	δ_H	δ
Aromatics				
Benzene	18.4	0	2.0	18.5
Naphthalene	19.2	2.0	5.9	20.2
Phenol	18.0	5.9	14.9	24.1
Pyridine	19.0	8.8	5.9	21.8
Toluene	18.0	1.4	2.0	18.2
Xylene	17.6	1.0	3.1	17.9
Paraffins				
n-Pentane	14.5	0	0	14.5
n-Hexane	14.9	0	0	14.9
n-Heptane	15.3	0	0	15.3
n-Octane	15.5	0	0	15.5
Alcohols				
Ethanol	15.8	8.8	19.4	26.5
1-Pentanol	15.9	5.9	13.9	21.9
1-Heptanol	16.0	5.3	11.7	20.5
Other Notable Compounds				
Acetone	15.5	10.4	7.0	19.9
Carbon Dioxide	15.7	6.3	5.7	17.9
Ethyl Acetate	15.8	5.3	7.2	18.2
Ethylene Glycol	17.0	11.0	26.0	33.0
Water	15.5	16.0	42.3	47.8

water wetting the well casing and prolonging the effective production life span of the well between treatments. Alternatively, refinery distillates, such as aromatic naphtha distillate, may be used as solvents (Trbovich and King 1991; Hansen 2007; Khaleel et al. 2015).

7.2.2 MECHANICAL REMOVAL

The mechanical removal of asphaltenes uses mechanical instruments to physically dislodge and remove solid deposits within the wellbore. These interventions techniques are deployed from workover service rigs and, more often than not, require a halt in production. The method selected, down time required, and consequent loss of revenue is dependent on the depth, length, and severity of the affected region. However, the thorough cleaning that it can provide while minimizing the hazardous chemicals needed makes it one of the preferred remediation methods ("Remedial treatment for asphaltene precipitation" n.d.).

Wireline is a traditional mode for in-situ operations. A reinforced wireline capable of carrying electrical signals allows for the collection of data and delivery of commands to the tools downhole. This form of conveyor is still widely used for the cleaning of vertical wells. However, this process has, in large part, been replaced by coiled-tube (CT) processes. The advent of coiled tubing in the 1980s revolutionized downhole operations by allowing one continuous length of insulated steel tubing to be deployed from a spool up to 4,572 m (15,000 feet) downhole. This development drastically decreases assembly time for hydraulic process equipment and, in some cases, allows for remediation of live wells (Leising and Newman 1993; The Oilfield Glossary - Schlumberger Oilfield Glossary 2017).

Jetting, also known as *hydroblasting*, is the most common coiled-tube mechanical removal technique. Using coiled tubing to convey a high-pressure wash nozzle to the deposition region, a series of bites and subsequent sweeps serve to break up solids and draw them uphole. Advancement in this method has resulted in the development of a concentric-coiled-tube-conveyed tool with combined jetting and vacuuming capabilities. In many cases, cleanout fluids are also used to facilitate dissociation of solids. This method requires significant shutdown time and is only effective for cases in which the deposits are soft enough to be dislodged by turbulence. For hardened deposits, more rigorous mechanics are required (Li et al. 2010).

More arduous cleaning methods involve drilling, scraping, and brushing operations. The oilfield service industry has developed hydraulically or electrically powered tools, outfitted with any combination of bits, blades, and bristles to dislodge deposited solids from well casings. Rotary tools are delivered downhole via coiled tubing for horizontal wells or wireline for vertical wells. These devices require wells to be shut down to conduct cleaning operations.

The ease of access to surface piping allows for specialized launchers to introduce plug-like devices, known as *pigs*, into live pipelines. Driven downline by fluid flow, a pig's primary function is noninvasive, mitigative treatments. Much like the tools described in the previous section, pigs may be furnished with wire bristles and retractable scraper blades to remove build-up. Intelligent pigs have also been developed to perform a number of diagnostic tests, such as sonar-detection of solid accumulation and leak detection. Poorly maintained piping runs the risk of sticking a pig behind excessive build-up. To prevent such a problem, pigging is an operation best used routinely (Tiratsoo 1992).

Because of the glutinous nature of solids deposits, satisfactory results using mechanical removal techniques alone are difficult to attain. In fact, evidence suggests that mechanical scraping operations result in either no net increased or even loss in production in the majority of the randomly selected, North American subject wells (Trbovich and King 1991). Furthermore, precipitation and deposition of asphaltenes can occur in regions beyond the range of mechanical intervention capabilities. Within wells, solid deposition occurs indiscriminately, meaning that the valves and devices essential for safe and efficient well production face the risk of failure because of asphaltene deposition (Kokal et al. 1995). Depletion has been known to cause pressure within the reservoir to drop into the asphaltene instability region, which can potentially cause precipitation, plugging of the porous media, or wettability alteration. These issues cause considerable economic and safety concerns that must be addressed but cannot be done through mechanical intervention alone. The solution to these efficacy problems is the conjunctive use of solvents.

Remediation methods are expensive, time-consuming, and potentially hazardous processes whose benefit is best weighed in barrels. Conclusions on their place in the oil industry are best drawn from the past. Cenegy (2001) described a Venezuelan well that faced asphaltene deposition problems severe enough to cause well failure within 7 months of cleaning. Cleaning involved a 12 to 24-hour xylene solvent soak, spotted to the correct well depth and agitated by wireline techniques, and when needed, supplemented with mechanical methods, such as scraping. This process cost approximately $50,000 and a 2-day shut-in period was required. As a result, the well could continue operation, producing approximately 713,934 barrels per year. Assuming that the loss of revenue as a result of shut-in periods is accounted for in the previous production value, at $22 per barrel, annual revenue associated with this well was $15,706,548 (Cenegy 2001). By no means are remediation techniques a comprehensive solution to the asphaltene problem. They do, however, play a vital role in the integrated approach taken to minimize its effect on the industry.

7.3 BEST PRACTICES

The injection of chemical additives to prevent asphaltene deposition has shown mixed results in the field. A large variety of asphaltene inhibitors has been developed to stabilize asphaltenes and reduce asphaltene precipitation. However, there are still several unsolved issues related to the performance of chemical inhibition. First, the experimental conditions and results obtained in the lab are not always consistent with those in the producing tubing. Moreover, in some cases, the injection of asphaltene inhibitors is not effective, and it could even worsen the asphaltene deposition problem in the field (Melendez-Alvarez et al. 2016). Thus, it is crucial to revisit the conventional techniques that have been used to assess the performance of the chemical additives. In the following sections, the limitations of the current techniques to screen asphaltene deposition inhibitors will be addressed. Based on the novel understandings of the mechanism of inhibition, the future direction for developing a new generation of asphaltene deposition inhibitors will be proposed.

7.3.1 Improper Assessment of Conventional Asphaltene Inhibitors

As discussed in Section 7.1.1.2.1.1, the ADT has been widely used to assess the performance of conventional asphaltene inhibitors at ambient conditions. The ADT ranks the effectiveness of the chemical additives based on their ability to reduce the size of asphaltene aggregates, which in turn reduces the amount of asphaltene sedimentation (Juyal et al. 2012). However, this technique suffers from several limitations. First, the ADT is conducted at ambient conditions so that the functionality of the selected chemicals might not be the same as that in the high-pressure and high-temperature environment during the oil production process. Additionally, the ADT requires the addition of n-heptane to the oil sample, at a ratio of 40 parts of n-heptane per part of the oil. Aside from the significant amount of solvent required, this high dilution may not be representative of the conditions in which asphaltenes precipitate and aggregate in the real system. In this case, results obtained from the ADT might be misleading because the selected chemicals are tested under the experimental condition that induces the highest amount of precipitation or the largest sizes of asphaltene aggregates (Melendez-Alvarez et al. 2016). Moreover, the ADT measures normal gravimetric sedimentation over a period of time after diluting the oil with an asphaltene precipitant (Juyal et al. 2012). According to Vargas et al. (2010), asphaltene deposition occurs when the aggregated asphaltenes follow a diffusion mechanism to the surface of the tubing, where they stick and build up a deposit. The tendency for asphaltene to adsorb and deposit onto different surfaces is determined by the interactions among the asphaltene-rich phase, the asphaltene-lean phase, and the solid substrate (Gonzalez 2008). Thus, it is not necessarily true that the decrease in the amount of sedimentation detected by the ADT would lead to the reduction in the amount of asphaltene deposition on the metallic surface.

Recently, a method was developed to assess the performance of asphaltene inhibitors based on the NIR spectroscopy (Melendez-Alvarez et al. 2016). According to Figure 7.17a, results from this method show that the onset of asphaltene precipitation shifts from 22 vol% to 29 vol% of n-heptane after the addition of 2,000 ppm of a commercial inhibitor. The shift of the onset may be misinterpreted as an increase of the stability of asphaltenes in the oil caused by the action of the inhibitor. However, the indirect method (as discussed in Section 3.4), a more sensitive approach to detect the onset of asphaltene precipitation, demonstrates that the asphaltene dispersant neither shifts the actual onset of asphaltene precipitation nor reduces the amount of precipitated asphaltenes. The results are presented in Figure 7.17b. Once asphaltenes are destabilized and start precipitating, the sizes of the particles are within the range of 100–400 nm. However, the detection limit of the methods based on microscopy or NIR light scattering is about 1 μm (Tavakkoli et al. 2015; Melendez-Alvarez et al. 2016). Therefore, the shift of onset observed in Figure 7.17a is misleading because the precipitated asphaltenes are already present before they can be detected. Because the dispersant slows aggregation down, the time for the first detection of these particles is delayed in the presence of the dispersant (Melendez-Alvarez et al. 2016). This is a major disadvantage of the techniques based on optical microscopy or NIR light scattering to assess the performance of asphaltene inhibitors.

The SDS is another technique that is used by laboratories to evaluate the stability of asphaltenes. Unlike the ADT and NIR spectroscopy, the SDS can analyze live oil samples at HPHT conditions. The SDS results presented in Figure 7.10 show that there is a shift in the asphaltene stability curve upon the addition of dispersants, implying that the chemical additives can modify the thermodynamic behavior of asphaltenes (Khaleel et al. 2015). As discussed, dispersants work to reduce the rate of asphaltene aggregation so that the system pressure drops further before the particles reach the minimum size to be detected by the SDS. The recorded AOP of a chemically treated system may appear to be lower than that of an untreated system, but in fact asphaltenes are already precipitated before they can be detected by the SDS (Khaleel et al. 2015). Therefore, the shift of the onset of asphaltene precipitation obtained from the SDS results might be caused by the insufficient sensitivity of the instrument and the reduced rate of asphaltene aggregation after the injection of the dispersants (Melendez-Alvarez et al. 2016).

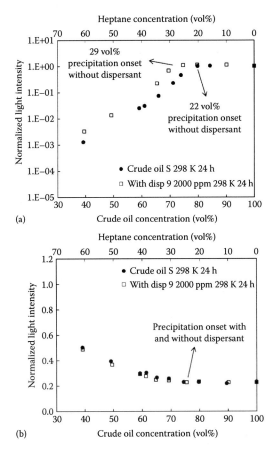

(a)

(b)

FIGURE 7.17 Results of the (a) direct method and (b) indirect method for crude oil S with 2,000 ppm of Dispersant 9, after mixing with *n*-heptane and aging for 24 h at 298 K. (Reprinted from Melendez-Alvarez, A. A. et al., *Fuel*, 179, 210–220, 2016. With permission.)

Chemical additives selected based on the mechanism of dispersion are only proved to prevent or slow down the asphaltene aggregation process. Therefore, asphaltene inhibitors evaluated by the conventional techniques such as ADT, SDS, and NIR spectroscopy might not be necessarily effective on the prevention of actual asphaltene deposition on the metallic surface. That is why the effectiveness of conventional asphaltene dispersants on the mitigation of asphaltene deposition has been questioned by different researchers (Akbar and Saleh 1989; Juyal et al. 2010; Gon and Fouchard 2016). Gon and Fouchard (2016) investigated the effectiveness of a commercial inhibitor on asphaltene deposition in their capillary deposition unit. They obtained a 20% increase in the asphaltene deposition rate after the addition of the commercial inhibitor. A similar phenomenon was observed in the organic solid deposition and control (OSDC) test. Juyal et al. (2010) reported that two asphaltene inhibitors increased the amount of deposition by 16.7% and 99.0%, respectively, compared to the condition without chemical treatment.

The unexpected results of the dispersants from the deposition test can be explained by the mechanism of asphaltene precipitation and deposition proposed by Vargas et al. (2010). According to Figure 1.6, the destabilized asphaltenes can lead to the formation of primary particles. These particles can further aggregate and form micro-aggregates, which can follow a diffusion mechanism to the surface of the oilfield pipelines, where they can stick and build up the deposit (Vargas et al. 2010). The aggregation and deposition of asphaltene micro-aggregates are two competing phenomena. Most of the conventional asphaltene dispersants could preserve small asphaltene aggregates, thereby inducing more asphaltene deposition according to the proposed mechanism (Vargas et al. 2010). Akbarzadeh et al. (2012) also concluded that asphaltene particles no longer tend to deposit after growing to a certain size. Other studies show that the deposition of asphaltenes is dominated by the Brownian motion of the nano-sized particles (Eskin and Ratulowski 2015). These findings suggest that there is an existence of critical size range of asphaltene aggregates favoring the deposition process. Thus, it is crucial to investigate the effect of inhibitor dosage on the performance of asphaltene inhibitors. According to the experimental and simulation studies conducted by Barcenas et al. (2008), higher dosages of the inhibitor significantly worsen the inhibition efficacy. The self-association of the inhibitor itself in the bulk phase might decrease the adsorption of the inhibitor molecules on the asphaltene surface (Barcenas et al. 2008; Rogel 2011). Additionally, the improper dosage of asphaltene inhibitors might induce the tendency of interactions between the inhibitor molecules and the metal surface. Consequently, asphaltene aggregates could have more frequent interactions with the surface to build up the deposit (Akbarzadeh et al. 2012).

Clearly, conventional techniques such as the ADT, the SDS, and the NIR spectroscopy are not as effective in screening asphaltene deposition inhibitors. In other words, strong asphaltene dispersants are not necessarily effective asphaltene deposition inhibitors. A case study revealing the relationship between the dispersive performance of the asphaltene inhibitor and the inhibitor's ability to reduce asphaltene deposition will be discussed in Case Study #10 in Chapter 8. The improper assessments would not only neglect chemicals targeting the actual asphaltene deposition problem, but also provide misleading results on the prevention of asphaltene deposition. Therefore, it is critical to conduct reliable deposition experiments to assess the performance of inhibitors on the prevention of actual asphaltene deposition.

7.3.2 NEW GENERATION OF ASPHALTENE INHIBITORS

Based on the novel understandings of asphaltene inhibition, a new generation of chemical additives targeting the actual deposition process needs to be developed. Instead of dispersing asphaltene aggregates, one of the possible mechanisms of inhibition is through aging disruption. According to Vargas et al. (2014), the precipitated phase of asphaltenes is a meta-stable phase that retains the physical softness of asphaltenes because of the presence of saturates, aromatics, and resins.

This meta-stable phase could be rearranged over time while the structure tries to reach a more stable configuration that causes some of the lighter components present in the precipitated phase to be expelled. The transition from the softer precipitated phase into the solid-like aged phase is irreversible and accelerated at high temperatures. To effectively mitigate asphaltene deposition, new inhibitors could be designed with the ability to disrupt the formation of compact and solid-like asphaltenes. In specific, the next generation inhibitors can be developed to have sufficient interactions with the asphaltene cores to prevent the sticking and stacking of asphaltenes during the aging process. They might have certain chemical structures to allow for strong and abundant π- π interactions with the aromatic rings in the asphaltene cores. Also, a certain length of aliphatic chains might be useful to block further interlocking of the asphaltene cores. The length of the alkyl chains should be carefully designed to maximize the efficiency of the chemical. Disrupting the mechanism of aging may allow asphaltenes to retain their softness, and as a result, the tendency to build up hard and compact deposits along the oilfield tubings could be significantly reduced. Consequently, the soft asphaltene deposits may be easily removed by the shear forces in the production lines.

Another potential approach to design a new generation of asphaltene inhibitors is through the mechanism of electrostatic interactions. Because of the presence of various functional groups and metal ions in the structure, asphaltenes are believed to possess certain surface charges. According to Lichaa and Herrera (1975), asphaltene particles can migrate to the electrode on the application of an electric field, indicating that controlling surface charges of the asphaltenes could be effective to prevent asphaltene deposition. Different studies agree that asphaltene particles are not likely to deposit after growing above a certain size (Vargas et al. 2010; Akbarzadeh et al. 2012; Eskin and Ratulowski 2015). Thus, the development of electro-responsive asphaltene flocculants to attract asphaltenes might be a future approach to control asphaltene deposition. In contrast to the conventional asphaltene dispersants, the new flocculants are able to increase the size of asphaltene aggregates by forming the large flocculant-asphaltene complexes through the electrostatic interactions. The flocculated asphaltenes may have less tendency to diffuse to the surface of the pipelines; instead, they could be remaining in the bulk phase. At the surface facilities, the flocculated asphaltenes could be easily separated from the crude oils.

Therefore, new mechanisms of inhibition, such as aging disruption and electrostatic interaction, may become the future direction for the development of next generation chemicals targeting the actual deposition process. Moreover, integrated approaches for simultaneous mitigation of multiple flow assurance problems could be of great commercial interest. As an example, ongoing efforts to address corrosion and asphaltene deposition at the same time are described in Section 9.2.4.

Additionally, a better knowledge of the molecular nature and morphology of the deposited asphaltenes can provide valuable insights on the design of the next generation asphaltene inhibitors. To achieve that, deposits should be collected from the representative deposition experiments and then characterized by different techniques such as scanning electron microscope, X-ray powder diffraction (XRD),

Fourier transform infrared spectroscopy (FTIR), elemental analysis, and so on. Because different oil samples have distinct compositions, the application of certain asphaltene inhibitor might not be universal. According to the analysis of the deposits collected from the RealView deposition cell, the variation of the heteroatom class distribution shows the difference in the mechanism of inhibition (Juyal et al. 2010). Thus, the characterization of the deposits can help researchers unveil the relationship between the properties of asphaltenes (aromaticity, molecular weight, polarity, etc.) and the choice of the asphaltene inhibitor. Consequently, the correct asphaltene inhibitor can be readily selected to solve specific asphaltene deposition problems based on the properties of asphaltenes. Moreover, important information about the inhibitor-surface interactions can be obtained from the analysis of the deposits.

According to previous studies, the critical size of the asphaltene aggregates plays an important role on their tendency to build up asphaltene deposition. It is crucial to elucidate the relationship between the size of the asphaltene aggregates and the dosage of inhibitor. The improper amount of injected chemicals could be detrimental to the flow assurance efforts. Therefore, the dosage of the chemical additives should be carefully controlled to avoid the generation of certain sizes of asphaltene aggregates favoring asphaltene deposition. Furthermore, an appropriate way to measure the size of asphaltene aggregates in the crude oil should be developed to investigate the correlation between the critical particle size and the asphaltene deposition rate.

REFERENCES

Abrahamsen, E. L. 2012. Organic flow assurance: Asphaltene dispersant/inhibitor formulation development through experimental design, University of Stavanger, Norway. Retrieved from http://brage.bibsys.no/xmlui/handle/11250/182547.

Akbar, S. H., and A. A. Saleh. 1989. A comprehensive approach to solve asphaltene deposition problem in some deep wells. In Society of Petroleum Engineers. doi:10.2118/17965-MS.

Akbarzadeh, K., D. Eskin, J. Ratulowski, and S. Taylor. 2012. Asphaltene deposition measurement and modeling for flow assurance of tubings and flow lines. *Energy & Fuels* 26 (1): 495–510. doi:10.1021/ef2009474.

Akbarzadeh, K., A. Hammami, A. Kharrat, D. Zhang, S. Allenson, J. Creek, S. Kabir et al. 2007. Asphaltenes—problematic but rich in potential. *Oilfield Review* 19 (2): 22–43.

Atta, A. M., and A. M. Elsaeed. 2011. Use of rosin-based nonionic surfactants as petroleum crude oil sludge dispersants. *Journal of Applied Polymer Science* 122 (1): 183–192. doi:10.1002/app.34052.

Barcenas, M., P. Orea, E. Buenrostro-González, L. S. Zamudio-Rivera, and Y. Duda. 2008. Study of medium effect on asphaltene agglomeration inhibitor efficiency. *Energy & Fuels* 22 (3): 1917–1922. doi:10.1021/ef700773m.

Boukherissa, M., F. Mutelet, A. Modarressi, A. Dicko, D. Dafri, and M. Rogalski. 2009. Ionic liquids as dispersants of petroleum asphaltenes. *Energy & Fuels* 23 (5): 2557–2564. doi:10.1021/ef800629k.

Bouts, M. N., R. J. Wiersma, H. M. Muijs, and A. J. Samuel. 1995. An evaluation of new aspaltene inhibitors; Laboratory study and field testing. *Journal of Petroleum Technology* 47 (9): 782–787. doi:10.2118/28991-PA.

Brandt, H. C. A., E. M. Hendriks, M. A. J. Michels, and F. Visser. 1995. Thermodynamic modeling of asphaltene stacking. *The Journal of Physical Chemistry* 99 (26): 10430–10432.

Broseta, D., M. Robin, T. Savvidis, C. Féjean, M. Durandeau, and H. Zhou. 2000. Detection of asphaltene deposition by capillary flow measurements. *SPE/DOE Improved Oil Recovery Symposium*. Tulsa, OK: Society Petroleum Engneers. Retrieved from https://www.onepetro.org/conference-paper/SPE-59294-MS.

Carbognani, L. 2001. Dissolution of solid deposits and asphaltenes isolated from crude oil production facilities. *Energy & Fuels* 15 (5): 1013–1020. doi:10.1021/ef0100146.

Cenegy, L. M. 2001. Survey of successful world-wide asphaltene inhibitor treatments in oil production fields. In Society of Petroleum Engineers. doi:10.2118/71542-MS.

Chang, C. L., and H. S. Fogler. 1994a. Stabilization of asphaltenes in aliphatic solvents using alkylbenzene-derived amphiphiles. 1. Effect of the chemical structure of amphiphiles on asphaltene stabilization. *Langmuir* 10 (6): 1749–1757. doi:10.1021/la00018a022.

Chang, C. L., and H. S. Fogler. 1994b. Stabilization of asphaltenes in aliphatic solvents using alkylbenzene-derived amphiphiles. 2. Study of the asphaltene-amphiphile interactions and structures using fourier transform infrared spectroscopy and small-angle X-ray scattering techniques. *Langmuir* 10 (6): 1758–1766. doi:10.1021/la00018a023.

Chaogang, C., J. Guo, N. An, B. Ren, Y. Li, and Q. Jiang. 2013. Study of asphaltene deposition from Tahe crude oil. *Petroleum Science* 10 (1): 134–138. doi:10.1007/s12182-013-0260-y.

De Boer, R. B., Klaas Leerlooyer, M. R. P. Eigner, and A. R. D. Van Bergen. 1995. Screening of crude oils for asphalt precipitation: Theory practice and the selection of inhibitors. *SPE Production & Facilities* 10 (1): 55–61.

Eskin, D., and J. Ratulowski. 2015. Regarding the role of the critical particle size in the asphaltene deposition model. *Energy & Fuels* 29 (11): 7741–7742. doi:10.1021/acs.energyfuels.5b01915.

Gon, S., and D. M. Fouchard. 2016. Modified asphaltene capillary deposition unit: A novel approach to inhibitor screening. *Energy & Fuels* 30 (5): 3687–3692.

Gonzalez, D. L. 2008. Modeling of asphaltene precipitation and deposition tendency using the PC-SAFT equation of state. Rice University. Retrieved from https://scholarship.rice.edu/handle/1911/22140.

Gonzalez, F. A. 2015. Personal Communication. Asphaltene deposition economic impact. Reservoir performance global community of practice Lead BP, March.

Hammami, A., C. H. Phelps, T. Monger-McClure, and T. M. Little. 2000. Asphaltene precipitation from live oils: An experimental investigation of onset conditions and reversibility. *Energy Fuels* 14 (1): 14–18.

Hansen, C. M. 2007. *Hansen Solubility Parameters: A User's Handbook*. 2nd ed. Hoboken, NJ: CRC Press.

Hildebrand, J. H., and R. L. Scott. 1964. *The Solubility of Nonelectrolytes*. New York: Dover Publications.

Hoepfner, M. P. 2013. Investigations into asphaltene deposition, stability, and structure, University of Michigan. Retrieved from http://hdl.handle.net/2027.42/100081.

Hosseini, A., E. Zare, S. Ayatollahi, F. M. Vargas, W. G. Chapman, K. Kostarelos, and V. Taghikhani. 2016. Electrokinetic behavior of asphaltene particles. *Fuel* 178: 234–242. doi:10.1016/j.fuel.2016.03.051.

Hu, Y. F., and T. M. Guo. 2005. Effect of the structures of ionic liquids and alkylbenzene-derived amphiphiles on the inhibition of asphaltene precipitation from CO_2-injected reservoir oils. *Langmuir* 21 (18): 8168–8174. doi:10.1021/la050212f.

Ibrahim, H. H., and R. O. Idem. 2004. Interrelationships between asphaltene precipitation inhibitor effectiveness, asphaltenes characteristics, and precipitation behavior during *n*-heptane (light paraffin hydrocarbon)-induced asphaltene precipitation. *Energy & Fuels* 18 (4): 1038–1048. doi:10.1021/ef0340460.

Jorda, R. M. 1966. Paraffin deposition and prevention in oil wells. *Journal of Petroleum Technology* 18 (12): 1605–1612. doi:10.2118/1598-PA.

Juyal, P., V. Ho, A. Yen, and S. J. Allenson. 2012. Reversibility of asphaltene flocculation with chemicals. *Energy & Fuels* 26 (5): 2631–2640. doi:10.1021/ef201389e.

Juyal, P., A. T. Yen, R. P. Rodgers, S. Allenson, J. Wang, and J. Creek. 2010. Compositional variations between precipitated and organic solid deposition control (OSDC) asphaltenes and the effect of inhibitors on deposition by electrospray ionization Fourier transform ion cyclotron resonance (FT-ICR) mass spectrometry. *Energy & Fuels* 24 (4): 2320–2326. doi:10.1021/ef900959r.

Karambeigi, M. A., M. Nikazar, and R. Kharrat. 2016. Experimental evaluation of asphaltene inhibitors selection for standard and reservoir conditions. *Journal of Petroleum Science and Engineering* 137: 74–86. doi:10.1016/j.petrol.2015.11.013.

Kelland, M. A. 2009. *Production Chemicals for the Oil and Gas Industry.* 2nd ed. Boca Raton, FL: CRC Press.

Khaleel, A., M. Abutaqiya, M. Tavakkoli, A. A. Melendez-Alvarez, and F. M. Vargas. 2015. On the prediction, prevention and remediation of asphaltene deposition. In Society of Petroleum Engineers. doi:10.2118/177941-MS.

Khaleel, A., and F. M. Vargas. 2017. Mitigation of asphaltene deposition by dead oil/maltene reinjection. In Preparation.

Khvostichenko, D. S., and S. I. Andersen. 2009. Electrodeposition of asphaltenes. 1. Preliminary studies on electrodeposition from oil–heptane mixtures. *Energy & Fuels* 23 (2): 811–819. doi:10.1021/ef800722g.

Kokal, S. L., and S. G. Sayegh. 1995. Asphaltenes: The cholesterol of petroleum. *Middle East Oil Show.* Society of Petroleum Engineers. Retrieved from https://www.onepetro.org/conference-paper/SPE-29787-MS.

Kraiwattanawong, K., H. S. Fogler, S. G. Gharfeh, P. Singh, W. H. Thomason, and S. Chavadej. 2009. Effect of asphaltene dispersants on aggregate size distribution and growth. *Energy & Fuels* 23 (3): 1575–1582. doi:10.1021/ef800706c.

Kuang, J., A. Melendez-Alvarez, J. Yarbrough, M. Garcia-Bermudes, M. Tavakkoli, D. S. Abdallah, and F. M. Vargas. 2017. Novel way to assess the performance of asphaltene dispersants on the prevention of asphaltene deposition.

Lauer, R. S. 2007. The use of high performance polymeric coatings to mitigate corrosion and deposit formation in pipeline applications. *CORROSION 2007*, Tennessee, TN.

Leising, L. J., and K. R. Newman. 1993. Coiled-tubing drilling. *SPE Drilling & Completion* 8 (4): 227–232. doi:10.2118/24594-PA.

León, O., E. Rogel, J. Espidel, and G. Torres. 2000. Asphaltenes: Structural characterization, self-association, and stability behavior. *Energy & Fuels* 14 (1): 6–10. doi:10.1021/ef9901037.

Leontaritis, K. J. et al. 1989. Asphaltene deposition: A comprehensive description of problem manifestations and modeling approaches. *SPE Production Operations Symposium.* Society of Petroleum Engineers. Retrieved from https://www.onepetro.org/conference-paper/SPE-18892-MS.

Li, J., J. Misselbrook, and M. Sach. 2010. Sand cleanouts with coiled tubing: choice of process, tools and fluids. *Journal of Canadian Petroleum Technology* 49 (8): 69–82. doi:10.2118/113267-PA.

Lichaa, P. M., and L. Herrera. 1975. Electrical and other effects related to the formation and prevention of asphaltene deposition problem in Venezuelan crudes. In Society of Petroleum Engineers. doi:10.2118/5304-MS.

Liu, X., T. Chen, P. Chen, H. Montgomerie, T. H.Hagen, B. Wang, and X. Yang. 2012. Understanding mechanisms of scale inhibition using newly developed test method and developing synergistic combined scale inhibitors. In Society of Petroleum Engineers. doi:10.2118/156008-MS.

Melendez-Alvarez, A. A., M. Garcia-Bermudes, M. Tavakkoli, R. H. Doherty, S. Meng, D. S. Abdallah, and F. M. Vargas. 2016. On the evaluation of the performance of asphaltene dispersants. *Fuel* 179: 210–220. doi:10.1016/j.fuel.2016.03.056.

The Oilfield Glossary - Schlumberger Oilfield Glossary. 2017. Retrieved from http://www.glossary.oilfield.slb.com/.

Remedial Treatment for Asphaltene Precipitation. (n.d). In PetroWiki. Retrieved from http://petrowiki.org/Remedial_treatment_for_asphaltene_precipitation.

Rodriguez, D. L. G. 2008. Modeling of asphaltene precipitation and deposition tendency using the PC-SAFT equation of state. Rice University. Retrieved from https://scholarship.rice.edu/handle/1911/22140.

Rogel, E. 2011. Effect of inhibitors on asphaltene aggregation: A theoretical framework. *Energy & Fuels* 25 (2): 472–481. doi:10.1021/ef100912b.

Sachanen, A. N. et al. 1945. Chemical constituents of petroleum. Retrieved from http://agris.fao.org/agris-search/search.do?recordID=US201300655727.

Schantz, S. S., and W. K. Stephenson. 1991. Asphaltene deposition: Development and application of polymeric asphaltene dispersants. In Society of Petroleum Engineers. doi:10.2118/22783-MS.

Smith, D. F., G. C. Klein, A. T. Yen, M. P. Squicciarini, R. P. Rodgers, and A. G. Marshall. 2008. Crude oil polar chemical composition derived from FT–ICR mass spectrometry accounts for asphaltene inhibitor specificity. *Energy & Fuels* 22 (5): 3112–3117. doi:10.1021/ef800036a.

Straub, T. J., S. W. Autry, and G. E. King. 1989. An investigation into practical removal of downhole paraffin by thermal methods and chemical solvents. Oklahoma City, OK: Society of Petroleum Engineers. doi:10.2118/18889-MS.

Tavakkoli, M., M. R. Grimes, X. Liu, C. K. Garcia, S. C. Correa, Q. J. Cox, and F. M. Vargas. 2015. Indirect method: A novel technique for experimental determination of asphaltene precipitation. *Energy & Fuels* 29 (5): 2890–2900. doi:10.1021/ef502188u.

Thermo Fisher Scientific. 2017. Toluene (Material Safety Data Sheet). Retrieved from https://www.fishersci.com/shop/msdsproxy?productName=T326P4&productDescription=TOLUENE.

Tiratsoo, J. N. H. 1992. *Pipeline Pigging Technology*. Houston, TX: Gulf Professional Publishing.

Trbovich, M. G., and G. E. King. 1991. Asphaltene deposit removal: Long-lasting treatment with a co-solvent. In Society of Petroleum Engineers. doi:10.2118/21038-MS.

Vargas, F. M., J. L. Creek, and W. G. Chapman. 2010. On the development of an asphaltene deposition simulator. *Energy & Fuels* 24 (4): 2294–2299. doi:10.1021/ef900951n.

Vargas, F. M., M. Garcia-Bermudes, M. Boggara, S. Punnapala, M. Abutaqiya, N. Mathew, S. Prasad, A. Khaleel, M. Al Rashed, and H. Al Asafen. 2014. On the development of an enhanced method to predict asphaltene precipitation. In Offshore Technology Conference. doi:10.4043/25294-MS.

Vilas Boas Favero, C., A. Hanpan, P. Phichphimok, K. Binabdullah, and H. S. Fogler. 2016. Mechanistic investigation of asphaltene deposition. *Energy & Fuels* 30 (11): 8915–8921. doi:10.1021/acs.energyfuels.6b01289.

Wang, J., C. Li, L. Zhang, G. Que, and Z. Li. 2009. The properties of asphaltenes and their interaction with amphiphiles. *Energy & Fuels* 23 (7): 3625–3631. doi:10.1021/ef801148y.

Turbidity. (n.d.). In Wikipedia. Retrieved from https://en.wikipedia.org/wiki/Turbidity.

Zhang, L., G. Yang, J. Q. Wang, Y. Li, L. Li, and C. Yang. 2014. Study on the polarity, solubility, and stacking characteristics of asphaltenes. *Fuel* 128: 366–372. doi:10.1016/j.fuel.2014.03.015.

8 Case Studies and Field Applications

M. I. L. Abutaqiya, C. Sisco, J. Kuang, P. Lin,
F. Wang, M. Tavakkoli, and F. M. Vargas

CONTENTS

8.1 ASPHALTENE PRECIPITATION FROM CRUDE OILS

8.1.1 CASE STUDY #1: SCREENING ASPHALTENE INSTABILITY IN RESERVOIR SAMPLES

This case study demonstrates how a few simple experiments conducted in the laboratory can provide insight on whether asphaltene deposition is likely to occur in the field or not. Crude oil sample SB was received from a well with declining productivity, which, allegedly, was the result of asphaltene deposition either in the wellbore or near-wellbore region. This sample had a density of 867 kg/m^3 and a viscosity of 60.5 cP at ambient conditions.

The crude oil sample was first centrifuged at 10,000 rpm (revolutions per minute) for 20 minutes to remove sediments and water, and then the Indirect Method (Section 3.3) was used to detect asphaltene precipitation at 0.1 MPa and 300 K (Tavakkoli et al. 2015). Two precipitants, n-pentane and n-heptane, were added to the SB sample at various ratios. After an aging time of 24 hours, the samples were centrifuged again at 10,000 rpm for 20 minutes to settle unstable asphaltenes. After centrifugation, 0.5 cm^3 of the supernatant liquid was extracted and diluted with 3 cm^3 of toluene. The absorbance was measured using a spectrophotometer at 700 nm, and the effect of dilution was removed from the absorbance data (Section 3.3). The plots in Figure 8.1 show the corrected absorbance data for SB at 700 nm. The sudden drop in absorbance that occurs between 70 vol% and 80 vol% n-pentane is an indication of asphaltene precipitation.

Because of the high amount of n-alkane that is required to induce precipitation, one may conclude that the asphaltenes present in the crude oil sample are highly stable, which suggests that the decline in the well productivity may not be associated with asphaltene deposition. After several months of further investigation, the oil operator concluded that the problem was caused by the migration of fines from the deep reservoir toward the near-wellbore region.

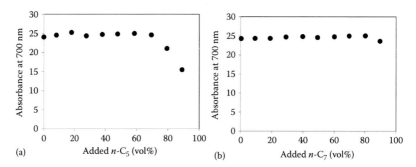

(a)   Added n-C$_5$ (vol%)   (b)   Added n-C$_7$ (vol%)

FIGURE 8.1 Results of the indirect method after correction of the dilution effect for the crude oil sample SB at 0.1 MPa and 300 K diluted with (a) n-pentane and (b) n-heptane and aged for 24 hours.

8.1.2 CASE STUDY #2: MODELING ASPHALTENE PRECIPITATION IN RESERVOIR FLUIDS WITH GAS INJECTION

Gas injection is one of several strategies that an operator uses to increase oil recovery. Under normal pressure depletion (primary production), only 5–20% of the original oil in place (OOIP) can be produced (Blunt et al. 1993). Miscible gas injection (which occurs when the injection gas and reservoir fluid are at sufficiently high pressures to form a single phase) favors an increased mobility of the oil by lowering its density and viscosity. Injection gases can include carbon dioxide, nitrogen, or hydrocarbon gases depending on their availability and vicinity to downstream processing units. Although it is true that the injection of gases at miscible conditions may reduce density and viscosity of the reservoir fluid, at the same time, they may also increase the likelihood of asphaltene precipitation, which may lead to deposition and blockage of the wellbore and adverse alteration of the near-wellbore region properties. Therefore, a full understanding of the effect of gas injection on both asphaltene stability and oil mobility is crucial for successful implementation of this operation. This case study portrays the application of the perturbed chain-statistical association fluid theory equation of state (PC-SAFT EOS; Section 4.2.3.2) to predict the effect of hydrocarbon and nonhydrocarbon gas injection on asphaltene precipitation at reservoir conditions.

The properties of the crude oils modeled in this case study are shown in Table 8.1. The flashed gas and flashed liquid compositions for each crude are shown in Tables 8.2 and 8.3, respectively. Crudes S1, B7, and U1 properties are taken from the work of Punnapala and Vargas (2013). Crude S14 properties are taken from the work of Vargas et al. (2014). PC-SAFT simulation parameters for each crude are shown in Table 8.4.

TABLE 8.1
PVT Properties of the Crude Oils Studied

Property	Crude S1	Crude B7	Crude U1	Crude S14
STO M (kg/kmol)	191	212.9	212	181
STO density (kg/m³)	823	846.4	867	829
GOR (Sm³/m³)	142	286	38	63
Reservoir temperature (K)	394.3	394.3	373.2	397.0
Bubble pressure (MPa)	14.4	27.4	5.8	7.6
Density at saturation (kg/m³)	—	610.5	764	692
Saturates, wt%	75.56	44.9	59.40	68.61
Aromatics, wt%	20.08	46.3	22.57	22.31
Resins, wt%	4.13	6.7	13.97	6.46
Asphaltenes, wt%	0.21	2.1	1.73	0.53

GOR, gas-to-oil ratio; M, molecular weight; STO, stock tank oil.

TABLE 8.2
Compositional Analysis of Flashed Gas for the Crude Oils Studied

	M	Flashed Gas Composition (mol%)			
		Crude S1	Crude B7	Crude U1	Crude S14
N_2	28.01	0.29	0.12	0.34	2.47
CO_2	44.01	3.59	4.91	3.29	6.33
H_2S	34.08	0.00	7.03	0.00	0.22
C_1	16.04	59.72	61.18	44.04	45.65
C_2	30.07	12.98	7.13	16.23	13.52
C_3	44.10	10.35	5.56	18.00	13.93
iC_4	58.12	2.30	1.52	2.84	3.52
nC_4	58.12	4.91	3.53	8.23	7.23
iC_5	72.15	1.69	1.63	2.14	2.30
nC_5	72.15	1.92	2.02	2.69	2.38
C_6	86.18	0.52	2.35	1.55	1.49
C_{7+}		1.73	3.02	0.65	0.96
M of C_{7+}		91.85	99.56	90.76	94.57

M, molecular weight.

TABLE 8.3
Compositional Analysis of Flashed Liquid for the Crude Oils Studied

	M	Flashed Liquid Composition (mol%)			
		Crude S1	Crude B7	Crude U1	Crude S14
N_2	28.01	0.00	0.00	0.00	0.00
CO_2	44.01	0.04	0.00	0.00	0.00
H_2S	34.08	0.00	0.00	0.00	0.00
C_1	16.04	0.21	0.00	0.00	0.00
C_2	30.07	0.46	0.03	0.22	0.25
C_3	44.1	1.83	0.19	1.65	1.10
$i-C_4$	58.12	1.12	0.15	0.62	0.77
$n-C_4$	58.12	3.81	0.58	3.20	2.63
$i-C_5$	72.15	2.68	0.68	2.05	2.34
$n-C_5$	86.18	3.76	1.14	3.62	3.40
$n-C_6$	86.117	3.30	3.22	6.15	6.66
$i-C_6$	86.117	3.14	0.00	0.46	0.00
M-cyclo-C_5	84.16	0.00	0.73	0.87	0.00
c-C_5	70.13	0.06	0.00	0.00	0.00
C_7	100.20	0.00	4.67	5.57	8.38
$n-C_7$	100.20	2.97	0.00	0.00	0.00
$i-C_7$	100.20	3.06	0.00	0.00	0.00

(*Continued*)

TABLE 8.3 (*Continued*)
Compositional Analysis of Flashed Liquid for the Crude Oils Studied

		Flashed Liquid Composition (mol%)			
	M	Crude S1	Crude B7	Crude U1	Crude S14
c-C_7	84.16	1.41	0.00	0.00	0.00
Benzene	78.11	0.33	0.00	0.20	0.00
Ethyl-benzene	106.17	0.00	0.65	0.59	0.00
C_8	114.13	0.00	6.57	5.93	9.00
n-C_8	114.231	2.46	0.00	0.00	
i-C_8	114.231	3.80	0.00	0.00	
c-C_8	98.19	0.99	1.24	0.75	
Toluene	92.14	1.04	0.98	1.02	
n-C_9	128.258	1.98	0.00	0.00	
i-C_9	128.258	3.52	0.00	0.00	
c-C_9	112.21	0.44	0.00	0.00	
Xylenes	106.17	1.47	2.27	0.84	
n-C_{10}	142.28	1.79			
i-C_{10}	142.28	2.96			
c-C_{10}	126.24	0.29			
C_{10} aromatics	120.19	1.66			
n-C_{11}	156.31	1.64			
i-C_{11}	156.31	2.18			
c-C_{11}	140.27	0.24			
C_{11} aromatics	134.08	1.07			
Plus fraction (mol%)		44.29	76.9	66.26	65.47
M of plus fraction		302.23	247.24	275.55	227.65
Overall M of STO		191.02	212.90	212.38	180.60

M, molecular weight.

8.1.2.1   Injection of Hydrocarbon Gas

Hydrocarbon gas injection is usually performed to maintain reservoir pressure and, thus, maintain production. This gas typically originates as an associated gas which is coproduced with oil or it may be produced separately from surrounding gas wells. Before undertaking gas injection for enhanced oil recovery (EOR), oil producers first perform flow assurance studies on the live oil sample to understand the potential occurrence and severity of asphaltene precipitation and deposition problems as a function of gas amount and composition. These high-pressure, high-temperature (HPHT) experiments, however, are expensive, time-consuming, and may be subject to significant and, sometimes unknown, experimental uncertainties (Section 3.2). A modeling method based on PC-SAFT EOS can be used to design and analyze such experiments, and hence, reduce the need of extensive experimental work and

TABLE 8.4

Simulation Parameters for the Crude Oils Studied

	Component	γ	M (g/mol)	z_i (mol%)	m	σ (Å)	ε/k (K)	M (g/mol)	z_i (mol%)	m	σ (Å)	ε/k (K)	
Crude S1	N_2	—	28.01	0.17	1.21	3.31	90.96	28.01	0.10	1.21	3.31	90.96	Crude B7
	CO_2	—	44.01	2.09	2.07	2.79	169.21	44.01	3.70	2.07	2.79	169.21	
	H_2S	—	34.08	0.00	1.652	3.07	227.34	34.08	5.30	1.652	3.07	227.34	
	C_1	—	16.04	34.83	1.00	3.70	150.03	16.04	46.30	1.00	3.70	150.03	
	C_2	—	30.07	7.57	1.61	3.52	191.42	30.07	5.40	1.61	3.52	191.42	
	C_3	—	44.10	6.04	2.00	3.62	208.11	44.10	4.20	2.00	3.62	208.11	
	Heavy gas	0.000	67.20	7.62	2.57	3.75	229.38	74.70	10.30	2.76	3.77	232.63	
	Saturates	0.000	176.1	34.16	5.37	3.91	250.36	193.2	18.00	5.81	3.92	251.56	
	A+R	0.080	256.1	7.53	7.18	3.97	270.33	264.6	6.60	5.84	4.09	345.75	
	Asphaltenes	0.358	2818.5	0.01	57.86	4.23	350.40	2183.9	0.004	45.09	4.22	349.97	
Crude U1	N_2	—	28.0	0.10	1.21	3.31	90.96	28.01	0.91	1.21	3.31	90.96	Crude S14
	CO_2	—	44.0	0.90	2.07	2.79	169.21	44.01	2.34	2.07	2.79	169.21	
	H_2S	—	34.08	0.00	1.652	3.07	227.34	34.08	0.08	1.652	3.07	227.34	
	C_1	—	16.0	12.50	1.00	3.70	150.03	16.04	16.84	1.00	3.70	150.03	
	C_2	—	30.1	4.60	1.61	3.52	191.42	30.07	4.99	1.61	3.52	191.42	
	C_3	—	44.1	5.10	2.00	3.62	208.11	44.10	5.14	2.00	3.62	208.11	
	Heavy gas	0.000	65.4	5.10	2.53	3.74	228.45	66.29	6.60	2.55	3.74	228.89	
	Saturates	0.000	177.9	52.90	5.42	3.91	250.48	158.18	49.45	4.91	3.89	248.81	
	A+R	0.690	310.4	18.60	5.71	4.22	410.87	257.68	13.65	5.67	4.12	367.33	
	Asphaltenes	0.292	2690.4	0.10	61.53	4.19	334.21	5000	0.012	100.8	4.25	353.88	

increasing the consistency of the results. Figure 8.2 summarizes a case in which PC-SAFT was used to simulate the effect of gas injection on asphaltene precipitation. The simulation parameters were tuned to the live oil experiments for Crude U8 (off-shore field). Crude U8 is the fluid that was characterized in Section 4.4.1 and its properties and simulation parameters can be found therein. The results for the gas injection at 10 mol%, 20 mol%, and 30 mol% cases are purely predictive, and the gas composition is provided in the figure caption. The excellent agreement between the PC-SAFT predictions and the HPHT experiments confirms the reliability of this EOS for designing the experiments and validating the consistency of the results.

Figure 8.3 shows another example from an on-shore field (Crude S1).

The results shown in Figures 8.2 and 8.3 confirm the benefit of using PC-SAFT model in aiding in the design and implementation of EOR strategies. By comparing Figures 8.2 and 8.3, Crude U8 is more susceptible to asphaltene precipitation problems than Crude S1 because the asphaltene phase envelope shifts significantly upward (while the amount of methane in both injection gases is about the same).

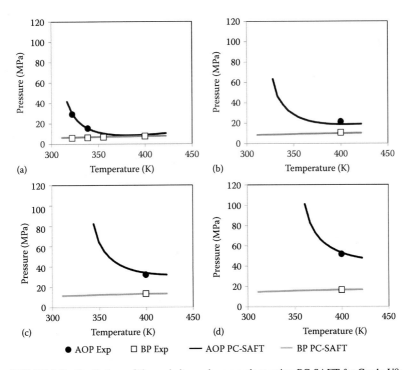

FIGURE 8.2 Predictions of the asphaltene phase envelope using PC-SAFT for Crude U8: (a) Live oil, (b) live oil + 10% gas injection, (c) live oil + 20% gas injection, and (d) live oil + 30% gas injection. Composition of injection gas (mol%) is N_2: 0.83; H_2S: 0.49; CO_2: 4.96; C_1: 73.54; C_2: 11.34; C_3: 5.19; and C_{4+}: 3.65.

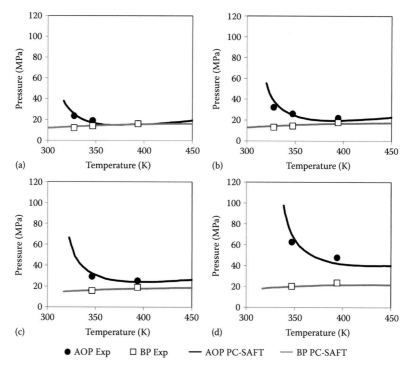

FIGURE 8.3 Predictions of the asphaltene phase envelope using PC-SAFT for Crude S1: (a) Live oil + 5%, (b) live oil + 10% gas injection, (c) live oil + 15% gas injection, and (d) live oil + 30 gas injection. Composition of injection gas (mol%) is N_2: 0.398; H_2S: 0.00; CO_2: 3.89; C_1: 71.31; C_2: 11.91; C_3: 7.22; and C_{4+}: 5.26. (Adapted from Punnapala, S. and Vargas, F. M., *Fuel*, 108, 417–429, 2013. With permission.)

Therefore, gas injection in Crude U8 should be performed with caution and may also require coinjection of asphaltene inhibitors.

PC-SAFT modeling can also help in designing experiments for flow assurance studies with different gas injection conditions. With only two or three experiments, which can be performed on live oil sample, the model can calibrated and used to design further experiments. Using PC-SAFT, a good estimate of the asphaltene onset pressure (AOP) can be readily obtained for live oil and gas blends, which helps to ensure that the experiments are conducted under adequate conditions, reducing the cost and time needed to complete these studies.

8.1.2.2 Injection of Nonhydrocarbon Gas

In addition to hydrocarbon gas injection, nonhydrocarbon gases like CO_2, N_2, and H_2S may be used in EOR. CO_2 injection is the most common type of nonhydrocarbon gas injection (Blunt et al. 1993) because of its excellent solvent properties and relative abundance. H_2S and N_2 may sometimes be injected separately or combined with hydrocarbon gases. N_2 and CO_2 gases are known to induce the precipitation of asphaltenes under miscible conditions. Because N_2 is a much stronger asphaltene precipitant, however, it may be responsible for the largest extent of asphaltene precipitation not only in the production tubing, but likely at the point of first contact deep in the reservoir. H_2S, on the other hand, because of its high solubility parameter value often acts as an asphaltene stabilizer. However, because of its toxicity and corrosive nature, it should be handled with extreme care both in the laboratory and in field applications.

To qualitatively understand the effect of nonhydrocarbon gas injection on the asphaltene phase stability, PC-SAFT is used to generate asphaltene phase envelopes for Crude U8 with different gases. The results are shown in Figure 8.4a; N_2 and CO_2 injection are both shown to induce asphaltene precipitation by shifting the onset curve upward compared to the live oil curve. H_2S injection shifts the asphaltene curves downward, meaning that asphaltenes are stabilized. Figure 8.4b shows the simulation results for CO_2 injection to Crude B7. Asphaltene simulation parameters (γ and M) for Crude B7 were tuned to match the AOP at 20 mol% injection; live oil AOP data is not available because the live crude is stable. PC-SAFT predictions for 30 mol% CO_2 injection match the experimental data very well.

Figure 8.5a–d shows the simulation results and experimental data for Crude U1. The simulation results demonstrate the effect of gas injection on the asphaltene onset and bubble pressures. In these plots, the area above the AOP curve represents the region where asphaltenes are stable. The area below the AOP curve and above the bubble pressure curve is where the asphaltenes may precipitate. The closer the conditions are to the bubble curve, the larger the driving force for asphaltene precipitation. Although, it is true that asphaltene precipitation is also

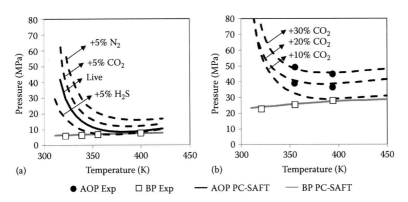

FIGURE 8.4 Effect of nonhydrocarbon gas injection on the asphaltene phase envelope of (a) Crude U8 and (b) Crude B7. (Adapted from Punnapala, S. and Vargas, F. M., *Fuel*, 108, 417–429, 2013. With permission.)

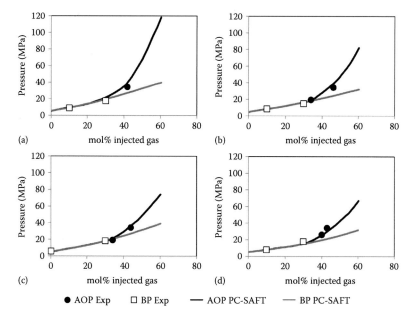

FIGURE 8.5 Asphaltene phase envelope using PC-SAFT for Crude U1 at T = 373 K with injection of (a) HC+N$_2$ injection (90:10 mol%), (b) HC + H$_2$S (96.4:3.6 mol%), (c) HC+CO$_2$ (80:20 mol%), and (d) HC+CO$_2$+N$_2$ (67:22:11 mol%). The hydrocarbon gas composition (mol%) is N$_2$: 0.40; H$_2$S: 0.19; CO$_2$: 2.44; C$_1$: 74.87; C$_2$: 14.36; C$_3$: 6.20; and C$_{4+}$: 1.54. (Adapted from Punnapala, S. and Vargas, F. M., *Fuel*, 108, 417–429, 2013. With permission.)

possible below the bubble point (and above what is known as the *lower asphaltene onset point*, or *LAOP*), because production takes place from high pressure and temperature (i.e., reservoir) to lower pressure and temperature (i.e., wellhead) the point of maximum risk of asphaltene precipitation and subsequent deposition is at the bubble point. If no deposition is observed near the bubble point pressure, beyond this point one is unlikely to find severe asphaltene deposition problems. Depressurization below the AOP increases the amount of asphaltene precipitation because the light components, such as CO$_2$, methane, ethane, and propane, are highly compressible, and therefore the fluid expands on depressurization, which results in a poor solvency power for asphaltenes.

The ability of PC-SAFT model to capture the asphaltene onset and saturation pressures as a function of temperature and on the addition of various concentrations of hydrocarbon and nonhydrocarbon gases is a reflection of the reliability of the model. For this reason, PC-SAFT EOS may provide valuable assistance in the design and implementation of strategies for EOR and asphaltene deposition mitigation.

8.1.3 Case Study #3: Detection of Experimental Discrepancies on Asphaltene Onset and Bubble Pressures

The most commonly used experimental method for detecting AOP of reservoir fluids are near-infrared (NIR) spectroscopy and high-pressure microscopy (HPM; Section 3.2). The insufficient sensitivity of the instrument, along with challenges associated to sample preparation, relatively slow kinetics of asphaltene aggregation, and nonequilibrium stages of depressurization, can significantly affect the final results of the experiments. This case study shows an example where PC-SAFT EOS can be used to detect experimental inconsistencies found in laboratory reports.

The experimental results of the flow assurance study conducted on Crude S14 are shown in Table 8.5. The flow assurance study was conducted on the live oil sample at three temperatures and three gas injection conditions. The hydrocarbon gas composition (mol%) is N_2: 0.65; H_2S: 0.031; CO_2: 6.73; C_1: 60.57; C_2: 13.70; C_3: 10.19; and C_{4+}: 8.13.

Inspection of the experimental results in Table 8.5 shows that there might be two main experimental issues. The AOP for the live oil at 333.2 K is much higher than the AOP for the 5% gas injection case, and the 10% gas injection case has a higher AOP value than the 20% case. These results are unusual because hydrocarbon gas injection is expected to increase the AOP value and increase the amount of asphaltene precipitation. The 10% and 20% gas injection cases also show similar bubble pressure values, which is unexpected. On injection of a gas that contains a high concentration of methane, one would expect the bubble pressure to increase as a result of the higher volatility of the mixture. Although, it is evident that there are some inconsistencies in the data reported in Table 8.5, one cannot readily pinpoint which one is the conflicting result.

To use the PC-SAFT model to assist in the investigation of this problem, at least two reliable experiments are required to provide inputs to optimize the asphaltene-simulation parameters (γ and M), as described in Section 4.3.1.3. To identify the most suitable experimental points for model calibration one should refer to the raw data of

TABLE 8.5
Experimental Results for the AOP and BP of Crude S14 at Different Temperatures. Pressures Reported in MPa

Temperature (K)	Live oil		+5% gas		+10% gas		+20% gas	
	AOP	BP	AOP	BP	AOP	BP	AOP	BP
397.0	N.D.[a]	7.1			29.8	13.9	25.6	14.1
333.2	21.9	6.0	18.1	6.7				
310.9	25.3	5.5						

Source: Vargas, F. M. et al., On the development of an enhanced method to predict asphaltene precipitation, Offshore Technology Conference, doi:10.4043/25294-MS, 2014. With permission.

[a] N.D., Not Detected.

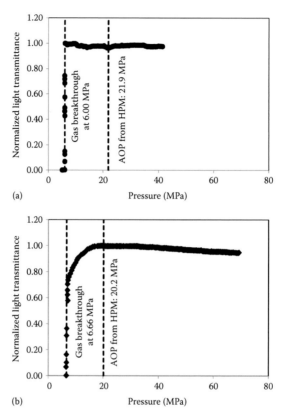

(a)

(b)

FIGURE 8.6 Near-infrared plot for Crude S14 (a) live oil and (b) with 5% hydrocarbon gas injection at 333.2 K.

NIR light transmittance versus pressure. Figure 8.6 shows the NIR plots for live oil and live oil with 5% of gas at 333.2 K. The clear and sustained visual dip in the NIR plot that is indicative of the AOP does not occur for the live oil case, and it was also reported that the HPM showed few solid particles precipitating at the reported onset (21.9 MPa). If a true onset is detected, then the number and size of the asphaltene particles is expected to increase as pressure decreases, decreasing the light transmittance at pressures below AOP. However, this was not observed for the live oil case at 333.2 K. The reported AOP value for live oil at 333.2 K is likely incorrect. The clear and sustained dip in NIR is observed for the 5% case, as shown in Figure 8.6b; thus, this point was chosen as the second point for optimization of the asphaltenes simulation parameters. After choosing the appropriate AOP data points for model calibration (i.e., AOP for live oil at 310.9 K and AOP for 5% injection at 333.2 K), the model can be calibrated and used to predict the precipitation of asphaltenes under various conditions. The simulation parameters are shown in Table 8.4 and the modeling results are shown in Figure 8.7.

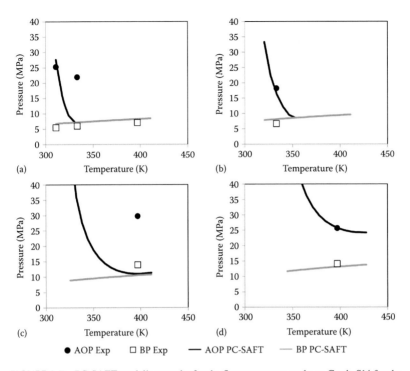

FIGURE 8.7 PC-SAFT modeling results for the flow assurance study on Crude S14 for the cases of (a) live oil, (b) 5% gas injection, (c) 10% gas injection, and (d) 20% as injection.

The experimental value of AOP for the live oil at 333.2 K, which was thought to be questionable from analysis of the experimental results, shows a significant deviation from PC-SAFT predictions. The model predicts that asphaltenes are at the threshold of instability at 333.2 K, supporting the experimental observation that asphaltene particles were clearly visible only near the bubble point, leading to the conclusion that the reported AOP at 333.2 K for the live oil is likely incorrect.

The simulation results for the 10% injection case (Figure 8.7c) show a significant deviation from experimental values of AOP and BP at 397.0 K, whereas the simulation results for the 20% case (Figure 8.7d) show an excellent agreement. Therefore, the experimental results for the 10% case are also deemed inconsistent. After the service laboratory repeated the experiments for 10% gas injection, it was possible to confirm that the first experiment was indeed flawed. Thus, once again, PC-SAFT EOS, in conjunction with a minimum set of reliable experimental data, can be a powerful tool to design and validate expensive and time-consuming high temperature AOP experiments.

8.1.4 Case Study #4: Effect of Oil Commingling and Oil-Based Mud

This case study investigates the effect of oil commingling and oil-based mud contamination on the asphaltene precipitation tendency. Gonzalez et al. (2007) and Panuganti et al. (2013) applied PC-SAFT EoS to study the effect of oil commingling and oil-based mud on asphaltene precipitation tendency. The simulation results discussed in the next section are based on the work of these authors.

8.1.4.1 Effect of Oil Commingling

Oil commingling refers to the practice of producing oil and gas from different zones through the same pipeline. It is a common subsea design consideration to reduce the number of flow lines, and thus, reduce capital and operating costs. The commingled oils can have different densities, gas-to-oil ratio (GOR), and viscosities meaning that the relative amounts of each crude in the blend play a significant role in determining the commingled oil properties. Of particular importance is the tendency of the commingled oil to precipitate asphaltenes. It is possible that asphaltenes are stable in two reservoir fluids, and yet, become unstable when they are commingled. An example is the commingling of heavy black oils with condensates. Condensates usually have a negligible amount of asphaltenes and do not tend to precipitate asphaltenes. Black oils have a relatively high amount of asphaltene but also have a high content of other fractions such as aromatics and resins, which can effectively stabilize asphaltenes in the oil. However, when mixed, condensates may act as an asphaltene precipitant to the black oil.

The effect of oil commingling of two oils Crude D and Crude E is shown in Figure 8.8 (Panuganti et al. 2013). The properties of these crudes are proprietary and unavailable for publication. As can be seen from Figure 8.8, the crudes do not show any asphaltene precipitation issues at their corresponding reservoir conditions. Crude E is expected to become unstable if operated at Crude D reservoir conditions. The commingling of both crudes at 50:50 volume ratio, however, results in a

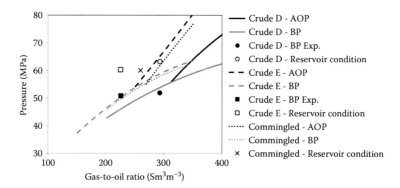

FIGURE 8.8 Effect of commingling of Crude 6 and Crude 7 on the asphaltene stability at 338.7 K. Comingling volume ratio is 50:50. (Adapted from Panuganti, S. R. et al., *Fluid Phase Equilib.*, 359, 2–16, 2013. With permission.)

crude oil that is stable at the corresponding condition. This implies that asphaltene deposition problems are not expected for the commingled oil at 260 Sm³/m³, which is in agreement with field observations.

8.1.4.2 Effect of Oil-Based Mud Contamination

Oil-based mud (OBM) is commonly injected during drilling operations to increase borehole stability, provide backup pressure, and reduce friction and wear of the drill bit. It consists of components in the range C_8–C_{34} but mainly by paraffinic C_{11}–C_{18} components (Pedersen et al. 2014). OBM may contaminate the near-wellbore reservoir fluids, and thus, alter its properties and composition. If contaminated reservoir fluid samples are collected and analyzed during early stages of the field development, they may be unrepresentative of the actual reservoir fluid in the deep reservoir. Properties affected include saturation pressure, GOR, stock tank oil (STO) density, and oil formation factor. Usually, companies perform a *fingerprint* analysis on bottom-hole samples to ensure that it is not contaminated (Pedersen et al. 2014). In this case study, the effect of OBM contamination on asphaltene stability is investigated.

Figure 8.9 shows the effect of OBM contamination on the AOP, bubble pressure, and GOR. According to Figure 8.9, increasing the level of OBM contamination reduces the AOP. At first, this might seem counterintuitive because of its paraffinic nature, OBM could be expected to make asphaltenes more unstable in the crude oil (i.e., increase the AOP). However, other factors such as the dilution of much lighter components, which are even stronger asphaltene precipitants, can result in asphaltene stabilization upon OBM contamination.

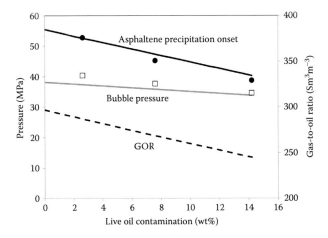

FIGURE 8.9 PC-SAFT simulation of the effect of mud contamination on the asphaltene onset, saturation pressure, and gas-to-oil at temperature = 354.3 K. (Adapted from Panuganti, S. R. et al., *Fluid Phase Equilib.*, 359, 2–16, 2013. With permission.)

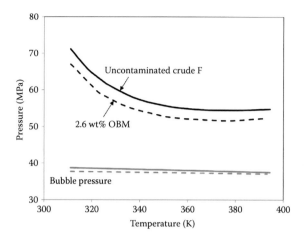

FIGURE 8.10 PC-SAFT modeling of asphaltene phase envelope for oil-based mud–contaminated sample. (Adapted from Panuganti, S. R. et al., *Fluid Phase Equilib.*, 359, 2–16, 2013. With permission.)

The predicted asphaltene onset and bubble pressure are in excellent agreement with available experimental data. The decreased GOR with OBM contamination is the main reason why asphaltenes become more stable. The effect of the OBM contamination on the asphaltene phase envelope is demonstrated in Figure 8.10.

8.1.5 CASE STUDY #5: EFFECT OF ASPHALTENE POLYDISPERSITY

Asphaltenes are a polydisperse mixture of the heaviest and most polarizable fraction in crude oils (Vargas et al. 2009). They are defined as a solubility class and can therefore contain a broad distribution of chemical structures and molecular sizes, extent of nano-clustering, and even physical appearance, depending on the subfraction of that solubility class that is under consideration. For example, asphaltenes that are precipitated with *n*-pentane (i.e., alkane with strongest asphaltene precipitation power at ambient pressure), also known as n-C_{5+} or simply C_{5+} asphaltenes, constitute the broadest asphaltene distribution at ambient pressure.

C_{5+} asphaltenes may be different in shape and structure than C_{7+} asphaltenes. Treating the asphaltenes as a polydisperse distribution of pseudo-components—instead of the single pseudo-component that is commonly used in the literature—may provide more realistic modeling results.

Tavakkoli et al. (2016) studied the effect of asphaltene polydispersity on phase-behavior modeling, laying out a robust method to predict asphaltene precipitation at high pressure and temperature with a model that was calibrated against experiments done on dead oil at ambient conditions. Using the Indirect Method (Section 3.4), the authors performed experiments to detect asphaltene precipitation and the amount precipitated for different precipitants. Asphaltene precipitation may be very slow, which results on a shift of the detection point at different aging times (Maqbool et al. 2009). For example, according to Figure 8.11a, the amount of *n*-heptane needed to observe the formation of asphaltene aggregates, can change from 40 vol% to 30 vol%, when the aging times varies from 1 hour to 1 month. Tavakkoli et al. (2016) concluded, fortunately, that although aging time is important in the lower precipitant concentration regions, the amount of precipitated asphaltenes at 90 vol% reaches an

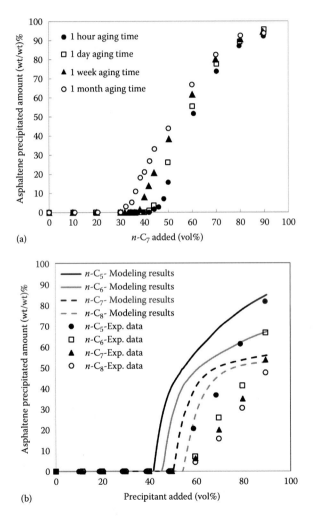

FIGURE 8.11   (a) Effect of aging time on the experimental results from the indirect method and (b) modeling results comparing PC-SAFT predictions to experimental results for 1-day aging time. (Adapted from Tavakkoli, M. et al., *Fluid Phase Equilib.*, 416, 120–129, 2016. With permission.)

equilibrium value in less than an hour. The convergence of precipitated asphaltene amount to a common point at 90 vol% of precipitant, which is nearly independent of the aging time, is shown in Figure 8.11a.

From these results one may conclude that the point at 90 vol% of precipitant reaches equilibrium in 1 hour or less; therefore, it can be used to calibrate the PC-SAFT model with confidence.

To represent asphaltene polydispersity, the authors chose a gamma distribution function with molecular weights ranging from a minimum of 1700 g/mol to a maximum of 30,000. Asphaltene parameters were then tuned to match the asphaltene precipitation amount with 90 vol% precipitant added, and the precipitants used were n-C_5, n-C_6, n-C_7, and n-C_8. Modeling results for the entire precipitation curve are shown in Figure 8.11b. The apparent discrepancy between the PC-SAFT predictions and experimental data might be explained by the kinetic effect. If the samples were given sufficient time to equilibrate—and it must be stressed that this can be an extraordinarily long time as reported by Maqbool et al. (2009)—it is likely that the experimental data would shift closer to the predicted precipitation curves.

The indirect method is currently suitable only for measurements at ambient pressure, and the application to HPHT is a topic of ongoing research. However, through the use of thermodynamic models, the parameters tuned to match experimental data collected at ambient conditions can be applied to HPHT conditions. The simulation results in Figure 8.12 show the effect of aging time on the fitting of simulation parameters. The curve labeled Eq, which stands for equilibrium, is obtained after fitting simulation parameters to precipitated asphaltenes at 90 vol% precipitant. The other

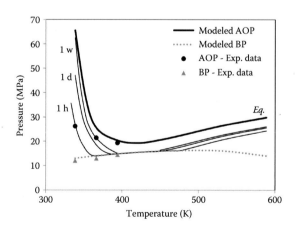

FIGURE 8.12 Modeling results of high-pressure and high-temperature asphaltene phase envelope using PC-SAFT showing the effect of aging time on the asphaltene phase envelope. (Adapted from Tavakkoli, M. et al., *Fluid Phase Equilib.*, 416, 120–129, 2016. With permission.)

curves were obtained by tuning simulation parameters to the asphaltene onset at the corresponding aging time 1 h: 1 hour aging; 1 d: 1 day aging; 1 w: 1 week aging). The results reveal that aging time has a strong effect on asphaltene phase behavior predictions at HPHT. The experimental data shown in Figure 8.12 were obtained from HPM and light-scattering techniques. However, because of the slow kinetics of asphaltene aggregation and the insufficient sensitivity of the solid detection system (SDS) and HPM instruments (Section 3.2), AOP measurements are likely underpredicted. Had equilibrium been achieved, asphaltenes would be in a more aggregated state and yield a higher onset pressure. The curve labeled *Eq* is expected to correspond to these higher pressures at each temperature.

Figure 8.12 also shows that the difference between the equilibrium curve predictions and the measured AOP decreases as temperature increases. This can be explained by taking into account the effect of temperature on aggregation kinetics. Higher temperatures lead to higher asphaltene mobility and sample equilibration occurs more readily than for lower temperatures. Therefore, it seems that within the time of the experiment and a temperature of 400 K, asphaltenes have approached the equilibrium state, and the reported AOP is closer to the true thermodynamic equilibrium state. For this reason, when a set of AOP experimental data is available at different temperatures, the AOP at the highest temperature may be considered more reliable because it is expected to be relatively closer to the equilibrium state than the values at lower temperatures.

8.1.6 CASE STUDY #6: COMPARISON BETWEEN DIFFERENT THERMODYNAMIC MODELS

Comparisons between the different thermodynamic models described in Section 4.2.3 are presented here with a focus on capturing accurately asphaltene phase behavior. There are difficulties in comparing directly the results from different researchers because of the various modeling approaches each chooses to adopt. However, general comparisons between different models frequently studied in the asphaltene modeling literature are discussed, such as oil characterization, experimental data required for tuning simulation parameters, capabilities and limitations of the approaches, and accuracy in predicting asphaltene phase behavior.

There are two types of thermodynamic models widely used for asphaltene phase behavior predictions: solution theories and EOS models. In solution models, based on regular solution theory and Flory-Huggins theory, oils are characterized as binary mixtures consisting of a bulk solvent with uniform properties and an asphaltene component. The main input parameters to solution models are molar volume and solubility parameter for both the solvent and asphaltene, and the solubility parameters are often tuned to match titration experiments at ambient condition. This simple binary mixture characterization method often leads to correspondingly large deviations from experimental observations. In addition, solution models can only be applied to liquid-liquid equilibrium, whereas an EOS model is capable of calculating both liquid-liquid and vapor-liquid equilibrium.

Because of their flexibility and the vast improvements in computing power, EOS models that were previously constrained by slow processors have surpassed solution models in popularity. The solution models, which require only a few easily measured parameters, lack the complexity to capture phase behavior at a wide range of temperature, pressure, and composition and are also capable of predicting only saturation properties. EOS models, despite requiring much more input information, can capture both saturation and non-saturation properties at a wide range of conditions.

Among the required input information for EOS models is a compositional characterization of the mixture under consideration. For crude oils, obtaining composition is non-trivial, and researchers have proposed several approaches for characterizing these complex mixtures. In addition to composition, each EOS requires component-specific parameters to quantify the intermolecular forces that describe phase behavior. Cubic EOS models, such as Soave-Redlich-Kwong (SRK) and Peng-Robinson (PR), are used widely in the oil industry to predict asphaltene precipitation and phase behavior because of their simplicity and speed (Vafaie-Sefti et al. 2003; Sabbagh et al. 2006). These models require knowledge of the critical temperature, critical pressure, and the acentric factor for each component. A volume correction may also be added to improve the liquid-density predictions (Péneloux et al. 1982).

The most common characterization method for obtaining all required input information (compositions and EOS parameters) for crude oil components is that proposed by Whitson (Whitson 1983). Cubic EOS models struggle to accurately predict phase behavior for large, complex molecules like resins and asphaltenes, especially facing problems in liquid-liquid calculations. Because asphaltene precipitation is modeled as a liquid-liquid equilibrium process, other EOS models should be considered to describe this phenomenon more accurately.

Advanced equations of state derived from statistical thermodynamic perturbation theory have recently gained wide acceptance for modeling asphaltene systems. Among the various models derived from perturbation theory, the PC-SAFT and cubic-plus-association (CPA) equations of state are the most popular for studying asphaltene precipitation. The CPA model combines the dispersion interactions described by a cubic EOS with the association term in SAFT. Some authors used SRK to represent dispersion interactions (Kontogeorgis et al. 1996), whereas others used PR EOS (Li and Firoozabadi 2010). The PC-SAFT has been shown to accurately predict the asphaltene precipitation onset without considering the association term (Gonzalez 2008; Vargas et al. 2009; Panuganti et al. 2013; Punnapala and Vargas 2013). This means that fewer tuning parameters are required to model crude oils with PC-SAFT than are required with CPA. In fact, Punnapala and Vargas (2013) reduced the number of simulation parameters required to characterize asphaltenes from three (m, σ, and ε/K) to two (γ and M) without the loss of accuracy in modeling asphaltene onset for crude oil blends with gas injection. Additionally, the speed that makes cubic EOS models attractive is lost with CPA because of the addition of an association term so that computational time for CPA is on par with PC-SAFT with association. However, because some experimental evidence reveals that the phase behavior of asphaltenes is dominated by dispersion forces and not polar interactions or hydrogen bonding (Boek et al. 2009; Czarnecki 2009; Sedghi and Goual 2010),

TABLE 8.6

Comparison between Different Thermodynamic Models

Model	Fundamental Theory	Computational Speed	Model Parameters	Capability
Solution Model	Regular solution theory; Flory-Huggins theory	High	δ, V	LLE
Cubic EOS	Physical distribution	High	$T_c, P_c, \omega,$ (Peneloux correction)	VLE+LLE
CPA	Physical distribution + association from thermodynamic perturbation theory	Low	$T_c, P_c, \omega,$ (Peneloux correction), association parameters $\kappa_{ij}, \varepsilon_{ij}$	VLE+LLE
PC-SAFT	Statistical association fluid theory	Low	m, σ, ε	VLE+LLE

CPA, Cubic plus association; EOS, equation of state; PC-SAFT, perturbed chain-statistical association fluid theory.

the association term may be dropped, making the PC-SAFT approach without association not only an accurate model, but also less computation intensive. Also, the PC-SAFT model without association requires fewer experimental data points, which in turn can significantly reduce the cost of expensive experimental studies. Table 8.6 presents a comparison between the different thermodynamic models for asphaltene precipitation.

Figure 8.13 presents the predicted asphaltene precipitation amount as a function of pressure by using Flory-Huggins solution model, PR EOS, and CPA EOS, respectively. Nor-Azlan and Adewumi (1993) combined an EOS and Flory-Huggins model to predict asphaltene precipitation from oil under pressure depletion. The molecular weight and solubility parameter of the asphaltene fraction were tuned using experimental precipitation data. The model overpredicts the asphaltene precipitation amount, as shown in Figure 8.13a because of the difficulty of evaluating solubility parameters at high pressure and temperature, but the trend is predicted well by the solution model. Asphaltene precipitation amounts predicted by PR and CPA are presented in Figure 8.13b. Szewczyk and Behar (1999) measured the amount of asphaltene dissolved in the liquid phase of a crude oil during pressure depletions at different temperatures and then applied PR EOS to predict asphaltene-precipitation amounts at various temperatures and pressures by fitting the asphaltene parameters to experimental data at 303 K. Li and Firoozabadi (2010) then used CPA EOS to describe the asphaltene phase behavior of the same oil sample, incorporating the association interactions between asphaltene molecules and resin molecules. The critical properties of the pseudo-components, including asphaltene, were taken from literature and the association energy between asphaltene and resin, ε_{AR}, was tuned as a function of temperature to match the measured upper precipitation-onset

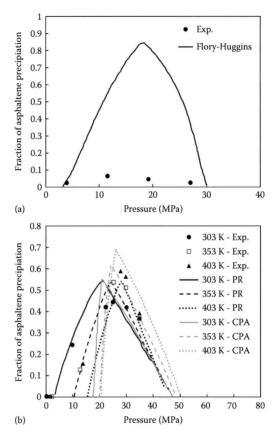

FIGURE 8.13 Asphaltene precipitation fraction as a function of pressure (a) for a crude oil at 373 K solid line predicted by Flory-Huggins solution model (From Nor-Azlan, N. and Adewumi, M. A., *SPE Eastern Regional Meeting*, Society of Petroleum Engineers, 1993. With permission.); (b) for a crude oil at 303 K, 353 K, and 403 K; black lines predicted by Peng-Robinson EOS (From Szewczyk, V. and Behar, E., *Fluid Phase Equilib.*, 158–160, 459–469, 1999); grey lines predicted by CPA EOS. (From Li, Z. and Firoozabadi, A., *Energy Fuels*, 24, 2956–2963, 2010. With permission.)

pressures at the highest and lowest temperatures. CPA yields satisfactory predictions on asphaltene precipitation amount above the bubble pressure, where PR especially struggled, and the apparent deviations below the bubble point is usually of little concern because asphaltene precipitation data measured below the bubble point is especially prone to error, so the accuracy or inaccuracy of predictions in this region is not easy to evaluate.

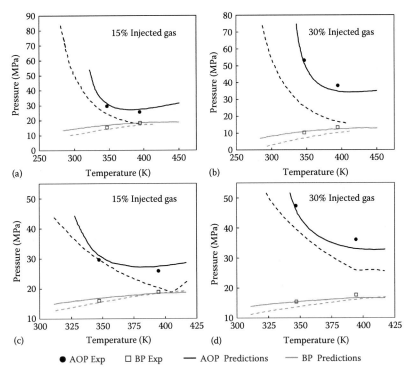

FIGURE 8.14 Asphaltene phase behavior of a crude oil under gas injections (a) + (b): after estimating parameters for 5 mol% of gas injection (solid line: PC-SAFT; dash line: SRK) (Adapted from Panuganti, S. R. et al., *Fluid Phase Equilibria.*, 359, 2–16, 2013.); (c) + (d): after estimating parameters for 5 mol% of gas injection (solid line: PC-SAFT; dash line: CPA). (Adapted from AlHammadi, A. A. et al., *Energy Fuels*, 29, 2864–2875, 2015. With permission.)

Comparisons between SRK, CPA, and PC-SAFT on asphaltene phase behavior predictions of a crude oil sample under gas injection are presented in Figure 8.14. Simulation parameters are fit to measured bubble pressures and AOPs at different temperatures with 5 mol% gas injection for SRK with Peneloux correction, PC-SAFT (Panuganti et al. 2013) and CPA (AlHammadi et al. 2015). The predictions from SRK show large deviations from experimental data under a high amount of gas injection, as shown in Figures 8.14a and 8.14b. Figures 8.14c and 8.14d show the results from the same oil with CPA and PC-SAFT. The predictions from CPA are improved compared to those from SRK. However, the liquid density is not accurately predicted by CPA, which leads to the deviation from experimental data at high gas injections. The predictions of CPA can be enhanced by adding a volume correction term to the model. More parameters are required for such cases, and the computation time may increase. PC-SAFT outperforms other equations of state model on asphaltene instability predictions at high gas injections by a single set of parameters fit to the experimental data at 5 mol% gas injection as shown in Figure 8.14. The crude oil is characterized based on the flashed gas and the STO. PC-SAFT parameters of asphaltene are tuned to the experimental onset data and asphaltene precipitation is accurately predicted at various temperatures, pressures, and compositions. PC-SAFT is capable of accurately predicting not only equilibrium properties but also derivative properties, such as isothermal compressibility, compared to CPA (AlHammadi et al. 2015). PC-SAFT also shows excellent predictions on PVT experiments, such as GOR, B_o, and composition results in DV.

8.2 ASPHALTENE DEPOSITION IN PIPELINES AND POROUS MEDIA

8.2.1 CASE STUDY #7: FIELD APPLICATIONS OF THE ASPHALTENE DEPOSITION TOOL

The asphaltene deposition simulator (ADEPT) introduced in Section 6.1.2 is applied here to field cases without parameter tuning. Because ADEPT first assumes constant deposition rate, a modified pseudo-transient simulator with changed deposition rate is presented. The results from these two simulation tools is also discussed by Kurup et al. (2012).

8.2.1.1 Sensitivity Analysis of Scaling Factor

The scaling factor (ScF), which was introduced in Section 6.1.2, plays an important role in extending the deposition rate from laboratory to field data, and boundary layer thickness is needed to calculate ScF. For fluids flowing through the pipeline, there are three different kinds of boundary layers. In order of decreasing thickness, these are the momentum boundary layer, laminar sublayer, and mass transfer boundary layer. These three different types of boundary layers are used to calculate the ScF. Equation 8.1 shows how to calculate momentum boundary layer:

$$\delta_{mom} = 62.7 D_t Re^{-7/8} \tag{8.1}$$

Equation 8.2 shows how to calculate laminar sublayer (Eskin et al. 2010)

$$\delta_{lam} = 5 \frac{\rho_f}{\mu_f u^*} \tag{8.2}$$

Equation 8.3 demonstrates how to calculate mass-transfer boundary layer (Deen 1998)

$$\delta_{mt} = \frac{D_t}{Sh} \text{ where } Sh = 0.023 Re^{0.8} Sc^{1/3} \tag{8.3}$$

When asphaltene deposition takes place in a production tubing, an increase in the pressure drop may occur as a result of the diameter reduction. In the example reported by Kurup et al. (2012) the field operator detected a frictional pressure drop

TABLE 8.7
Frictional Pressure Drop Predictions

	Frictional Pressure Drop (psi)	
	Smooth Deposit	Rough Deposit
K_d **based on**	$f = 0.013$	$f = 0.018$
Momentum boundary layer	502	700
Laminar boundary layer	424	605
Mass transfer boundary layer	372	519

Source: Kurup, A. S. et al., *Energy Fuels*, 26, 5702–5710, 2012. With permission.

of 648 psi. Kurup et al. (2012) used the Darcy-Weisbach formula (Whitaker 1968) to calculate the frictional pressure drop:

$$\Delta P_{friction} = f \frac{L}{D_t} \frac{U^2}{2g} \rho g \qquad (8.4)$$

Rough and smooth deposits were considered in their study. The results for the calculated frictional pressure drop for smooth deposit and rough deposit are shown in Table 8.7. It was found that the measured field pressure drop is close to the calculated frictional pressure drop when the deposit is considered as rough and the laminar sublayer is applied as the boundary layer.

In their study, Kurup et al. (2012) used ADEPT to study the effect of Reynold's number (Re) to the deposition profile. It was found that when oil flow rate increases, the thickness of deposition profile decreases. Based on the deposition profile, the mass of deposited asphaltene can be calculated. From their simulation results, it is observed that doubling Re causes the reduction of deposited asphaltene by 41%.

A sensitivity analysis was also performed to study the effect of the parameters, such as Re, the viscosity of crude oil, and particle size, on the ScF. Three different capillary deposition constants, two different particle sizes, two different values of viscosity, and three different Re were analyzed. The results of this sensitivity analysis are shown in Figure 8.15a–c. From Figure 8.15a–c, there are several phenomena that can be observed. First, when the particle size changes from 2 nanometers to 2 micrometers, the scaling factor decreases. Second, when the value of capillary deposition constant increases from 0.0013 to 0.13, the effect for Re is less important. Thus, when the capillary deposition constant is 0.13, changing Re will not affect the scaling factor. Third, as the Re increases, the scaling factor decreases. In other words, when Re increases, the scaled deposition constant will decrease because of the decreased ScF. This behavior was also observed in microchannel experiments and in Haskett and Tartera (1965). Last but not least, the ScFs of different viscosities are compared. It is found that as the viscosity increases, the ScF decreases. In other words, the oil that has higher viscosity should present a lower deposition rate.

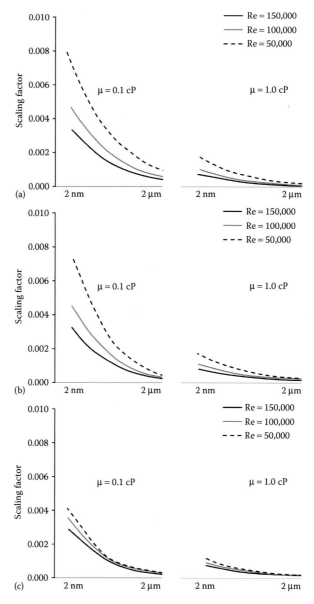

FIGURE 8.15 Sensitivity of the deposition scale-up factor. Deposition constant = (a) 0.0013 (b) 0.013, and (c) 0.13. (Adapted from Kurup, A. S. et al., *Energy Fuels*, 26, 5702–5710, 2012. With permission.)

8.2.1.2   Comparison of Steady-State Simulator and
Pseudo-Transient Simulator

The simulation results mentioned previously are performed by a steady-state simulator where the deposition rate is assumed constant. Nonetheless, in real cases, when the asphaltene starts to build up in the pipe, it may cause the cross-sectional area of the pipe to decrease and the flow rate of oil to increase. The deposited asphaltene may also increase the frictional pressure drop and change the pressure profile of the pipe. The buildup of asphaltenes might also change the temperature profile. Because of the changes in pressure and temperature, the thermodynamic stability of asphaltenes in oil may also change. Therefore, to study the effect of asphaltenes buildup, a pseudo-transient simulator is required. This simulator will recalculate the flow rate based on the decreased pipeline diameter because of deposited asphaltenes. It will also recalculate the pressure profile by adding the additional frictional pressure drop caused by the deposited asphaltene. The solubility of asphaltene in the oil along the axial length will be updated based on the modified pressure profile.

Figure 8.16a shows the results from the pseudo-transient simulator and steady-state simulation for deposit buildup times of 9, 22, and 35 days. The results calculated from steady-state simulator are represented by dotted lines and the results obtained from pseudo-transient simulator are drawn in solid lines. It is shown that for a short build up time, the deposit thickness is both at around 0.1 in for two simulators. For a longer period, such as 22 days, the difference of deposition thickness is not much. However, the deposition profile is a little different from steady-state and pseudo-steady state simulators. When the buildup time becomes longer, it seems that the difference of the thickness of deposit between pseudo-transient and steady-state simulator becomes more obvious. For 35 days, it is clear that the deposit thickness calculated from the pseudo-transient simulator is thinner compared to the steady-state simulator. It means that deposition rate will decrease because of the buildup of asphaltene deposit layer. This situation is also observed in the field and reported by Haskett and Tartera (1965). The reduced diameter of the pipe increases the velocity of oil, which in turn reduces the amount of additional deposited asphaltene. The other observation is that when the deposit buildup occurs, the position of the highest thickness in the deposition profile moves to upstream of the flow. Figure 8.16b demonstrated the variation of asphaltene solubility along with the axial length for different times. When deposit buildup increases, the frictional pressure drop increases, which modifies the pressure profile. The updated pressure profile may cause the solubility curve of asphaltene to shift. This phenomenon has also been reported by Haskett and Tartera (1965). Although the pseudo-transient simulator considers the effect of the deposit buildup, it is computationally expensive. Moreover, it seems that when the duration is short, the results from the steady-state simulator and from the pseudo-transient simulator are similar. For quick estimates, the steady-state simulator can be used.

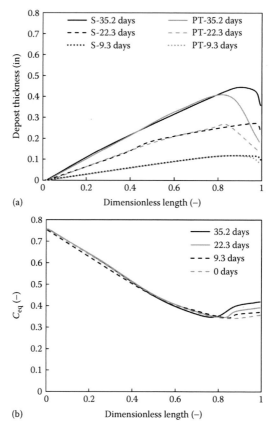

(a)

(b)

FIGURE 8.16 (a) Comparison of deposit thickness from Pseudo-transient simulator and steady-state simulator. (b) Comparison of asphaltene solubility (C_{eq}) with deposit buildup. (Adapted from Kurup, A. S. et al., *Energy Fuels*, 26, 5702–5710, 2012. With permission.)

8.2.2 Case Study #8: Effect of Corrosion on Asphaltene Deposition

Corrosion and asphaltene deposition are among the most severe problems in the wellbore during the oilfield production process. Corrosion can cause detrimental material degradation, and asphaltene deposition can lead to significant reduction of the well productivity. Although these two problems tend to occur simultaneously, mitigation strategies are usually the result of independent studies. The possible relationship between corrosion and asphaltene deposition or the interaction between corrosion and asphaltene inhibitors is not fully understood. This case study is focused on understanding the effect of corrosion on the asphaltene deposition tendency. Furthermore, a new strategy for the mitigation of corrosion-induced asphaltene deposition is presented. The experimental results are taken from the work of Sung et al. (2016).

Corrosion of carbon steel surfaces can lead to the formation of rust (iron oxides) and higher extent of surface roughness. To study the effects of corrosion products on the asphaltene deposition tendency, deposition experiments were conducted in the packed bed column (Section 5.1.1.4) filled with carbon steel spheres under three different scenarios: (a) noncorroded spheres, (b) corroded spheres with the rust left on, and (c) precorroded spheres with rust removed. The deposition experiments were performed using 5 wt% model oil P at ambient conditions. The volume ratio of oil and n-heptane was maintained at 30/70, with a total flow rate of 9 cm^3/h. The amount of deposits was collected after 6 hours. The mass of deposition on the three types of carbon steel surfaces was quantified and compared. Figure 8.17 shows the surface of the original spheres (before corrosion), the corroded spheres (with rust), and the sphere with higher surface roughness (after corrosion and without rust). Images were taken using a HIROX KH8700 3D Digital Microscope.

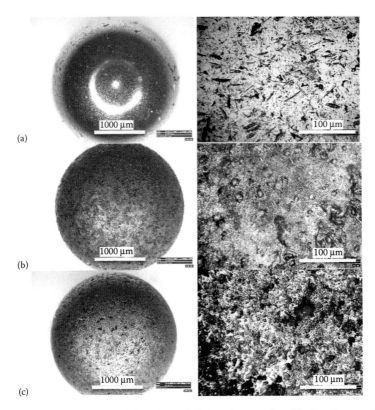

FIGURE 8.17   The surface of the spheres (a) before the corrosion, (b) after the corrosion (with rust), and (c) after the corrosion (without rust).

TABLE 8.8

Amount of Deposited Asphaltenes on the Original Spheres, on the Corroded Spheres with Rust, and on the Corrode Spheres without Rust

	Deposited Asphaltenes (% of Total Infused Asphaltenes)		
	Original Spheres	**Corroded Spheres (with Rust)**	**Corroded Spheres (without Rust)**
Spheres	23.4 ± 1.4	33.3 ± 1.0	33.6 ± 1.5

According to Table 8.8, the amount of asphaltenes deposited on the corroded spheres with and without the presence of the rust is 42.3% and 43.6% more than that on the original spheres, respectively. Thus, a larger amount of asphaltene deposits is observed on the spheres with higher surface roughness compared to that on the original spheres. This phenomenon can be explained by the increased total surface area of the spheres which creates more surface interactions with asphaltenes in the bulk phase. Thus, the corroded oilfield pipelines might worsen the asphaltene deposition problem because of the increased number of sites for asphaltenes to deposit. Besides the surface roughness of the metal, the presence of iron oxides on the surface seems to facilitate the deposition of asphaltenes. It might be associated with the interactions between iron oxides and asphaltenes. Wang et al. (2014) reported that significant amount of iron oxides is observed in the compositional analysis of deposited materials obtained from the production lines. Additionally, the amount of asphaltenes adsorbed on iron minerals like hematite is considerably more than those on other minerals (Carbognani et al. 1999). According to Murgich et al. (2001), low hydrogen-to-carbon (H/C) ratio and heteroatom contents of asphaltenes can promote the adsorption on the iron oxide mineral. Furthermore, Wang and Guidly (1994) determined that ferric ions on the mineral surface are potential adsorption sites for polar components such as asphaltenes in the crude oil. Therefore, the corrosion of the production tubing, which results in an increased surface roughness and the presence of iron compounds, may aggravate the asphaltene deposition problem.

Sung et al. (2016) further investigated the effect of ferric ions on the asphaltene-deposition tendency. Based on the results presented in Table 8.9, the injection of

TABLE 8.9

Amount of Deposited Materials on the Spheres for the Experiment without the Presence of an Aqueous Phase and the Experiment with the Presence of FeCl₃ Solution

	Deposited Materials (% of Total Infused Asphaltenes)		
		With FeCl$_3$ Solution	
	With Emulsified Water	**Ferric Ions Included**	**Ferric Ions Excluded**
Spheres	24.9 ± 1.5	36.2 ± 1.4	20.7 ± 1.5

20,000 ppm of ferric ions induces 45.4% more deposition compared to the experiment with only emulsified water. A similar phenomenon is observed by Wang et al. (2014) in their capillary deposition flow loop. However, the deposited materials might contain both asphaltenes and iron. To obtain the actual amount of asphaltene in the deposited material, the ethylenediaminetetraacetic acid solution (EDTA) was used to chelate the ferric ions present in the deposited phase (Sung et al. 2016). EDTA is a hexadentate ligand that can be ionized to form a stable complex with metal ions. According to Table 8.9, the resulting asphaltene deposition after excluding the mass of iron is 16.9% less than that in the experiment with only emulsified water.

Based on these experimental results, ferric ions in the bulk phase have a high affinity to the asphaltene molecules. Ferric ions may actually induce the flocculation of asphaltenes and produce large aggregates that might be transported along the tubing reducing the amount of deposited material; however, there is also a risk that these iron-asphaltene complexes are attracted to the metal surface, in which case the amount of deposit may actually increase. Analysis of a sample of the deposit collected from the wellbore can reveal important information about the mechanisms of the formation of such deposit. Different researchers have showed that larger asphaltene aggregates have less tendency to deposit (Vargas 2009; Eskin et al. 2011; Tavakkoli et al. 2014), but the presence of iron ions in the precipitated phase can also modify the particle/surface interaction, which may result on an increased deposition tendency.

The injection of a chelating agent, such as EDTA, can effectively bind to ferric ions and reduce their interaction with asphaltenes. The use of EDTA to mitigate the iron-induced asphaltene deposition was reported by Sung et al. (2016). Table 8.10, summarizes some of the results obtained from their packed bed column experiments, where asphaltene deposition is taken place in the presence of an aqueous ferric solution, with and without EDTA.

According to Table 8.10, the addition of the EDTA solution significantly decreases the amount of asphaltene deposition on the carbon steel spheres by 63.3%, indicating that EDTA can effectively mitigate the iron-induced asphaltene deposition on the metal surface. However, the experiments also revealed that the amount of deposited asphaltenes on the polytetrafluoroethylene (PTFE) surface significantly increases in

---

TABLE 8.10

Amount of Deposited Asphaltenes on the Spheres and the PTFE Column for the Experiments with the $FeCl_3$ Solution and with the $FeCl_3$ and EDTA Solutions

| | Deposited Asphaltenes (% of Total Infused Asphaltenes) | |
	With $FeCl_3$ Solution	With $FeCl_3$ and EDTA Solutions
Spheres	20.7 ± 1.5	7.6 ± 0.5
PTFE column	4.7 ± 0.1	12.8 ± 1.2
Total	25.4 ± 1.6	20.4 ± 1.7

EDTA, ethylenediaminetetraacetic acid; PTFE, polytetrafluoroethylene.

the presence of EDTA, which was an unexpected result. According to the literature, asphaltenes do not tend to deposit on a PTFE surfaces (Gonzalez 2008; Tavakkoli et al. 2014). Nevertheless, the iron-EDTA complex seems to have a much higher affinity to the PTFE surface. The experiments were repeated multiple times, and consistent results were obtained. Thus, one of the lessons learned from these experiments is that when multiple chemicals are coinjected to address multiple problems, such as corrosion and asphaltene deposition, a number of studies must be conducted not only to guarantee the thermal stability and the compatibility between the different chemicals, but also between the chemicals, the different fractions of the oil, and the interaction with the metal and nonmetal surfaces, if any.

8.2.3 CASE STUDY #9: ASPHALTENE DEPOSITION IN POROUS MEDIA

In Section 6.2.3, a novel modeling method for asphaltene deposition in porous media was presented along with some results of experiments conducted in mircofluidic devices. This case study is focused on the application of the modeling method to interpret the results obtained from microfluidic devices and core flood experiments.

8.2.3.1 Asphaltene Deposition in Microchannels

Figure 8.18a shows the structure of the microchannel studied in this work. The white circles show the cylindrical obstacles and the black regions are the flow paths. Figure 8.18b displays the velocity profiles in the microchannel obtained using the Lattice-Boltzmann method (LBM) (Section 6.2.3).

The simulation results are not only in good agreement with the deposited asphaltene amount but also match well with the deposit shape. Figure 8.18c shows the shape of deposited asphaltene from the micromodel experiment. The fluid flows from left to right. White circles are the obstacles in the microchannel and the dark areas represent the deposited asphaltene. Asphaltene deposition is observed in the form of conical shapes upstream of the different obstacles. Figure 8.18d shows the simulation results for the shape of deposited asphaltene from the LBM. The color scale represents the fraction of deposited asphaltene, which ranges from red for the highest amount to blue where no deposition occurs.

Figure 8.18e compares the amount of deposited asphaltene simulated against the experimental values obtained from the microfluidic deposition tests using 20 vol% crude oil C and 80 vol% n-heptane at flow rate of 60, 100, and 200 μL/min, respectively. The red circles represent experimental data, which were obtained from image analysis, and the blue line represents the simulation results. The values of the surface deposition coefficient resulted in a reasonable match between the modeling results and the experimental data. The coefficients of surface deposition for 60 (μL/min), 100 (μL/min) and 200 (μL/min) are 4.44×10^{-5}, 1.69×10^{-5}, and 8.24×10^{-6}, respectively.

The entrainment coefficient is kept constant (3.6916×10^{-3}) for all three different flow rates.

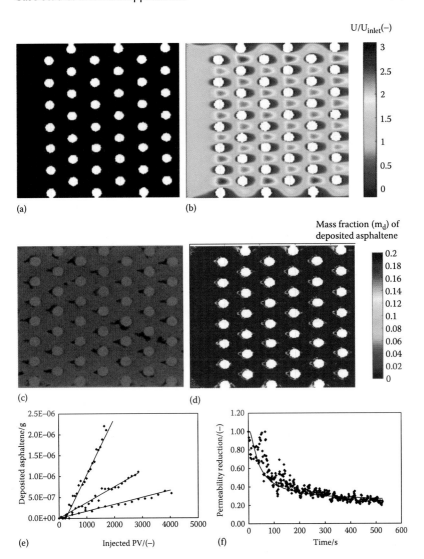

FIGURE 8.18 Results of asphaltene deposition for micromodel: (a) Structure of micromodel, (b) velocity profile of micromodel, (c) deposition profile from the experiment, (d) deposition profile from Lattice-Boltzmann method simulation, (e) asphaltene deposition amount at different flow rates from experiment and simulation, and (f) permeability reduction of micromodel from experiment and simulation.

Permeability reduction can be correlated to the amount of deposited asphaltene using the equation proposed by Gruesbeck and Collins (1982):

$$\frac{K}{K_0} = \exp(-\Upsilon\sigma^4) \tag{8.5}$$

K and K_0 represent the permeability with and without asphaltene deposit, respectively, σ is the volume fraction of the deposited asphaltene to the pore volume, Υ is a nondimensional characteristic constant for a given porous medium. Figure 8.18f shows the permeability reduction as a function of time for the case of 100 μL/min. According to this figure, the correlation works well after tuning to the available experimental data. The measured pressure drop across the porous medium is recorded and converted into permeability reduction using Darcy's law. In Figure 8.18f, the blue circles are the experimental data and the solid blue line is the simulation result. The value of Υ was a 3900.

From both modeling and experiment results, it is observed that the amount of deposited asphaltene decreases when the flow rate increases. In other words, when the flow rate increases, the surface deposition coefficient decreases. This means that by increasing the flow rate, the tendency of surface deposition tend to decrease. There are two possible reasons for this phenomenon. (1) Asphaltenes have a less tendency to deposit onto the surface when the flow rate increases. At higher flow rates, more participated asphaltene particles will be carried away before they attach to the surface. (2) The entrainment mechanism has a greater effect when the flow rate increases. This trend can also found in (Eskin et al. 2010, 2011, 2012; Akbarzadeh et al. 2012). They mention that this could be the result of the shear stress effect, for which they propose Equation 8.6. In this correlation, they used three parameters to correlate the shear stress and deposition mass flux.

$$q_a = k_{sr} \cdot q_0 \tag{8.6}$$

q_0 is the deposition mass flux in the absence of shear removal;
k_{sr} is the shear removal term; and
q_a is the actual deposition mass flux.

$$k_{sr} = \left(a + \frac{b}{\tau_w} \right)^n \tag{8.7}$$

We simplify their correlation by neglecting the parameter a to obtain Equation 8.8.

$$k_{sr} = \left(\frac{b}{\tau_w} \right)^n \tag{8.8}$$

It is found that b and n can be set to 0.0036545 and 0.592155, respectively.

8.2.3.2 Fluid Flow and Deposition Modeling in Real Porous Media

Figure 8.19a shows the three-dimensional image obtained with a Hirox 3D Microscope of the Berea sandstone core plug (high permeability) which was

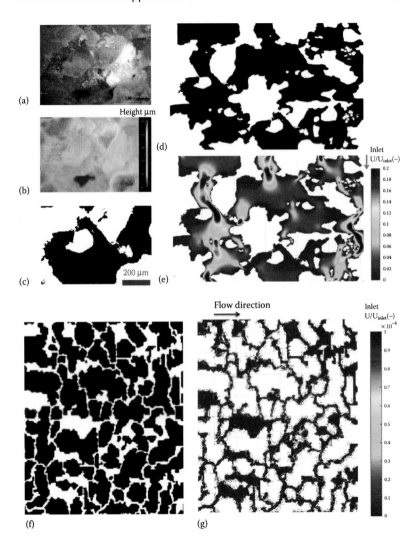

FIGURE 8.19 Results of asphaltene deposition for real porous media: (a) Three-dimensional microscopy of porous media, (b) color code map for height of porous media, (c) binary file of porous media, which can be imported directly into LBM code, (d) binary file of reconstructed porous media, (e) velocity profile of reconstructed porous media, (f) microscopic pore structure of a Berea sandstone (Reprinted from Boek, E. S. and M. Venturoli, M., *Comput. Math. Appl.*, 59, 2305–2314, 2010. With permission.), and (g) LBM simulated asphaltene deposition profile for the Berea sandstone.

used for a core flood test. The resolution is 0.4 μm and the length of this figure is 611.2 μm. Figure 8.19b shows the depth analysis of the porous media. The red and blue colors represent high and low depth, respectively. By using image processing, Figure 8.19b can be converted into Figure 8.19c, which then can be directly imported into the LBM code to calculate the fluid flow pattern for this particular geometry. The black color in Figure 8.19c represents the open space available for fluid flow, whereas the white areas correspond to the location of the solid phases.

Several micrographs have been taken from the high permeability core plug and stitched into a bigger porous medium, as depicted in Figure 8.19d. The obstacles are represented in white and the pore space is in black. The fluid flow of this porous media is calculated by the LBM code, and the results are presented in Figure 8.19e. The velocity profile represents a normalized velocity with respect to the inlet velocity. The color scale goes from blue for zero velocity to red for the maximum velocity. However, using this kind of method to reconstruct porous media would get an overestimated porosity. Therefore, this pore structure is not used to simulate asphaltene deposition because the porosity is too high.

In one more study, the porous medium is obtained from the work of Boek and Venturoli (2010). They display a two-dimensional micromodel of Berea stone. This micromodel is made at Schlumberger Cambridge Research. The structure of this micromodel is based on a thin section of Berea sandstone rock. The porosity of this structure for Berea stone is much closer to the porosity measured than the one obtained from three-dimensional microscopy.

Figure 8.19f shows the pore structure for Berea sandstone core plug from Boek and Venturoli (2010) with white area representing void pore space and the black area representing the grains. This novel two-dimensional deposition model proposed in this work can also be applied to simulate the core flood experiments. Some results are obtained for the core flood test. The surface deposition coefficient, a, the entrainment coefficient, b, and the constant of the porous medium, Υ, can be tuned to match available experimental data obtained from the core flood test. The entrainment coefficient is the same as the one obtained for the microchannel experiments (3.6916×10^{-3}). The surface deposition coefficient is 12×10^{-5}, and the constant of the porous medium (Υ) is 45. Figure 8.19g shows the predicted normalized asphaltene deposition profile. The color code of the deposition profile goes from zero deposition (blue color) to maximum deposition (red color). In this case, the parameters were tuned to the results of permeability reduction data from experiments and predictions are done for the heterogeneous medium shown in Figure 8.19g.

When the proposed modeling method is used in conjunction with core flooding experiments, one can correlate the permeability reduction for different stages of asphaltene precipitation and deposition in the porous medium. The modeling results showed a reasonable agreement with the experimental data obtained from the experiments. Figure 8.20 compares the modeling results for permeability reduction to the experimental results obtained from a certain core flood test. Blue circles represent the experimental data and the blue line shows the simulation results.

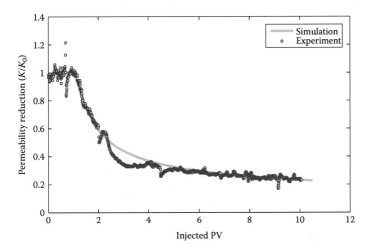

FIGURE 8.20   Permeability reduction by asphaltene deposition for the core flood test.

8.3   MITIGATION STRATEGIES FOR ASPHALTENE DEPOSITION

8.3.1   CASE STUDY #10: ANALYSIS OF THE POOR PERFORMANCE OF SOME ASPHALTENE DISPERSANTS TO REDUCE ASPHALTENE DEPOSITION

As discussed in Chapter 7, multiple approaches have been used to assess the performance of asphaltene inhibitors. Conventional techniques, such as asphaltene dispersion test (ADT), NIR spectroscopy, and SDS, have been widely used with the objective to identify the best asphaltene inhibitors. The advantages and disadvantages of the different methods were highlighted in Section 7.3.1.

Because mixed results have been reported in the oil fields after the injection of chemical additives, it is crucial to investigate the effectiveness of the conventional approaches to screen asphaltene dispersants. Moreover, the possible relationship between strong asphaltene dispersion and mitigation of actual deposition should be determined. This case study focuses on the discrepancies observed in some asphaltene inhibitors that present high dispersion efficiency but a poor performance in reducing actual asphaltene deposition, based on the experiments conducted in microfluidic devices and packed bed column.

Lin et al. (2014) used the best asphaltene dispersant selected from 15 chemical additives based on the results of ADT, and they investigated the performance of that chemical on the prevention of asphaltene deposition in a microfluidic device. The porous media are designed and fabricated as homogenous pore networks of posts with different geometries. In this case, the posts are circular with a diameter and smallest pore-throat spacing of 125 μm. In their experiment, the total flow rate of the fluid mixture of oil and n-heptane was fixed at 60 μL/min. Figure 8.21 shows the

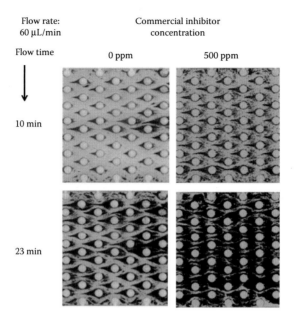

FIGURE 8.21 Deposition profiles from the mixture of oil and *n*-heptane with and without 500 ppm of the commercial inhibitor in the microchannels at 10 and 23 minutes from the start of the experiment. (Reprinted from Lin, Y. J. et al., *2015 AIChE Spring Meeting & 11th Global Congress on Process Safety*, Galveston, TX, 2014. With permission.)

results of the deposition experiments after 10 and 23 minutes of flowing the mixture with and without 500 ppm of the selected dispersant. According to the microscopic images, the injection of 500 ppm of the asphaltene inhibitor significantly increased the amount of asphaltene deposition with respect to the control experiment with no chemical added. Therefore, one can conclude that the conventional techniques of assessing asphaltene inhibitors based on the mechanism of dispersion might not be appropriate to select effective chemicals targeting the actual deposition problem. Instead, actual deposition experiments must be conducted.

To further investigate the relationship between asphaltene dispersion and deposition, a systematic study was proposed by Melendez-Alvarez et al. (2016).

To examine the dispersive performance of conventional asphaltene dispersants on the mitigation of asphaltene deposition on the metallic surface, the effects of Dispersants 8, 9, and 15 on the prevention of asphaltene aggregation and asphaltene deposition were evaluated by the direct method and by the multisection deposition test apparatus, respectively. First, the dispersive performance efficiency of the dispersants was determined at 298 K and 1 h aging time in crude oil S (Melendez-Alvarez et al. 2016). The amount of deposition on the spheres was then obtained from the experiment conducted in the multisection deposition test apparatus with and without the addition of the corresponding asphaltene dispersant at 298 K. The relationship between the

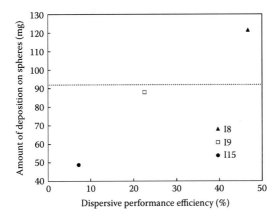

FIGURE 8.22   Results of the relationship between the dispersive performance efficiency of Dispersants 8, 9, and 15 at 70 ppm and the corresponding amount of deposition on the spheres for 70% *n*-heptane in crude oil S.

dispersive performance efficiency of Dispersants 8, 9, and 15 at 70 ppm and the corresponding amounts of deposition on spheres in crude oil S are presented in Figure 8.22. The amount of deposition from the untreated crude oil S is represented by the dotted line. Under the current experimental conditions, Dispersant 8 was the most effective asphaltene dispersant (DPE = 46.9%), followed by Dispersant 9 (DPE = 22.7%) and Dispersant 15 (DPE = 7.3%) (Melendez-Alvarez et al. 2016). From the three tested chemicals, the most effective dispersant (Dispersant 8) turned out to be the worst asphaltene deposition inhibitor. Crude oil S treated with 70 ppm of Dispersant 8 produced 31.7% more deposition on spheres than the untreated crude oil. In comparison, the same dosage of the weakest asphaltene dispersant (Dispersant 15) reduced the amount of deposition on spheres by 46.7%. Thus, if the asphaltene deposition inhibitors are only evaluated by their ability to disperse asphaltene aggregates, effective chemicals to prevent asphaltene deposition might be easily overlooked, and the operator may end up qualifying chemicals for field trials that may worsen the asphaltene deposition problem.

The effect of the dispersing power of Dispersants 8, 9, and 15 on the prevention of asphaltene deposition was further assessed in crude oil S9. According to the results obtained from the direct method and from the multisection deposition test apparatus, the relationship between the dispersive performance efficiency of the dispersants at 60 ppm and the corresponding amount of deposition on spheres in crude oil S9 is presented in Figure 8.23.

In crude oil S9, Dispersant 8 was again the most-effective asphaltene dispersant (DPE = 66.5%), followed by Dispersant 15 (DPE = 51.7%) and Dispersant 9 (DPE = 42.2%). Under the current experimental conditions, the least-effective asphaltene dispersant (Dispersant 9) generated 16.0% more deposition on spheres than the untreated crude oil. Although Dispersant 8 was ranked as the best asphaltene dispersant in both crude oils, its performance on the prevention of deposition was different.

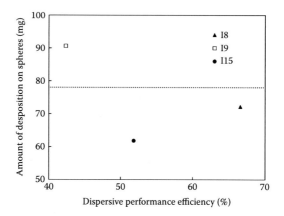

FIGURE 8.23 Results of the relationship between the dispersive performance efficiency of Dispersants 8, 9, and 15 at 60 ppm and the corresponding amount of deposition on the spheres for 70% *n*-heptane in crude oil S9.

Dispersant 8 decreased the amount of deposition on spheres by 7.7% in crude oil S9. Dispersant 15 was not the most powerful dispersants in either crude oil S or crude oil S9; however, it was the most effective chemical to reduce asphaltene deposition on the metallic surface. Specifically, the addition of Dispersant 15 in crude oil S and crude oil S9 reduced the amount of deposition on spheres by 46.7% and 20.5%, respectively.

The assessment of the performance of conventional asphaltene dispersants on asphaltene deposition with the packed bed column provides clear indication that there is no direct relationship between good asphaltene dispersion and an effective deposition reduction. Additionally, chemicals that work well in one fluid might not work and can even worsen the deposition problems in other fluids.

With this case study, one may conclude that conventional techniques such as ADT, SDS, and NIR spectroscopy are not effective in screening asphaltene deposition inhibitors. Experimental evidence from the comparison of dispersion tests and actual deposition experiments in microfluidic devices and packed bed column demonstrates that the improper assessments of chemical additives provide misleading results that could significantly hinder the strategies to reduce asphaltene deposition in the field. Representative deposition experiments should be conducted to screen the most effective chemical additives. Inhibitors developed based on the mechanism of dispersion might not work as expected; thus, a new generation of inhibitors should be developed based on other mechanisms of inhibition as it will be discussed in Section 9.2.4.

8.4 FINAL REMARKS

This chapter provides a series of case studies to better understand the methods to predict asphaltene precipitation and deposition and some of the challenges associated to the development of strategies for asphaltene deposition mitigation. The modeling and experimental techniques that were presented in Chapters 3 to 6, provide the foundation and the tools for a proper analysis and interpretation of laboratory experiments and field data. Computer modeling using PC-SAFT EOS can be effectively used for the development of EOR scenarios because it can successfully capture the effect of gas injection on the asphaltene phase envelope. In addition, PC-SAFT can be also used for detecting experimental inconsistencies in asphaltene precipitation experiments. Modeling of asphaltene deposition in the production tubing can be done using ADEPT, whereas in porous media, methods such as LBM can be implemented.

Corrosion products can affect the tendency of asphaltene deposition in wellbores and transportation pipelines because corrosion can lead to a higher extent of surface roughness and a higher concentration of iron oxides, which may further induce asphaltene deposition. Also, it is experimentally determined that the existence of ferric ions in the bulk phase can stabilize asphaltenes and deteriorate the asphaltene deposition problem. A novel approach of using the EDTA solution to mitigate iron-induce asphaltene precipitation was presented. However, more asphaltene deposition on the PTFE-coated tubing was observed. Thus, different approaches to the prevention of asphaltene deposition should be carefully selected to avoid the incompatibility problems.

The conventional asphaltene inhibitors are usually developed to disperse asphaltene aggregates. Nevertheless, it is not necessarily true that the chemicals showing strong dispersive performance can reduce asphaltene deposition. According to the deposition experiments in the microfluidic devices and the packed bed column, the most powerful asphaltene dispersants selected by the ADT may worsen the deposition problem. Thus, conventional assessment of asphaltene inhibitors based on the mechanism of dispersion might not be appropriate to screen the chemicals targeting the actual asphaltene deposition problem.

REFERENCES

Akbarzadeh, K., D. Eskin, J. Ratulowski, and S. Taylor. 2012. Asphaltene deposition measurement and modeling for flow assurance of tubings and flow lines. *Energy & Fuels* 26 (1): 495–510. doi:10.1021/ef2009474.

AlHammadi, A. A., F. M. Vargas, and W. G. Chapman. 2015. Comparison of cubic-plus-association and perturbed-chain statistical associating fluid theory methods for modeling asphaltene phase behavior and pressure–volume–temperature properties. *Energy & Fuels* 29 (5): 2864–2875. doi:10.1021/ef502129p.

Blunt, M., J. Fayers, and F. Orr. 1993. Carbon dioxide in enhanced oil recovery. *Energy Conversion and Management* 34 (9–11): 1197–1204.

Boek, E. S., and M. Venturoli. 2010. Lattice-Boltzmann studies of fluid flow in porous media with realistic rock geometries. *Computers & Mathematics with Applications* 59 (7): 2305–2314. doi:10.1016/j.camwa.2009.08.063.

Boek, E. S., D. S. Yakovlev, and T. F. Headen. 2009. Quantitative molecular representation of asphaltenes and molecular dynamics simulation of their aggregation. *Energy & Fuels* 23 (3): 1209–1219. doi:10.1021/ef800876b.

Carbognani, L., M. Orea, and M. Fonseca. 1999. Complex nature of separated solid phases from crude oils. *Energy & Fuels* 13 (2): 351–358.

Czarnecki, J. 2009. Stabilization of water in crude oil emulsions. Part 2. *Energy & Fuels* 23 (3): 1253–1257. doi:10.1021/ef800607u.

Deen, W. M. 1998. *Analysis of Transport Processes.* New York: Oxford University Press.

Eskin, D., J. Ratulowski, K. Akbarzadeh, and S. Andersen. 2012. Modeling of asphaltene deposition in a production tubing. *AIChE Journal* 58 (9): 2936–2948.

Eskin, D., J. Ratulowski, K. Akbarzadeh, and S. Pan. 2011. Modelling asphaltene deposition in turbulent pipeline flows. *The Canadian Journal of Chemical Engineering* 89 (3): 421–441.

Eskin, D., J. Ratulowski, K. Akbarzadeh, S. Pan, and T. Lindvig. 2010. Modeling asphaltene deposition in oil transport pipelines. In *Proceedings of the 7th International Conference on Multiphase Flow (ICMF)*, Tampa, FL.

Gonzalez, D. L. 2008. Modeling of asphaltene precipitation and deposition tendency using the PC-SAFT equation of state. Rice University. Retrieved from http://hdl.handle.net/1911/22140.

Gonzalez, D. L., G. J. Hirasaki, J. Creek, and W. G. Chapman. 2007. Modeling of asphaltene precipitation due to changes in composition using the perturbed chain statistical associating fluid theory equation of state. *Energy & Fuels* 21 (3): 1231–1242. doi:10.1021/ef060453a.

Gruesbeck, C., and R. E. Collins. 1982. Entrainment and deposition of fine particles in porous media. *Society of Petroleum Engineers Journal* 22 (6): 847–856. doi:10.2118/8430-PA.

Haskett, C. E., and M. Tartera. 1965. A practical solution to the problem of asphaltene deposits-Hassi Messaoud field, Algeria. *Journal of Petroleum Technology* 17 (4): 387–391. doi:10.2118/994-PA.

Kontogeorgis, G. M., E. C. Voutsas, I. V. Yakoumis, and D. P. Tassios. 1996. An equation of state for associating fluids. *Industrial & Engineering Chemistry Research* 35 (11): 4310–4318. doi:10.1021/ie9600203.

Kurup, A. S., J. Wang, H. J. Subramani, J. Buckley, J. L. Creek, and W. G. Chapman. 2012. Revisiting asphaltene deposition tool (ADEPT): Field application. *Energy & Fuels* 26 (9): 5702–5710.

Li, Z., and A. Firoozabadi. 2010. Cubic-plus-association equation of state for asphaltene precipitation in live oils. *Energy & Fuels* 24 (5): 2956–2963. doi:10.1021/ef9014263.

Lin, Y. J., M. Tavakkoli, N. T. Mathew, Y. F. Yap, J. C. Chai, A. Goharzadeh, F. M. Vargas, and S. L. Biswal. 2014. Probing asphaltene deposition using microfluidic channels. In *2015 AIChE Spring Meeting & 11th Global Congress on Process Safety*, Galveston, TX.

Maqbool, T., A. T. Balgoa, and H. S. Fogler. 2009. Revisiting asphaltene precipitation from crude oils: A case of neglected kinetic effects. *Energy & Fuels* 23 (7): 3681–3686. doi: 10.1021/ef9002236.

Melendez-Alvarez, A. A., M. Garcia-Bermudes, M. Tavakkoli, R. H. Doherty, S. Meng, D. S. Abdallah, and F. M. Vargas. 2016. On the evaluation of the performance of asphaltene dispersants. *Fuel* 179: 210–220. doi:10.1016/j.fuel.2016.03.056.

Murgich, J., E. Rogel, O. Leon, and R. Isea. 2001. A molecular mechanics density functional study of the absorption of fragments of asphaltenes and resins on the (001) surface of Fe_2O_3. *Petroleum Science and Technology* 19 (3–4): 437–455.

Nor-Azlan, N., and M. A. Adewumi. 1993. Development of asphaltene phase equilibria predictive model. In *SPE Eastern Regional Meeting*. Society of Petroleum Engineers. Retrieved from https://www.onepetro.org/conference-paper/SPE-26905-MS.

Panuganti, S. R., M. Tavakkoli, F. M. Vargas, D. L. Gonzalez, and W. G. Chapman. 2013. SAFT model for upstream asphaltene applications. *Fluid Phase Equilibria* 359: 2–16. doi:10.1016/j.fluid.2013.05.010.

Pedersen, K. S., P. L. Christensen, and J. A. Shaikh. 2014. *Phase Behavior of Petroleum Reservoir Fluids.* 2nd ed. London: CRC Press.

Péneloux, A., E. Rauzy, and R. Fréze. 1982. A consistent correction for Redlich-Kwong-Soave volumes. *Fluid Phase Equilibria* 8 (1): 7–23. doi:10.1016/0378-3812(82)80002-2.

Punnapala, S., and F. M. Vargas. 2013. Revisiting the PC-SAFT characterization procedure for an improved asphaltene precipitation prediction. *Fuel* 108: 417–429. doi:10.1016/j.fuel.2012.12.058.

Sabbagh, O., K. Akbarzadeh, A. Badamchi-Zadeh, W. Y. Svrcek, and H. W. Yarranton. 2006. Applying the PR-EoS to asphaltene precipitation from n-alkane diluted heavy oils and bitumens. *Energy & Fuels* 20 (2): 625–634. doi:10.1021/ef0502709.

Sedghi, M., and L. Goual. 2010. Role of resins on asphaltene stability. *Energy & Fuels* 24 (4): 2275–2280. doi:10.1021/ef9009235.

Sung, C. A., M. Tavakkoli, A. Chen, and F. M. Vargas. 2016. Prevention and control of corrosion-induced asphaltene deposition. In Offshore Technology Conference. doi:10.4043/27008-MS.

Szewczyk, V., and E. Behar. 1999. Compositional model for predicting asphaltenes flocculation. *Fluid Phase Equilibria* 158–160: 459–469. doi:10.1016/S0378-3812(99)00107-7.

Tavakkoli, M., A. Chen, and F. M. Vargas. 2016. Rethinking the modeling approach for asphaltene precipitation using the PC-SAFT equation of state. *Fluid Phase Equilibria* 416: 120–129.

Tavakkoli, M., M. R. Grimes, X. Liu, C. K. Garcia, S. C. Correa, Q. J. Cox, and F. M. Vargas. 2015. Indirect method: A novel technique for experimental determination of asphaltene precipitation. *Energy & Fuels* 29 (5): 2890–2900. doi:10.1021/ef502188u.

Tavakkoli, M., S. R. Panuganti, F. M. Vargas, V. Taghikhani, M. R. Pishvaie, and W. G. Chapman. 2014. Asphaltene deposition in different depositing environments: Part 1. Model oil. *Energy & Fuels* 28 (3): 1617–1628. doi:10.1021/ef401857t.

Vafaie-Sefti, M., S. A. Mousavi-Dehghani, and M. Mohammad-Zadeh. 2003. A simple model for asphaltene deposition in petroleum mixtures. *Fluid Phase Equilibria* 206 (1–2): 1–11. doi:10.1016/S0378-3812(02)00301-1.

Vargas, F. M., Garcia-Bermudes, M. Boggara, S. Punnapala, M. I. L. Abutaqiya, N. T. Mathew, S. Prasad, A. Khaleel, M. H. Al Rashed, and H. Y. Al Asafen. 2014. On the development of an enhanced method to predict asphaltene precipitation. In Houston, TX: Offshore Technology Conference. doi:10.4043/25294-MS.

Vargas, F. M. 2009. *Modeling of Asphaltene Precipitation and Arterial Deposition.* Houston, TX: Rice University.

Vargas, F. M., D. L. Gonzalez, J. L. Creek, J. Wang, J. Buckley, G. J. Hirasaki, and W. G. Chapman. 2009. Development of a general method for modeling asphaltene stability. *Energy & Fuels* 23 (3): 1147–1154. doi:10.1021/ef800666j.

Wang, F. H. L., and L. J. Guidly. 1994. Effect of oxidation-reduction condition on wettability alteration. *Society of Petroleum Engineers.* doi: 10.2118/20504-PA.

Wang, J., T. Fan, J. S. Buckley, and J. L. Creek. 2014. Impact of water cut on asphaltene deposition tendency. *Offshore Technology Conference.* doi:10.4043/25411-MS.

Whitaker, S. 1968. *Introduction to Fluid Mechanics.* Malabar, FL: Krieger Publishing Company:

Whitson, C. H. 1983. Characterizing hydrocarbon plus fractions. *Society of Petroleum Engineers Journal* 23 (4): 683–694. doi:10.2118/12233-PA.

9 Conclusions, Current Research, and Future Directions

S. Enayat, F. Lejarza, R. Doherty, A. T. Khaleel,
C. Sisco, J. Kuang, M. I. L. Abutaqiya,
N. Rajan Babu, M. Tavakkoli, and F. M. Vargas

CONTENTS

9.1   LESSONS LEARNED

Different subjects were covered throughout this book. Physical properties, chemical structure, and composition of crude oils and asphaltenes were discussed in Chapter 2. In Chapters 3 to 6, the mechanisms of asphaltene precipitation and deposition were explained, as were several methods to model and describe asphaltene phase behavior in the wellbore and reservoir. Moreover, different experimental techniques for measuring the onset of asphaltene precipitation at various conditions were discussed. In addition, several methods for conducting asphaltene deposition tests at both ambient and high-pressure, high-temperature (HPHT) conditions were presented. Furthermore, different techniques and strategies for asphaltene deposition prevention, mitigation, and remediation were discussed in Chapter 7. Finally, Chapter 8 reviewed several case studies to apply the knowledge and tools provided throughout the book with the intention of addressing asphaltene deposition problems. In Chapter 9, each discussed topic is summarized, and concluding remarks are provided.

9.1.1   CRUDE OIL AND ASPHALTENE CHARACTERIZATION

Crude oil and asphaltene characterization has always been an area of concern in the oil and gas industry. Specifically, asphaltene characterization is an important topic because of the challenges that asphaltene represents in crude oil production, such as emulsion stabilization, asphaltene precipitation, and deposition, which can ultimately affect the production of oil. This is translated into losses of millions for the industry. Determining crude oil properties and characteristics is an important step in the development of cost-effective strategies for anticipation and remediation of these problems. Important factors that affect the flow rate of crude oil and its quality are density, viscosity, molecular weight, and water content. One of the most important factors that contribute to the behavior of crude oil at various stages during its extraction and processing is its chemical composition. Because of the high complexity of this hydrocarbon mixture, the determination of its individual components is impractical. Different techniques have been developed to integrate several components of crude oil into groups divided according to their boiling temperature, their chemical structure, or their polarity. The current methods have some limitations; they can be lengthy and consume a lot of solvents, and they can be expensive, or inaccurate. The challenge in this area is the development of methods, that are fast and affordable and whose results are reproducible and provide a reasonable representation of the crude oil composition.

Additional challenges are faced when dealing with asphaltenes. Better techniques for determining asphaltene composition would lead to greater success when modeling their behavior or designing a chemical to prevent their aggregation and deposition. Asphaltenes are highly polarizable complex polyaromatic molecules that contain heteroatoms and metals and can form nanoclusters. As a result of this structural complexity, several experimental techniques are necessary to describe the physiochemical properties of asphaltenes. The developed techniques cover the analysis of their functional groups by Fourier transform infrared (FTIR), Raman, and nuclear magnetic resonance (NMR) spectroscopy, the compositional analysis by elements, the determination of the thermal decomposition patterns, and the analysis of their molecular weight distribution. The results of these methods can be used in the calculation of other properties such as asphaltenes aromaticity, which is necessary for the modeling of asphaltenes phase behavior.

Aromaticity is a key property to understand the stability of crude oil, and it varies for the different fractions in the asphaltene polydispersed distribution. Different alkanes can be used to separate different asphaltene fractions from this distribution. Light paraffinic solvent, such as *n*-pentane, precipitate a larger fraction of the distribution. The results obtained by different methods show that fractions might vary in their heteroatoms content and molecular weight.

One of the challenges caused by asphaltene deposition in the oil reservoir is wettability alteration of the rock surfaces. Oil recovery significantly decreases once water-wet surfaces are altered to oil-wet. Wetting properties of mineral surfaces depend on different parameters such as physical and chemical properties of the solid surfaces, crude oil properties, temperature, pressure, and brine composition. Among different crude oil subfractions, asphaltenes and resins significantly affect the wettability of the rock surfaces. Asphaltenes have a greater impact compared to resins subfraction because asphaltene molecules have more polar functional groups and a higher tendency to aggregate (Buckley 1998). Contact angle experiments are often used to investigate the wettability behavior of mineral surfaces at various conditions. Among different proposed methodologies for contact angle measurements, the most widely used are the captive bubble method and the dual-drop dual-crystal technique (Xu 2005).

9.1.2 EXPERIMENTAL TECHNIQUES FOR DETERMINATION OF ASPHALTENE PRECIPITATION AND DEPOSITION

A number of experimental techniques have been developed to detect the onset of asphaltene precipitation at reservoir conditions, which includes: gravimetric technique, acoustic resonance technology (ART), light scattering technique (LST), high-pressure microscopy (HPM), HPHT filtration, and quartz crystal resonator (QCR) technique. Among these techniques, LST and HPM have become the industry's standard to measure asphaltene precipitation onset. However, because of the high cost associated with these experiments, researchers investigated pairing experiments at ambient conditions using dead oil with thermodynamic models to predict asphaltene phase behavior. Among the different proposed techniques, the indirect method was found to surpass the other approaches by sensitivity, applicability, and effectiveness. The indirect method is sensitive enough to detect the precipitation of particles that are as small as 100 nm. It is also applicable to crude oils with low or high asphaltene content, light or dark in color. This technique can also be used to quantify the amount of asphaltene precipitated (Tavakkoli et al. 2015). Additionally, the effect of kinetics on asphaltene precipitation onset was addressed using the indirect method (Tavakkoli et al. 2016). As per the work presented by Maqbool et al. (2009), they concluded that the true thermodynamic onset needed for thermodynamic models must be obtained from long-term experiments. Measuring the true thermodynamic onset is essential because the quality of the predictions obtained from the thermodynamic model will depend on the experimental data used to tune the parameters of the model. Tavakkoli et al. (2016) showed that the amount of asphaltene precipitated at

90 vol% of precipitant was not a function of time. Thus, at this amount of precipitant, the system has reached thermodynamic equilibrium, and thermodynamic models must be tuned to this value.

It is generally accepted that asphaltene precipitation is a necessary but insufficient step to form asphaltene deposition. Extensive research has been performed to study the asphaltene precipitation process; however, the mechanism of asphaltene deposition has not been fully elucidated. The development of representative experimental tools to investigate asphaltene deposition is of great significance to improve the accuracy of current asphaltene deposition simulators and provide a better understanding of the prevention and remediation of asphaltene deposition. In Chapter 5, current experimental techniques on the determination of asphaltene deposition in the wellbores and near-wellbore regions are reviewed and compared. Based on the deposition studies in different setups, the effects of water, electrolytes, corrosion, and internal surface coatings on asphaltene deposition are discussed.

The most commonly used experimental setups for the determination of asphaltene deposition in wellbores and pipelines are the capillary flow loop, the quartz crystal microbalance, the RealView deposition cell, and the packed bed column. The capillary flow loop is capable of estimating the amount and distribution of deposition by the pressure drop across the capillary. The deposits can also be recovered and quantified. However, in some cases, the mass of deposits obtained from pressure drop data and mass balance calculations are not consistent. This technique is always limited to study dead oil samples under laminar flow regime. The quartz crystal microbalance, with dissipation experiments, is highly sensitive to detect the adsorption of asphaltenes on different surfaces by consuming a small amount of sample, whereas the limited amount of adsorbed asphaltenes cannot well represent the actual asphaltene deposition process. The RealView deposition cell is feasible to simulate the thermal and hydrodynamic conditions in the production lines. In this setup, asphaltene deposition is induced by depressurizing the live oil sample. However, it requires a large amount of sample and the measurement is costly. The packed bed column is a newly developed apparatus to investigate asphaltene deposition by direct mass balance using a small amount of oil sample. The axial deposition profile can be directly obtained, and the physical and chemical properties of the deposits can be easily characterized. The packed bed column still needs future improvements on the determination of asphaltene deposition in live oil samples under the turbulent flow conditions.

Core-flooding tests and microfluidic experiments for determination of asphaltene deposition in porous media were explained in Section 5.1.2. Core-flooding tests have the feasibility of investigating asphaltene deposition in a real rock sample at HPHT conditions. However, a core-flooding experiment is usually complicated and costly and requires a large amount of oil sample. A microfluidic device, on the other hand, allows easy visualization and determination of asphaltene deposition by using a small volume of the oil sample. Nevertheless, current microfluidic experiments are performed at relatively low pressures, which are not comparable to the field conditions. Also, deposition in the two-dimensional structure may not be representative of that in the near-wellbore region. Finally, because of the high linear flow velocities in a microfluidic device, the results may overestimate the formation damage caused by asphaltene deposition.

9.1.3   MODELING METHODS FOR PREDICTION OF ASPHALTENE PRECIPITATION AND DEPOSITION

Understanding asphaltene thermodynamics—especially the prediction of saturation curves and asphaltene precipitation amounts for various temperature and pressure conditions—is the first step in designing remediation strategies for asphaltene deposition. Thermodynamics governs the equilibrium distribution of components and phases of a crude oil mixture, which determine whether asphaltenes are in a single-stable phase or if they have split into a second liquid phase, making them candidates for depositing on the pipe walls. The models commonly used to describe asphaltene thermodynamics were introduced in Chapter 4 and perturbed chain form of the statistical association fluid theory (PC-SAFT) received most of our attention because of its capability in describing accurately the thermodynamics of heavy molecules that are present in large quantities in crude oil systems.

There is still ongoing work in asphaltene thermodynamics focused on refining the thermodynamic models used for predicting crude oil phase behavior; though PC-SAFT was the only model of the SAFT family discussed in detail in this work, researchers are still proposing modifications of varying complexity to the SAFT framework to capture certain physical forces that PC-SAFT might not model accurately.

The more active area of research for asphaltene thermodynamics is on fluid characterization methods. The PC-SAFT characterization approach of Panuganti et al. (2012) was covered in detail. In this approach, the stock tank oil (STO) is modeled as a mixture of only three components (saturates, aromatics + resins, and asphaltenes), and yet the asphaltene onset curves generated using PC-SAFT with this characterization approach are consistent with the onsets measured experimentally. However, the lumping together of a broad solubility class like asphaltenes is problematic for modeling accurately both precipitation onsets and precipitation amounts throughout the asphaltene unstable region. In the modeling approach of Panuganti, the asphaltene component parameters are fit to high-pressure asphaltene onset pressure (AOP) data. The implication of this approach is that the asphaltene component parameters will be fit to the asphaltenes that drive precipitation (i.e., the heaviest, most unstable fraction that precipitates first) and neglect the more stable asphaltenes that precipitate further into the asphaltene unstable region. This approach is suitable if only the AOP is of interest, but it is often the case that the more useful information is the amount of asphaltene that precipitates as a function of temperature and pressure, not just when they become thermodynamically unstable.

To obtain more reliable results on precipitation amount, the Panuganti approach was modified by Tavakkoli et al. (2016) to consider a polydisperse distribution of the asphaltene fraction of the oil. In this approach, the most unstable asphaltenes occupy the high molecular weight end of the asphaltene distribution and the more stable asphaltenes occupy the low molecular weight end of the distribution. The PC-SAFT parameters are fit to asphaltene precipitation amounts determined by the indirect method instead of to high-pressure AOP data. Obtaining asphaltene precipitation data by the indirect method is both significantly cheaper than high-pressure data, and the nature of the experiments are such that thermodynamic equilibrium is nearly guaranteed whereas the kinetic effect plays some nonnegligible role in the measurement of high-pressure AOP by conventional techniques.

Although the prediction and analysis of possible asphaltene precipitation are extremely critical, they are only part of the solution. It is equally important to predict the transport of precipitated asphaltenes and identify their deposition tendencies. Asphaltene deposition is a complex process that not only depends on pressure, temperature, and composition of the fluid, but also depends on flow regime, surface type, and particle-surface interactions. Therefore, a predictive tool to estimate the amount and profile of asphaltene deposited along the wellbore and pipelines as well as in porous media is required.

Modeling of asphaltene deposition in both wellbore and porous media was described in Chapter 6. A detailed review of the modeling methods adopted for predicting asphaltene deposition over the years was presented. The asphaltene deposition tool (ADEPT), which is based on capillary deposition tests, has proven to be a good method to evaluate the location and magnitude of the maximum deposit thickness, thereby, providing valuable support for flow assurance risk assessment and production optimization (Kurup et al. 2012). Also, modeling asphaltene aggregation by employing population balance model, rather than a simple chemical reaction mechanism and thereby evaluating particle size distribution, has been one of the noteworthy developments (Maqbool et al. 2011). Modeling of asphaltene deposition in porous media by using Lattice-Boltzmann method (LBM) based on results from microchannel and core-flood experiments was also introduced. Although several important works have been done, it should be noted that there is still a requirement to develop a computationally efficient simulator that can predict asphaltene precipitation, aggregation, and deposition on a single platform.

9.1.4 ASPHALTENE MITIGATION AND REMEDIATION STRATEGIES

Asphaltene deposition in the wellbores and pipelines can significantly reduce the oilfield productivity. In some extreme cases, a wellbore can be completely plugged. Millions of dollars are spent to prevent and remediate asphaltene deposition every year. Extensive research has been conducted to develop cost-effective approaches to the mitigation of asphaltene deposition.

Strategies such as chemical treatment by asphaltene inhibitors, control of operating conditions, and internal surface coating have been attempted to prevent the formation of asphaltene deposition in the wellbores. Chemical additives have been designed and developed based on multiple mechanisms. Dispersion, as the most commonly used mechanism, is being used to stabilize asphaltenes by reducing the size of asphaltene aggregates. Additionally, aging prevention can inhibit the sticking and stacking tendency of the asphaltene aggregates. Another mechanism such as electrostatic interactions aims to flocculate asphaltenes through the sites containing surface charges. Different laboratory techniques have been used to assess the performance of the chemical additives. Asphaltene dispersion test, turbidity measurements, direct spectroscopy, and solid detection system are being used to evaluate the performance of chemicals on the prevention of asphaltene precipitation and aggregation. More advanced techniques such as capillary deposition flow loop, RealView deposition cell, and packed bed column screen chemicals based on their effectiveness on the mitigation of actual asphaltene deposition. Beside chemical inhibition, the control of the pressure and temperature along the wellbore can potentially reduce the occurrence or the magnitude of asphaltene deposition by defining the well's production profile. However, the development of a cost-effective solution to this problem is still under investigation. A novel approach to control asphaltene deposition by reinjection of maltenes or dead oil can effectively stabilize asphaltenes in the wellbore. Furthermore, effective internal surface coatings on the prevention of asphaltene deposition can be selected based on the Hamaker constants, which predict the sticking and spreading tendencies of asphaltenes on different materials.

In addition to the prevention strategies, solvent wash and mechanical removal are the most widely used methods to clean the deposited asphaltenes in the wellbores and pipelines. The solvent wash is always accomplished by injecting aromatic organic liquids that can redissolve the deposited asphaltenes, and the mechanical removal is used to eliminate the formation of hard and large deposits along the oilfield pipelines. However, it is acknowledged that these remediation strategies are usually expensive, time-consuming, or potentially hazardous to the environment.

As discussed in Chapter 7, asphaltene inhibitors that are evaluated based on conventional approaches might worsen the asphaltene deposition problem. In other words, strong asphaltene dispersants are not necessarily effective asphaltene deposition inhibitors. Best practices are needed to improve the current strategies on the prevention and remediation of asphaltene deposition. For example, the real impact of chemicals should be evaluated by representative deposition experiments, so that a new generation of asphaltene inhibitor targeting the mechanism of asphaltene deposition can be designed. Moreover, a cocktail of compatible inhibitors should be developed to not only prevent asphaltene deposition but also to control corrosion and other flow assurance problems.

9.1.5 CASE STUDIES AND FIELD APPLICATIONS

The modeling and experimental techniques presented in this book have shown promising capabilities for assessing and solving various real case studies.

The first case study showed how a simple screening for asphaltene stability using the indirect method can give results that are consistent with field observations. The indirect method was applied to a crude oil with an a priori knowledge that the field does not suffer asphaltene deposition problems. Using n-C_5 and n-C_7 as precipitants for the indirect method, the crude oil showed strong resistance to asphaltene precipitation, implying that asphaltenes are stable and asphaltene deposition problems should now be expected.

The second case study investigated the effect of gas injection, with different types and amounts of gasses, on the asphaltene onset. PC-SAFT predictions showed an excellent agreement with experimental data for hydrocarbon and non-hydrocarbon gas injections. The capability of the PC-SAFT model to perform such accurate predictions makes it a valuable tool to aid oil operators in developing enhanced oil recovery (EOR) scenarios.

The third case study focused on the capability of PC-SAFT to validate the consistency of the data produced by pressure-volume-temperature (PVT) experiments and asphaltenes test at HPHT.

In the fourth case study, the effect of oil commingling from two different wells as well as oil-based mud (OBM) contamination was investigated. It was found that OBM contamination stabilizes asphaltenes because of the dilution of lighter hydrocarbons in the crude oils. These lighter hydrocarbons are the main precursors for asphaltene precipitation from a live oil sample. Therefore, diluting the light components has a more pronounced effect on asphaltene stability than the addition of OBM, which is made of heavy paraffins. PC-SAFT is capable of capturing the effect of OBM contamination and matching the experimental data.

In the fifth case study, the effect of asphaltene polydispersity is investigated from a modeling perspective. Treating asphaltenes as a polydisperse mixture yields better predictions for the asphaltene onset and amount precipitated. An important outcome of the case study is the effect of aging time on the measured asphaltene onset. It was found that the equilibrium onset curve predicted by PC-SAFT is closer to experimental data at higher temperatures than it is for data at lower temperatures. This is explained in terms of the temperature-dependence on aggregation kinetics. At higher temperatures, equilibrium is expected to be achieved faster than at lower temperatures. Therefore, a better agreement between the equilibrium curves from PC-SAFT is expected for experimental values at higher temperatures.

In the sixth case study, different thermodynamic models, including solution model and several equations of state, were compared against each other regarding their performance on asphaltene precipitation and deposition predictions. The fundamental theory, computational speed, oil characterization method, fitting parameters, and capabilities of each model were reported. Solution models work well for asphaltene instability predictions but not for precipitation amount, and the temperature and pressure effects are not well considered in such models as illustrated in a crude oil example. The cubic equation of state with volume correction term can predict asphaltene precipitation amount as a function of pressure. However, the asphaltene precipitation onset pressures, especially as a function of temperature, are poorly predicted. The advanced equation of state developed based on statistical thermodynamics, such as PC-SAFT, outperformed cubic equation of state in not only liquid density predictions but also asphaltene phase. PC-SAFT and cubic plus association (CPA) are both capable of providing reliable oil and asphaltene phase behavior predictions. Compared to PC-SAFT, CPA needs more fitting parameters. In addition, derivative parameters, such as isothermal compressibility, are better predicted by the PC-SAFT equation of state.

In the seventh case study, Lattice-Boltzmann method is used for modeling of the asphaltene deposition in porous media. A novel model of asphaltene deposition is proposed. This model not only matches well with the deposited asphaltene amount and permeability reduction but also the deposited shape in microchannel experiment. Unlike previous models that required up to 16 adjustable parameters, the proposed computer model in this project requires only 3 adjustable parameters, which are tuned to match the results from the microfluidic or core-flood experiments. This is also the first two-dimensional asphaltene deposition model in porous media, which can open a new perspective for deposition model.

In the eighth case study, the effect of corrosion on the tendency of asphaltene deposition is investigated in the packed bed column deposition setup. Corrosion can increase surface roughness and the concentration of iron compounds, which may further induce asphaltene deposition in the wellbore. Also, it is experimentally determined that the existence of ferric ions in the bulk phase can impact asphaltene stability. The use of ethylenediaminetetraacetic acid (EDTA) solution to mitigate iron-induced asphaltene deposition on the metallic surface is successfully developed. However, surprisingly more asphaltene deposition on the polytetrafluoroethylene-coated tubing is observed. Thus, different approaches to the prevention of asphaltene deposition should be carefully selected to avoid the incompatibility not only of different chemicals that are being injected but also of chemicals and surfaces.

In the ninth case study, ADEPT is applied to model the asphaltene deposition of subsea pipeline in a fully predictive way. The results predicted from ADEPT are compared to field observations. It was seen that there was a good agreement between the predicted pressure drop and those measured at the field. This case study also introduces pseudo-transient simulator to study the effect of deposited asphaltene buildup. It is found that when the deposit buildup is considered, the deposition rate will decrease. When the thickness of deposit buildup is small, the difference between ADEPT and pseudo-transient simulator is not large. Thus, for low deposition rate, ADEPT can have a good estimation.

In the tenth case study, the performance of asphaltene inhibitors using commercial techniques is evaluated. The conventional asphaltene inhibitors are usually developed to disperse asphaltene aggregates. However, it is not necessarily true that the chemicals showing strong dispersive performance can help reduce asphaltene deposition. According to the deposition experiments in the microfluidic devices, the most powerful asphaltene dispersant selected by the asphaltene dispersion test (ADT) worsen the deposition problem in the microchannels. Moreover, the deposition tests in the packed bed column prove that there is no direct relationship between good asphaltene dispersants and effective asphaltene deposition inhibitors. The development of representative deposition studies and a new generation of asphaltene deposition inhibitors are of utmost importance to control the asphaltene deposition problem.

9.2   CURRENT RESEARCH AND FUTURE DIRECTIONS

In the following sections, several unresolved issues related to the asphaltene deposition problem are discussed. The interrelation of asphaltene and other flow assurance problems such as wax and gas hydrates is presented. Moreover, development of a more sensitive and accurate experimental tool for measuring the onset of asphaltene precipitation at HPHT is discussed. Afterward, several strategies and methods are proposed to address the challenges in asphaltene phase behavior and deposition modeling. In addition, new mechanisms of asphaltene inhibition are introduced, which lead to the development of new generation of asphaltene inhibitors. Finally, the idea of separating and using asphaltene as a precursor for the potential development of novel carbon-based materials with several applications in different areas is expressed.

9.2.1 INTERRELATION OF ASPHALTENES AND OTHER FLOW ASSURANCE PROBLEMS

9.2.1.1 Interaction and Coprecipitation of Wax and Asphaltene

Most reservoir fluids contain waxes, which are normal paraffins, slightly branched paraffins, and also naphthalenes with long paraffinic chains (Pedersen et al. 2006). Like asphaltenes, waxes can precipitate out of solution, deposit in pipelines, and cause major flow assurance problems. This is particularly problematic in low-rate wells, where the oil has a greater residence time in the wellbore. The increase in flow time allows a greater heat loss, decreasing the oil temperature, which in turn can lead to wax precipitation and deposition (Weingarten and Euchner 1988). Wax appearance temperature (WAT), also known as cloud point, and wax content can be measured using different techniques such as microscopy, differential scanning calorimetry (DSC), pulsed NMR, microfluidic and rheometry techniques, among others (Zhao et al. 2015; Japper-Jaafar et al. 2016; Molla et al. 2016).

Although several articles in the literature explore the relationship between the phase behavior of asphaltenes and that of waxes, the interrelation of these two components and its effect on flow assurance of waxy crude oils is still a major area of research. Lei et al. (2014) investigated the effect of asphaltene phase transition on wax crystallization using DSC and rheological measurements. They observed that increased asphaltene aggregation increases the WAT, or in other words, induces wax crystallization (Lei et al. 2014). One of the hypotheses proposed to explain this observation is that the asphaltene aggregates can serve as nucleation sites for the formation of wax crystals, inducing wax precipitation (Lei et al. 2014). Another explanation for the coprecipitation of asphaltene and wax was suggested by Kriz and Andersen (2005). They proposed that saturated wax molecules exist in a subcooled state stabilized by the spatial interference of dispersed asphaltenes in crude oil. Once the asphaltenes start aggregating, the amount of dispersed asphaltenes decreases. Therefore, these spatial interactions are weakened, allowing the saturated wax molecules to come together and crystallize (Kriz and Andersen 2005).

Recent research by (Lei et al. 2016) demonstrated that the dispersion degree of asphaltenes can influence the process of wax deposition. They reported that when the concentration of asphaltene is below the asphaltene critical concentration, there is a greater degree of dispersion, which suppresses wax crystallization (Lei et al. 2016). Using cold finger experiments, they observed that increasing the concentration of dispersed asphaltene increases the rate of wax deposition at the surface. However, at concentrations higher than the asphaltene critical concentration, when asphaltenes begin to form aggregates, the rate of wax precipitation increases. Therefore, the concentration gradient of wax molecules between the bulk and deposition interface drops, which in turn causes a decrease in the amount of wax deposit. Hence, they concluded that the phase behavior of asphaltene has a significant effect on the deposition of wax in waxy crude oil systems (Lei et al. 2016).

Several studies have been conducted to investigate the effect of asphaltenes structure on the coprecipitation of wax in crude oils. Molina et al. (2017), for example, discovered that most influential parameters on increasing the crystallization temperature are the number of peri-condensed carbons, the ratio of peripheral carbons to aromatic carbons, and the aromaticity factor. They also reported that structural properties such as the length of aliphatic chains and number of naphthalene rings cause in a decrease of the WAT, but the concentration of asphaltenes has no apparent effect, which is contradictory to the study presented by Lei et al. (2016). Moreover, Molina et al. (2017) acknowledged that although some of their results agree with the available literature (Carbognani et al. 2000; Carbognani and Rogel 2003; Fang et al. 2012), some of the relationships they observed are also contradictory to what has been previously reported (Venkatesan et al. 2003; Alcazar-Vara and Buenrostro-Gonzalez 2011). This discrepancy further reinforces the need for a more thorough understanding of the molecular interactions between asphaltene and wax particles. This understanding should account for the polydisperse nature of both components within crude oil samples and the significantly different properties of crude oils from different regions.

Moreover, there are few articles in the literature that investigate the computational models developed to study the aggregation kinetics of asphaltene and wax during coprecipitation processes. One of the few available models is the one recently developed by Sun et al. (2017). They implemented a modified diffusion-limited aggregation (DLA) model to explain the aggregation mechanism between asphaltene and wax particles. They investigated the aggregation process and the geometry of the aggregates using three main parameters: sticking coefficient between the aggregates, the ratio of particle size, and particle concentration. Nevertheless, one of their model shortcomings is that it assumes asphaltene and wax particles are spheres of constant size. It also assumes that the aggregation process is solely driven by Brownian motion without the influence of any other intermolecular forces. Further experimental data and computer simulations are required for developing a more realistic model to predict asphaltene and wax coprecipitation process.

Although recently there have been many advances in understanding and predicting the coprecipitation and deposition of asphaltene and wax particles in crude oil, further investigations in this field are required to get a more fundamental comprehension of the interrelation of these two components of crude oil, which are responsible for major flow assurance problems.

9.2.1.2 Gas Hydrates Formation Induced by Asphaltene Precipitation

Gas hydrates are defined as crystalline compounds of water clathrate cages that enclose gas molecules under certain pressure and temperature conditions (Gao 2008; Daraboina et al. 2015). Methane (C_1) and carbon dioxide (CO_2) are commonly found in crude oils and have the proper molecular size to stabilize hydrate lattices. Similarly to asphaltene and wax deposition, hydrate formation may result in plugging pipelines and ultimately shutdown of production (Pedersen et al. 2006). There are several studies in the literature on gas hydrates; however, the interrelation between asphaltene and gas hydrates is poorly documented (Sum et al. 2010).

Understanding the chemical and physical phenomena involving systems containing both asphaltenes and gas hydrates is essential for preventing and remediating the flow assurance problems in these systems.

It has been well studied that increasing the gas-to-oil ratio (GOR) increases asphaltene instability. It was found that both the bubble pressure and the asphaltene onset pressure increase significantly with increasing GOR (AlHammadi et al. 2017). This is because light gas components are strong precipitants for asphaltenes; thus, asphaltenes in the oils with higher GOR become unstable at higher pressure. The effect of dissolved gas on asphaltene deposition was studied by simulating the asphaltene deposit thickness over the length of the wellbore, using ADEPT (AlHammadi et al. 2017). The results obtained indicate that the magnitude of the deposit thickness is not sensitive to GOR. The location of the deposit, however, shifts to a lower depth with higher pressure when GOR is increased, as predicted by the phase behavior results (AlHammadi et al. 2017). Although GOR can be related to the formation of hydrates, the interrelation between asphaltenes and hydrates has not been studied in sufficient depth.

A study was conducted by using Champion Technologies Hydrate Rocking Cell (CTHRC) apparatus, which is used to investigate flow assurance problems caused either by high fluid viscosities or solids formation (Gao 2008). This methodology consists of monitoring and analyzing the movement of a stainless steel ball inside a cell with a fluid (in this case oil with different chemical additives) at a set temperature and pressure. Sensors are used to detect the movement of the ball in the cell and report any obstructions. Gao (2008) used this technology to test an oil sample with high asphaltene content at different water contents. He also used different combinations of two inhibitors, one for asphaltenes and the other one for hydrates. He observed that when the oil had a high water content and was treated with inhibitors for both asphaltenes and hydrates, no obstructions were observed. However, when the crude oil sample with high water content was treated with the hydrate inhibitor only, the inhibitor could not effectively prevent hydrate blockage (Gao 2008). Gao (2008) repeated the latter experiment by removing all the water from the oil to prevent the formation of hydrates. Using the same composition of hydrate inhibitor as before, the ball in the CTHRC setup was able to move freely, indicating that the previous obstruction was not caused by asphaltenes. The author then concluded that asphaltene precipitation in the presence of the gas hydrates can result in agglomeration and cause blockage of pipelines. Thus, in a crude oil, which exhibits formation of asphaltene aggregates and hydrates, failure to remediate the asphaltene problem can jeopardize the hydrate prevention techniques.

The study presented earlier suggests that asphaltene precipitation induces gas hydrate formation (Gao 2008). However, it provides no insight into the mechanism by which these two problems are interrelated. Gao (2008) mentioned that the spectroscopy techniques would be needed to obtain more quantitative results regarding the interrelation of asphaltene precipitation and gas hydrates formation. Microscopy techniques, such as scanning electron microscope (SEM), would also be useful to look at the structure of these aggregates to better understand the flocculation mechanisms (Gao 2008). Moreover, it would be interesting to study the interrelation between hydrates and asphaltene deposition using deposition simulators such as the ones described in Section 6.1. Another experimental variable, which was not addressed in the study by Gao (2008), is the interaction between different chemical additives used to prevent different flow assurance problems.

9.2.2 EXPERIMENTAL DETERMINATION OF ASPHALTENE PRECIPITATION ONSET AT HIGH PRESSURE AND HIGH TEMPERATURE (INDIRECT METHOD AT HIGH PRESSURE AND HIGH TEMPERATURE)

With no doubt, the indirect method has shown surpass performance at ambient conditions because of its sensitivity. However, designing a HPHT unit will be beneficial in many ways. This will allow the detection of an onset that is closer to the true onset of precipitation because of the higher sensitivity of the indirect method compared to the commercially available techniques designed for HPHT conditions. It can also be used to validate the results obtained on the effect of inhibitors at ambient conditions (i.e. no shift on the onset point after addition of the chemicals). The performance of these chemicals highly depends on the conditions at which they are tested. Thus, mimicking reservoir conditions will give more reliable results on their performance and prove that chemical inhibitors cannot shift the AOP curve. Additionally, the effect of water, electrolyte, and iron on asphaltene precipitation can also be studied at HPHT conditions. The design will include the LST at HPHT, which is already commercially available, coupled with a HPHT centrifugation unit. This is an ongoing project and the technology will be offered in the near future.

9.2.3 CHALLENGES IN ASPHALTENE PRECIPITATION AND DEPOSITION MODELING

The saturates, aromatics, resins, and asphaltenes (SARA)–based characterization method with PC-SAFT has proven to be a reliable method for predicting HPHT asphaltene phase behavior. However, several aspects of the model can still be improved to further enhance its capabilities. The model was first developed to require SARA analysis as an input to tune the simulation parameters with no rigorous methodology available in the absence of SARA. Relying on SARA analysis as an input may lead to unrepresentative modeling results because of the high uncertainty in the current commercial technologies for quantifying SARA. Abutaqiya et al. (2017) proposed a method for eliminating SARA from crude oil characterization for PVT property predictions, but this method cannot be applied to asphaltenic crudes. The next advancements in asphaltene characterization should be able to handle asphaltene property predictions without requiring SARA. Additionally, reducing the number of tuning parameters for asphaltene modeling from two (molecular weight and aromaticity) to one (molecular weight or aromaticity) would offer significant time savings to the optimization routine.

Another limitation of the traditional SARA-based approach is that it requires AOP data as an input, so it is not completely predictive. Reliably obtaining these data can be a difficult task given the wide differences in asphaltene structures in different crudes even from the same reservoir. However, with the advances in experimental techniques to characterize asphaltene and understanding its behavior, the development of a predictive model is still in progress. Characterizing the asphaltene fraction based on its aromaticity and molecular weight is a topic of ongoing research.

Vargas et al. (2010) and Kurup et al. (2012) performed the modeling of asphaltene aggregation and deposition based on experimental results from capillary deposition tests, which helped in the development of ADEPT. ADEPT is a one-dimensional axial dispersion model based on capillary deposition experiments. Asphaltene aggregation is modeled as a simple second-order reaction mechanism. The difference between the concentration of asphaltene in the oil-precipitant mixture, and the thermodynamic equilibrium concentration of asphaltene in the oil provides the necessary driving force for asphaltene precipitation. A detailed description of the application of ADEPT simulator to laboratory scale capillary tests and real oilfield cases were shown in Sections 6.1.2 and 8.2.1, respectively. With asphaltene deposition experiments being performed in new geometries such as RealView cell and packed bed column, there is a need to modify ADEPT such that asphaltene deposition in these geometries can be modeled. This forms an important part of ongoing research.

In addition to the capillary deposition experiments (Wang et al. 2004; Kurup et al. 2012) asphaltene deposition has been studied in various instruments such as the RealView cell (organic solids deposition and control device [OSDC]) (Eskin et al. 2011; Akbarzadeh et al. 2012), packed bed columns (Vilas Bôas Fávero et al. 2016; Kuang and Vargas, 2017) and micromodels (Lin et al. 2016; He et al. 2017). Studying asphaltene deposition in these geometries has facilitated the investigation of the deposition mechanism under different flow regimes, shear rates (Eskin et al. 2011), temperatures ranging from ambient to reservoir conditions (Vargas et al. 2010; Eskin et al. 2011; Kurup et al. 2012), particle-surface interactions, aggregate formation and depletion (Eskin et al. 2011; Maqbool et al. 2011), and critical particle size (Vargas et al. 2010; Eskin et al. 2011). The capabilities and limitations of using a capillary tube, RealView cell and packed bed column to perform asphaltene deposition experiments at the laboratory scale were discussed in detail in Section 5.1.1 and microchannel experiments in Section 5.1.2.

Eskin et al. (2011) developed a deposition model based on the experimental results obtained from a RealView cell. The model describes the particle size distribution evolution in time in a Couette-Taylor device, and along a pipe based on population balance equations and simulates deposition based on particle transport to the wall. Vilas Bôas Fávero et al. (2016) measured asphaltene deposition using a packed bed apparatus and developed a mass transfer limited one-dimensional deposition model to explain the deposition of nanometer-sized unstable asphaltenes in the viscous flow regime.

As the geometry in which deposition tests are performed gets more complicated, the asphaltene deposition simulator can be developed by using computational fluid dynamics (CFD) simulation. The application of Lattice-Boltzmann method to model asphaltene deposition in porous media by using results from microchannel and core-flood experiments was shown in Section 8.2.3. Similarly, finite element and finite volume discretization methods can be used to solve the convection-diffusion equation (partial differential equation) that governs the transport of asphaltenes toward the wall surface and further deposition on the surface.

The ADEPT simulator which has submodels for asphaltene precipitation, aggregation, and deposition can be modified and adapted to suit the deposition mechanism in geometries other than a capillary tube. Instead of a one-dimensional axial dispersion equation in ADEPT, a three-dimensional convection-diffusion equation with precipitation, aggregation, and deposition mechanisms can be solved to predict asphaltene deposition in RealView geometry and packed bed column. The flow field needs to be investigated in such a geometry, before proceeding to the mass balance calculation of asphaltenes in the required control volume. One such example is shown in Figure 9.1. It shows the fluid ($\rho = 742$ kg/m^3, $\mu = 0.0012$ Pa.s) flow simulation results obtained for a RealView geometry, where the diameter of the inner cylinder is 14 mm and that of the outer cylinder is 28 mm and the height of the apparatus is 40 mm (Eskin et al. 2012). The inner cylinder is rotating at 1020 rpm (revolutions per minute). The results have been obtained from COMSOL Multiphysics™, which is a commercial software based on Finite element method (Figure 9.1).

The laboratory scale experimental data can be used to estimate parameters required in the deposition model for an oil sample that needs to be investigated for the possibility of deposition of asphaltenes in the wellbore. CFD simulation can also be performed to account for how the growing deposit affects the deposition flux, removal of asphaltene deposit from one location, and its subsequent deposition at other locations. It can also be employed to track the deposition front to quantify the amount of asphaltene deposited accurately, which are the aspects of ongoing work.

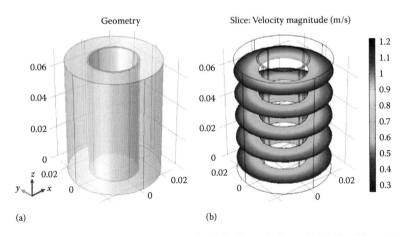

(a)   (b)

FIGURE 9.1   (a) Geometry of RealView cell and (b) slice velocity profiles in RealView cell.

9.2.4 DEVELOPMENT OF NEW GENERATION OF ASPHALTENE INHIBITORS

Conventional asphaltene inhibitors are developed based on their performance to stabilize asphaltenes by reducing the size of asphaltene aggregates. However, the injection of these chemicals has been showing mixed results in the oil fields. In some cases, the problem of asphaltene deposition is even worse after the chemical treatment. As discussed in Chapter 8 (Case Study #10), conventional techniques such as ADT, near-infrared (NIR) spectroscopy, and solid detection system (SDS) are improper to assess the performance of chemical additives on the prevention of actual asphaltene deposition. Chemicals developed based on the mechanism of dispersion might not be helpful on the inhibition of asphaltene deposition. Therefore, a new generation of asphaltene inhibitors should be developed based on other mechanisms of inhibition such as aging disruption and electrostatic interactions. Moreover, the development of integrated approaches to solving multiple problems simultaneously is of great significance to avoid the incompatibility issues associated with different chemical treatments.

According to the conceptual mechanism for asphaltene precipitation, aggregation, and aging proposed by Vargas et al. (2014), the aging of asphaltenes builds up the more compact and solid-like materials. This formation of the aged solid phase is irreversible, and it could be the main reason for the asphaltene deposition problem in the oil field (Vargas et al. 2014). Thus, potential asphaltene inhibitors targeting aging disruption may effectively prevent the formation of hard solid-like deposits. Alternatively, new asphaltene inhibitors can be developed based on the electrostatic interactions with the surface charges on the asphaltene particles. Different studies agree that asphaltene particles are not likely to deposit after growing to a certain size (Vargas et al. 2010; Akbarzadeh et al. 2012; Eskin and Ratulowski 2015). Instead of dispersing asphaltene aggregates, the new chemicals may act as electro-responsive flocculants to attract asphaltenes in the bulk phase. Thus, the larger asphaltene inhibitor complexes may have less tendency to stick and build up deposition on the oilfield pipelines.

Based on the new mechanisms of inhibition, integrated approaches can be developed to solve multiple problems in the production lines. Asphaltene deposition and corrosion are two of the most severe problems during the production of crude oils. Although corrosion and asphaltene deposition occur simultaneously, they are usually investigated independently. Two potential approaches that could simultaneously mitigate corrosion and asphaltene deposition are proposed. The first approach is to use the new generation of asphaltene inhibitors that would disrupt the aging of asphaltenes so that soft deposits are allowed to form on the surface of the wellbore. Shear forces caused by the oil flow are expected to create a thin film of soft protective asphaltene coating, which could provide some corrosion protection. Therefore, corrosion and asphaltene deposition could be controlled at the same time. Another approach is to design new electro-responsive asphaltene inhibitors/flocculants that are negatively charged to attract asphaltenes through various mechanisms, such as acid/base, polar, or London dispersion interactions. At the same time, an impressed current cathodic protection (ICCP) would be applied on the pipelines. Consequently, the charged asphaltene-inhibitor complexes could be repelled away from the negatively charged surface of the oilfield pipelines. Therefore, corrosion could be reduced by cathodic protection, and asphaltene deposition could be prevented by the application of electro-responsive inhibitors.

9.2.5 Separation and Utilization of Asphaltene for Production of Novel Materials

Asphaltene is considered as one of the major problems for oil and gas production, costing oil companies millions of dollars each year (Melendez Alvarez et al. 2016). The asphaltene deposition problem is not only limited to upstream operations, but it can also cause some operational difficulties in downstream and transportation specifically for heavy oils. Heavy oil is characterized with low American Petroleum Institute (API) gravity (API<20), high asphaltene content (>10 wt%), high carbon residue, and high viscosity. Efforts have been made to upgrade heavy oils and bitumen to more usable hydrocarbons through hydroconversion and reduction in viscosity. However, asphaltenes adsorption on the catalyst surface during the hydroconversion process has limited the ability to upgrade heavy oil (Al-Jabari and Husien 2007). This initiated the need to study asphaltene separation through adsorption. Multiple materials may be studied for the adsorption unit, such as magnetite and stainless steel. The high surface area associated with nano-sized adsorbents will be used. Finally, the magnetic properties of these adsorbents can be used to separate them from the deasphalted oil (maltenes). If the well is problematic, then the maltenes can be reinjected into the wellbore to reduce the amount of asphaltene precipitated as discussed in Section 7.1.2.2. If this is not the case, then it can be transported to refineries where it is processed more efficiently with the absence of asphaltenes.

The separated asphaltenes with no real market value could be used as a resource for the synthesis of carbon-based materials, which can potentially turn a problematic substance that has been causing damage to the economy of oil industries, into valuable and profitable products. These products can ultimately be used in various applications such as biological imaging and drug delivery (Wang and Hu 2014), Light-emitting diode (LEDs) technology, and photovoltaics (Li et al. 2011; Sarswat and Free 2015).

There have been many studies on the asphaltene behavior in the oilfield and many attempts have been made to prevent asphaltene precipitation and deposition. However, the knowledge about the asphaltene chemical structure is not sufficient. And more importantly, the potential commercial use of this fraction in the production of valued materials such as carbon-based nanoparticles has not been fully explored.

Carbon-based nanoparticles have drawn much attention because of their interesting and unique properties, such as high mechanical strength, good electrical and thermal conductivity (Li et al. 2009). These distinct physical properties have led the carbon nanoparticles to find their way in numerous applications such as electronics (Avouris et al. 2007), material science (Prasad et al. 2009), photovoltaic applications (Li et al. 2011), batteries and energy storage (Schuster et al. 2012), energy harvesting (Dai et al. 2012), biology and bio-imaging (Wang and Hu 2014), LEDs and quantum dots (Sarswat and Free 2015), and so on.

The currently available LEDs made from quantum dots have expensive rare earth heavy metals that are toxic and not environmental friendly. On the other hand, carbon-based fluorescent nanoparticles can be made that are less expensive, less toxic, and have better biocompatibility with respect to other conventional quantum dots such as cadmium selenide (Sarswat and Free 2015; Enayat et al. 2017b). Moreover, these materials can be potentially produced from very abundant and cheap resources such as asphaltenes. The potential market of quantum dots can significantly grow at a higher rate by substituting the currently commercialized heavy metal-based quantum dots with carbon dots while keeping the same quality and performance but lowering the overall cost.

Moreover, there are certain properties of asphaltenes that make them unique and exceptional with respect to other carbon-based nanoparticles. The nanoparticles derived from conventional carbon sources such as graphene, fullerene, and carbon nanotubes are just pure carbon substrates. In contrast, the asphaltene aggregates are already functionalized thanks to the presence of heteroatoms such as nitrogen, oxygen, sulfur, and trace metals like nickel and vanadium in the asphaltene molecular structure. The existence of these functional groups can potentially facilitate the functionalization step of the produced particles. This can be especially important when certain functionalization of the carbon-based particles is required depending on the desired application such as the production of photosensitive materials, energy storage, and photocatalysis reactions.

Currently, the asphaltene fraction is largely being employed in road construction. Also, it can be used as an additive to make water-repellent (hydrophobic) surfaces. Despite the current applications, the problem of asphaltene outweighs its advantages and a negative view toward them still remains. In this view, asphaltene is considered a major problem rather than a valuable material. Therefore, an extensive study on the asphaltene chemical and physical properties and identifying its potential use and application in different areas can ultimately change this notion.

The possibility of deriving nanoscale carbon-based chemical structures out of asphaltene was first proposed by Camacho-Bragado et al. (2002). In their study, asphaltene samples from Maya crude oil were investigated under high resolution transmission electron microscopy (TEM). As the samples were being observed under TEM, they came across fullerenic structures forming on the edge of the aggregates because of the electron irradiations during the microscopic analysis. However, once the fullerene onion structures are formed, they become unstable and ultimately disintegrate under further electron irradiation. Although this onion-shaped structure of asphaltene was induced by an electron beam, the authors claim that the beam current and irradiation conditions were much lower and milder than those typically used in the formation of fullerene onions. They also treated graphite under the same condition used for asphaltene and no onion structure formation was found. The authors conclude that the formation of the fullerenic structures was the result of the destruction of weak aliphatic chains during irradiation, which gives the rest of aromatic part of the asphaltene structure an opportunity to form a graphene-like structure.

In a different investigation, Wang et al. (2009) were able to synthesize carbon spheres from asphaltene with the size of 300–400 nm in diameter by a chemical vapor deposition (CVD) technique. The carbon spheres have potential applications in materials reinforcement for polymers and developing catalysts and lubricants (Xu et al. 2005). In this study, the reaction took place in a tube furnace as the argon gas carried the vaporized part of asphaltene into the hot zone of the furnace which was kept at 1237 K for 1 hour. During the pyrolysis, solid deposits of carbon-based microspheres were formed on the inner side of the tube furnace. Although this simple method for production of uniform microsphere carbon-based particles looks promising, based on the thermogravimetric analysis (TGA) the vaporized part of the asphaltene only accounts for the 5% of the initial sample (Enayat et al. 2017a). Therefore, even if we assume that all the vaporized fraction of the initial sample will turn into microspheres and deposit inside the furnace, the yield of the production using CVD will be only 5%.

Moreover, Danumah et al. (2011) investigated the potential production of carbon nanoparticles (CNP) from asphaltene. They cast a thin layer film of asphaltene on a carbon-coated copper grid, used for TEM studies, and heated the sample up to 473 K inside a TGA apparatus under nitrogen and oxygen atmosphere for 36 hours. The authors report the successful synthesis of CNPs with the average diameter of 20 nm. However, they use a dilute solution (0.0001 mg/cm^3) as the starting material for the production of the CNPs. Additionally, only one drop (10–15 µL) of the starting solution is cast on the copper grid for each experiment. A simple calculation reveals that only 10^{-9} g of CNPs can be synthesized at the end of the experiment, which makes this method to be quite impractical for a potential commercialization.

Furthermore, the potential application of asphaltene aggregates as the precursor for the synthesis of graphene was investigated in two different studies. First, Cheng et al. (2011) proposed a new technique for the production of multi-layered graphene sheets using a simple and inexpensive process similar to CVD. In this method, asphaltene samples are placed inside a covered ceramic crucible and heated by a natural gas burner. The thermal decomposition of the asphaltene generates fumes that are condensed on a secondary surface kept at 923 K. The synthesized graphene sheets are composed of mostly sp^2 carbon with relatively small defects (low D band to G band ratio in Raman). Next, Xu et al. (2013) were able to produce multilayered graphene sheets by catalytic carbonization of asphalt using vermiculite. The adsorption of asphalt by the expanded vermiculite followed by heating it at 973 K for an hour in argon flow resulted in the formation of 8–10 graphene layers and a width of tens of microns. The synthesized asphaltene-derived graphene sheets were tested as an anode in Li-ion batteries and showed slightly higher capacity (699.4 mAh/g at 50 mA/g) and much better cycling stability and conductivity (3327 $\pm$ 22 S/m) with respect to widely used reduced graphene oxide.

Most recently, water-soluble photoluminescent (PL) Carbon and graphene quantum dots (CQDs, GQDs) have been synthesized from petroleum coke and asphaltene by Wu et al. (2014) and Zhao et al. (2016), respectively. The quantum dots were prepared by oxidation of starting materials using strong acids such as HNO_3/H_2SO_4 followed by neutralization by aqueous ammonia. The final nitrogen doped GQDs show a relatively strong PL emission with a quantum yield of 16%–18% within a broad excitation range and the pH (acidity) of 2–12. The prepared GQDs were used on HeLa cells and demonstrated biocompatibility with the cell survival rate of more than 90% and effective fluorescent probe performance for cell imaging (Zhao et al. 2016).

Other than the production of carbon-based particles from asphaltene, Wu et al. (2015) have managed to take advantage of the asphaltenes' inherent properties, such as high thermal and mechanical stability. In two different studies, asphaltenes were functionalized and added as low-cost fillers to poly(styrene-butadiene-styrene) copolymer (SBS) and epoxy pre-polymer (bisphenol A diglycidyl ether) (Wu and Kessler 2015). In both cases, the rigid structure of the asphaltene enhanced the thermal stability and mechanical strength of the composite.

In another study, sulfonated asphalt based solid acid catalyst was synthesized by the carbonization of vegetable oil asphalt followed by the acidic digestion of the pyrolysis product by concentrated sulfuric acid under reflux. In a similar approach, Sulfonated multiwalled carbon nanotube (s-MWCNT) catalyst was prepared and along with the asphalt-based catalyst was used in a biodiesel production reaction. The results show that the much larger average pore diameter of asphalt-based catalyst makes it be more catalytic active, which ultimately leads to a higher biodiesel conversion, compared to the s-MWCNT catalyst (Shu et al. 2009).

According to Abujnah et al. (2016), asphaltenes can be used as organic semiconductors because asphaltenes can absorb light in a broad range of visible and NIR spectrum, have large conjugated system that is able to conduct charge, are inexpensive and naturally abundant, and have high thermal and mechanical stability (Wu and Kessler 2015). The potential absorption and the band gap difference between different fractions (from light to heavy) of asphaltenes can make them a good candidate to be used as donor and acceptor components of organic solar cells. Abujnah et al. reported the first real data on using asphaltenes as a sensitizer in dye-sensitized solar cells (DSSC) with the light to energy conversion efficiency of 1.8% (Abujnah et al. 2016). Further fractionation and functionalization of asphaltenes can improve their absorbance; hence, the efficiency of the solar cells, which can potentially lead to large-scale utilization of asphaltenes in the photovoltaic industry.

As it can be seen, recently, different researchers have started to notice the chemical and physical potential of the asphaltene fraction and have tried to employ and make use of this material in different areas of chemistry and material science. However, this research is still in its early stages and more extensive studies on asphaltene utilization are still needed.

9.3 FINAL REMARKS

Despite of all the efforts that people have devoted to the understanding, prediction, and mitigation of asphaltene deposition during oil production, to date this flow-assurance problem still persists. The cost associated to asphaltene deposition is enormous, and most of the solutions that are available in the market are usually implemented on a trial and error basis.

With this book we aim to provide a quick immersion into the complex problems of asphaltene precipitation and deposition and provide the foundation and state-of-the-art methodologies for its prediction and mitigation. Moreover, motivated by the idea that what is currently considered undesirable and problematic, asphaltenes may be converted into novel materials with highly specialized properties; we intend to inspire current and future scientists, engineers, and entrepreneurs to become actively involved in this area to accelerate the development of these technologies.

With the active participation of multidisciplinary teams, with members from industry and the academia, we steadily advance in establishing best practices to prevent and remediate the asphaltene deposition problem across the world. By combining rigorous experimental procedures and advanced modeling methods, along with the development of a new generation of asphaltene inhibitors, we are confident that we are on track to find the most cost-effective solutions that the oil industry urgently needs.

REFERENCES

Abujnah, R. E., H. Sharif, B. Torres, K. Castillo, V. Gupta, and R. R. Chianelli. 2016. Asphaltene as light harvesting material in dye-sensitized solar cell: Resurrection of ancient leaves. *Journal of Environmental & Analytical Toxicology* 06 (01): 345. doi:10.4172/2161-0525.1000345.

Abutaqiya, M. I. L., S. R. Panuganti, and F. M. Vargas. 2017. Efficient algorithm for the prediction of PVT properties of crude oils using the PC-SAFT EoS. *Industrial & Engineering Chemistry Research* 56: 6088–6102. doi:10.1021/acs.iecr.7b00368.

Akbarzadeh, K., D. Eskin, J. Ratulowski, and S. Taylor. 2012. Asphaltene deposition measurement and modeling for flow assurance of tubings and flow lines. *Energy & Fuels* 26 (1): 495–510. doi:10.1021/ef2009474.

Alcazar-Vara, L. A., and E. Buenrostro-Gonzalez. 2011. Characterization of the wax precipitation in Mexican crude oils. *Fuel Processing Technology* 92 (12): 2366–2374. doi:10.1016/j.fuproc.2011.08.012.

AlHammadi, A. A., Y. Chen, A. Yen, J. Wang, J. L. Creek, F. M. Vargas, and W. G. Chapman. 2017. Effect of the gas composition and gas/oil ratio on asphaltene deposition. *Energy & Fuels* 31 (4): 3610–3619. doi:10.1021/acs.energyfuels.6b02313.

Al-Jabari, M., and M. Husien. 2007. Review of adsorption of asphaltenes onto surfaces and its application in heavy oil recovery/upgrading and environmental protection. In. *Proceedings of the International Congress of Chemistry & Environment*, ICCE. Kuwait.

Avouris, P., Z. Chen, and V. Perebeinos. 2007. Carbon-based electronics. *Nature Nanotechnology* 2 (10): 605–615. doi:10.1038/nnano.2007.300.

Buckley, J. S. 1998. Wetting alteration of solid surfaces by crude oils and their asphaltenes. *Oil Gas Science and Technology* 53 (3): 303–312. doi:10.2516/ogst:1998026.

Camacho-Bragado, G. A., P. Santiago, M. Marin-Almazo, and M. Espinosa. 2002. Fullerenic structures derived from oil asphaltenes. *Carbon* 40: 2761–2766.

Carbognani, L., L. DeLima, M. Orea, and U. Ehrmann. 2000. Studies on large crude oil alkanes. II. Isolation and characterization of aromatic waxes and waxy asphaltenes. *Petroleum Science and Technology* 18 (5–6): 607–634. doi:10.1080/10916460008949863.

Carbognani, L., and E. Rogel. 2003. Solid petroleum asphaltenes seem surrounded by alkyl layers. *Petroleum Science and Technology* 21 (3–4): 537–556. doi:10.1081/LFT-120018537.

Cheng, I. F., Y. Xie, R. A. Gonzales, P. R. Brejna, J. P. Sundararajan, B. A. F. Kengne, D. E. Aston, D. N. McIlroy, J. D. Foutch, and P. R. Griffiths. 2011. Synthesis of graphene paper from pyrolyzed asphalt. *Carbon* 49 (8): 2852–28561. doi:10.1016/j.carbon.2011.03.020.

Dai, L., D. W. Chang, J. B. Baek, and W. Lu. 2012. Carbon nanomaterials for advanced energy conversion and storage. *Small (Weinheim an Der Bergstrasse, Germany)* 8 (8): 1130–1166. doi:10.1002/smll.201101594.

Danumah, C., A. J. Myles, and H. Fenniri. 2011. Graphitic carbon nanoparticles from asphaltenes. *MRS Proceedings* 1312: mrsf10-1312-jj05-26. doi:10.1557/opl.2011.133.

Daraboina, N., S. Pachitsas, and N. von Solms. 2015. Natural gas hydrate formation and inhibition in gas/crude oil/aqueous systems. *Fuel* 148: 186–190. doi:10.1016/j.fuel.2015.01.103.

Enayat, S., F. Lejarza, M. Tavakkoli, and F. M. Vargas. 2017a. Development of asphaltene coated melamine sponges for water-oil separation. In Preparation.

Enayat, S., M. Tavakkoli, and F. M. Vargas. 2017b. Development of water soluble and photo-luminescent carbon based nanoparticles from asphaltenes. In Preparation.

Eskin, D., J. Ratulowski, K. Akbarzadeh, and S. Andersen. 2012. Modeling of asphaltene deposition in a production tubing. *AIChE Journal* 58 (9): 2936–2948.

Eskin, D., J. Ratulowski, K. Akbarzadeh, and S. Pan. 2011. Modelling asphaltene deposition in turbulent pipeline flows. *The Canadian Journal of Chemical Engineering* 89 (3): 421–441.

Eskin, D., and J. Ratulowski. 2015. Regarding the role of the critical particle size in the asphaltene deposition model. *Energy & Fuels* 29 (11): 7741–7742. doi:10.1021/acs.energyfuels.5b01915.

Fang, L., X. Zhang, J. Ma, and B. Zhang. 2012. Investigation into a pour point depressant for Shengli crude oil. *Industrial & Engineering Chemistry Research* 51 (36): 11605–11612. doi:10.1021/ie301018r.

Gao, S. 2008. Investigation of interactions between gas hydrates and several other flow assurance elements. *Energy & Fuels* 22 (5): 3150–3153. doi:10.1021/ef800189k.

He, P., Y. J. Lin, M. Tavakkoli, J. Creek, J. Wang, F. M. Vargas, and S. L. Biswal. 2017. Effect of flow rates on the deposition of asphaltenes in porous media. In Preparation.

Japper-Jaafar, A., P. T. Bhaskoro, and Z. S. Mior. 2016. A new perspective on the measurements of wax appearance temperature: Comparison between DSC, thermomicroscopy and rheometry and the cooling rate effects. *Journal of Petroleum Science and Engineering* 147: 672–681. doi:10.1016/j.petrol.2016.09.041.

Kriz, P., and S. I. Andersen. 2005. Effect of asphaltenes on crude oil wax crystallization. *Energy & Fuels* 19 (3): 948–953. doi:10.1021/ef049819e.

Kuang, J., and F. M Vargas. 2017. Novel way to assess the performance of dispersants on the prevention of asphaltene deposition. In Preparation.

Kurup, A. S., J. Wang, H. J. Subramani, J. S. Buckley, J. L. Creek, and W. G. Chapman. 2012. Revisiting asphaltene deposition tool (ADEPT): Field application. *Energy & Fuels* 26 (9): 5702–5710.

Lei, Y., S. Han, and J. Zhang. 2016. Effect of the dispersion degree of asphaltene on wax deposition in crude oil under static conditions. *Fuel Processing Technology* 146: 20–28. doi:10.1016/j.fuproc.2016.02.005.

Lei, Y., S. Han, J. Zhang, Y. Bao, Z. Yao, and Y. Xu. 2014. Study on the effect of dispersed and aggregated asphaltene on wax crystallization, gelation, and flow behavior of crude oil. *Energy & Fuels* 28 (4): 2314–2321. doi:10.1021/ef4022619.

Li, H., C. Xu, N. Srivastava, and K. Banerjee. 2009. Carbon nanomaterials for next-generation interconnects and passives: Physics, status, and prospects. *IEEE Transactions on Electron Devices* 56 (9): 1799–1821. doi:10.1109/TED.2009.2026524.

Li, Y., Y. Hu, Y. Zhao, G. Shi, L. Deng, Y. Hou, and L. Qu. 2011. An electrochemical avenue to green-luminescent graphene quantum dots as potential electron-acceptors for photovoltaics. *Advanced Materials* 23 (6): 776–780. doi:10.1002/adma.201003819.

Lin, Y. J., P. He, M. Tavakkoli, N. T. Mathew, Y. Y. Fatt, J. C. Chai, A. Goharzadeh, F. M. Vargas, and S. L. Biswal. 2016. Examining asphaltene solubility on deposition in model porous media. *Langmuir* 32 (34): 8729–8734. doi:10.1021/acs.langmuir.6b02376.

Maqbool, T., A. T. Balgoa, and H. S. Fogler. 2009. Revisiting asphaltene precipitation from crude oils: A case of neglected kinetic effects. *Energy & Fuels* 23 (7): 3681–3686. doi:10.1021/ef9002236.

Maqbool, T., S. Raha, M. P. Hoepfner, and H. S. Fogler. 2011. Modeling the aggregation of asphaltene nanoaggregates in crude oil–precipitant systems. *Energy & Fuels* 25 (4): 1585–1596. doi:10.1021/ef1014132.

Melendez-Alvarez, A. A., M. Garcia-Bermudes, M. Tavakkoli, R. H. Doherty, S. Meng, D. S. Abdallah, and F. M. Vargas. 2016. On the evaluation of the performance of asphaltene dispersants. *Fuel* 179: 210–220. doi:10.1016/j.fuel.2016.03.056.

Molina V., D. E. Ariza León, and A. Chaves-Guerrero. 2017. Understanding the effect of chemical structure of asphaltenes on wax crystallization of crude oils from Colorado oil field. *Energy & Fuels* 31 (9): 8997–9005. doi:10.1021/acs.energyfuels.7b01149.

Molla, S., L. Magro, and F. Mostowfi. 2016. Microfluidic technique for measuring wax appearance temperature of reservoir fluids. *Lab on a Chip* 16 (19): 3795–3803. doi:10.1039/C6LC00755D.

Panuganti, S. R., F. M. Vargas, D. L. Gonzalez, A. S. Kurup, and W. G. Chapman. 2012. PC-SAFT characterization of crude oils and modeling of asphaltene phase behavior. *Fuel* 93: 658–669. doi:10.1016/j.fuel.2011.09.028.

Pedersen, K. S., P. L. Christensen, and J. A. Shaikh. 2006. *Phase Behaviour of Petroleum Reservoir Fluids*. London, UK: CRC Press.

Prasad, K. E., B. Das, U. Maitra, U. Ramamurty, and C. N. R. Rao. 2009. Extraordinary synergy in the mechanical properties of polymer matrix composites reinforced with 2 nanocarbons. *Proceedings of the National Academy of Sciences of the United States of America* 106 (32): 13186–13189. doi:10.1073/pnas.0905844106.

Sarswat, P. K., and M. L. Free. 2015. Light emitting diodes based on carbon dots derived from food, beverage, and combustion wastes. *Physical Chemistry Chemical Physics* 17 (41): 27642–27652. doi:10.1039/c5cp04782j.

Schuster, J., G. He, B. Mandlmeier, T. Yim, K. T. Lee, T. Bein, and L. F. Nazar. 2012. Spherical ordered mesoporous carbon nanoparticles with high porosity for lithium-sulfur batteries. *Angewandte Chemie* 51 (15): 3591–3595. doi:10.1002/anie.201107817.

Shu, Q., Q. Zhang, G. Xu, Z. Nawaz, D. Wang, and J. Wang. 2009. Synthesis of biodiesel from cottonseed oil and methanol using a carbon-based solid acid catalyst. *Fuel Processing Technology* 90 (7–8): 1002–1008. doi:10.1016/j.fuproc.2009.03.007.

Sum, A. K., L. E. Zerpa, J. L. Salager, C. A. Koh, and D. Sloan. 2010. Surface chemistry and gas hydrates in flow assurance. *Industrial & Engineering Chemistry Research* 50 (1): 188–197.

Sun, W., W. Wang, Y. Gu, X. Xu, and J. Gong. 2017. Study on the wax/asphaltene aggregation with diffusion limited aggregation model. *Fuel* 191: 106–113. doi:10.1016/j.fuel.2016.11.063.

Tavakkoli, M., A. Chen, and F. M. Vargas. 2016. Rethinking the modeling approach for asphaltene precipitation using the PC-SAFT equation of state. *Fluid Phase Equilibria* 416: 120–129.

Tavakkoli, M., M. R. Grimes, X. Liu, C. K. Garcia, S. C. Correa, Q. J. Cox, and F. M. Vargas. 2015. Indirect method: A novel technique for experimental determination of asphaltene precipitation. *Energy & Fuels* 29 (5): 2890–2900. doi:10.1021/ef502188u.

Vargas, F. M., M. Garcia-Bermudes, M. Boggara, S. Punnapala, M. I. L. Abutaqiya, N. T. Mathew, S. Prasad et al. 2014. On the development of an enhanced method to predict asphaltene precipitation. In Houston, TX: Offshore Technology Conference. doi:10.4043/25294-MS.

Vargas, F. M., J. L. Creek, and W. G. Chapman. 2010. On the development of an asphaltene deposition simulator. *Energy & Fuels* 24 (4): 2294–2299. doi:10.1021/ef900951n.

Venkatesan, R., J. A. Östlund, H. Chawla, P. Wattana, M. Nydén, and H. S. Fogler. 2003. The effect of asphaltenes on the gelation of waxy oils. *Energy & Fuels* 17 (6): 1630–1640. doi:10.1021/ef034013k.

Vilas Bôas Fávero, C., A. Hanpan, P. Phichphimok, K. Binabdullah, and H. S. Fogler. 2016. Mechanistic investigation of asphaltene deposition. *Energy & Fuels* 30 (11): 8915–8921. doi:10.1021/acs.energyfuels.6b01289.

Wang, J., J. S. Buckley, and J. L. Creek. 2004. Asphaltene deposition on metallic surfaces. *Journal of Dispersion Science and Technology* 25 (3): 287–298.

Wang, X., J. Guo, X. Yang, and B. Xu. 2009. Monodisperse carbon microspheres synthesized from asphaltene. *Materials Chemistry and Physics* 113 (2–3): 821–823. doi:10.1016/j.matchemphys.2008.08.053.

Wang, Y., and A. Hu. 2014. Carbon quantum dots: Synthesis, properties and applications. *Journal of Materials Chemistry C* 2 (34): 6921. doi:10.1039/C4TC00988F.

Weingarten, J. S., and J. A. Euchner. 1988. Methods for predicting wax precipitation and deposition. *SPE Production Engineering* 3 (01): 121–126. doi:10.2118/15654-PA.

Wu, H., and M. R. Kessler. 2015. Asphaltene: Structural characterization, molecular functionalization, and application as a low-cost filler in epoxy composites. *RSC Advances* 5 (31): 24264–24273. doi:10.1039/C5RA00509D.

Wu, H., V. K. Thakur, and M. R. Kessler. 2015. Novel low-cost hybrid composites from asphaltene/SBS tri-block copolymer with improved thermal and mechanical properties. *Journal of Materials Science* 51 (5): 2394–2403. doi:10.1007/s10853-015-9548-1.

Wu, M., Y. Wang, W. Wu, C. Hu, X. Wang, J. Zheng, Z. Li, B. Jiang, and J. Qiu. 2014. Preparation of functionalized water-soluble photoluminescent carbon quantum dots from petroleum coke. *Carbon* 78: 480–489. doi:10.1016/j.carbon.2014.07.029.

Xu, C., G. Ning, X. Zhu, G. Wang, X. Liu, and J. Gao. 2013. Synthesis of graphene from asphaltene molecules adsorbed on vermiculite layers. *Carbon* 62: 213–221. doi:10.1016/j.carbon.2013.05.059.

Xu, L., W. Zhang, Q. Yang, Y. Ding, W. Yu, and Y. Qian. 2005. A novel route to hollow and solid carbon spheres. *Carbon* 43 (5): 1090–1092. doi:10.1016/j.carbon.2004.11.032.

Xu, W. 2005. Experimental investigation of dynamic interfacial interactions at reservoir conditions. Louisiana State University. Retrieved from http://etd.lsu.edu/docs/available/etd-04112005-141253/.

Zhao, P., M. Yang, W. Fan, X. Wang, and F. Tang. 2016. Facile one-pot conversion of petroleum asphaltene to high quality green fluorescent graphene quantum dots and their application in cell imaging. *Particle & Particle Systems Characterization* 1–10. doi:10.1002/ppsc.201600070.

Zhao, Y., K. Paso, J. Norrman, H. Ali, G. Sørland, and J. Sjöblom. 2015. Utilization of DSC, NIR, and NMR for wax appearance temperature and chemical additive performance characterization. *Journal of Thermal Analysis and Calorimetry* 120 (2): 1427–1433. doi:10.1007/s10973-015-4451-1.

Index

Note: page numbers in *italics* refer to figures or tables.

Made in the USA
Coppell, TX
11 August 2021

60322524R00228